Air Pollution, Clean Energy and Climate Change

Air Pollution, Clean Energy and Climate Change

Anilla Cherian

This edition first published 2022

Registered Offices
John Wiley & Sons, Inc., 111 River Street, Hoboken, NJ 07030, USA
John Wiley & Sons Ltd, The Atrium, Southern Gate, Chichester, West Sussex, PO19 8SQ, UK

Editorial Office
9600 Garsington Road, Oxford, OX4 2DQ, UK

For details of our global editorial offices, customer services, and more information about Wiley products visit us at www.wiley.com.

Wiley also publishes its books in a variety of electronic formats and by print-on-demand. Some content that appears in standard print versions of this book may not be available in other formats.

Library of Congress Cataloging-in-Publication Data

Names: Cherian, Anilla, author.
Title: Air pollution, clean energy and climate change / Anilla Cherian.
Description: Hoboken, NJ : Wiley, 2022. | Includes
  bibliographical references and index.
Identifiers: LCCN 2021049302 (print) | LCCN 2021049303 (ebook) | ISBN
  9781119771586 (hardback) | ISBN 9781119771593 (adobe pdf) | ISBN
  9781119771609 (epub)
Subjects: LCSH: Air–Pollution. | Clean energy. | Climatic changes.
Classification: LCC TD883 .C4355 2022 (print) | LCC TD883 (ebook) | DDC
  363.739/2–dc23/eng/20211115
LC record available at https://lccn.loc.gov/2021049302
LC ebook record available at https://lccn.loc.gov/2021049303

Cover Design: Wiley
Cover Image: © Anilla Cherian, photo by Alison Sheehy Photography

Set in 9.5/12.5pt STIXTwoText by Straive, Pondicherry, India
Printed and bound by CPI Group (UK) Ltd, Croydon, CR0 4YY

C9781119771586_180322

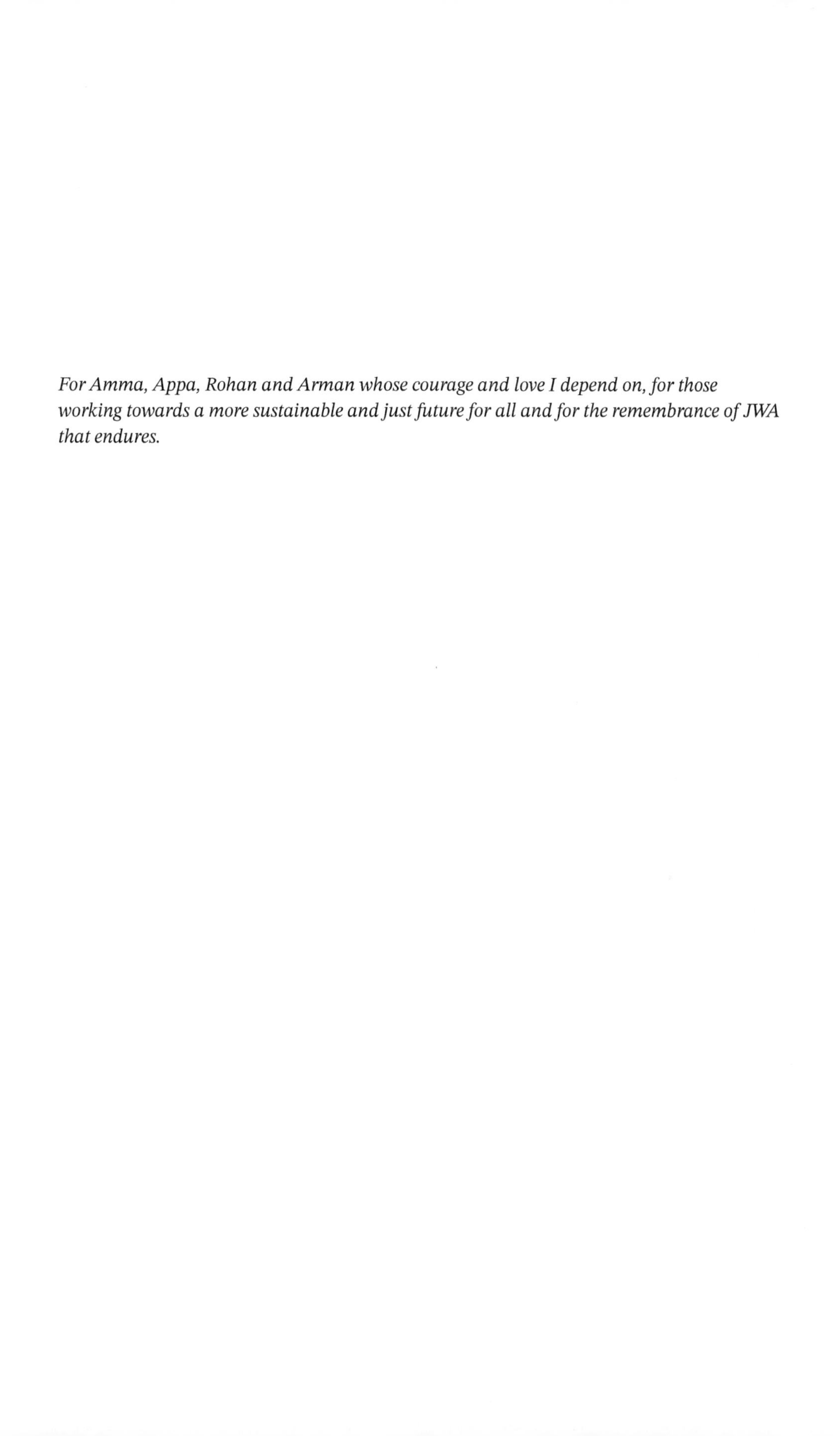

For Amma, Appa, Rohan and Arman whose courage and love I depend on, for those working towards a more sustainable and just future for all and for the remembrance of JWA that endures.

Contents

Preface

In the past few years, irrevocable losses have been experienced in the personal circles that each of us identifies as family/friends and in the intersecting circles that expand. This book has been written in the margins of grief, and in the midst of the real-time widow work of making sure my sons who tragically lost their beloved father somehow didn't shatter under the weight of sadness. But, writing a preface for a book that is inextricably interwoven with loss also comes with a sense of gratitude for the courage and loyalty of true friends, and for the tireless advocates who have spoken out for climate justice, especially those whose voices are no longer with us.

The unprecedented scope and scale of climatic impacts present a clear and present danger to our shared planet. Sadly, there is ample evidence that the immediate costs of climatic adversities will be felt more deeply by those most marginalized who lack safety nets and resilience measures to adapt to extreme climatic change. We now need to unequivocally acknowledge that the collective global failure to address climate change represents the largest inter-generational human rights violation of our time. Our collective failure to act conclusively to curb climate change condemns the poorest and most vulnerable among us who have contributed the least to the problem of greenhouse gas (GHG) emissions to suffer the most. It is also undeniably clear that poor and vulnerable lives will continue to be devastated if we ignore the costs of the largest environmental health risk – air pollution – facing some of the most populous cities in the world.

The sixth and most recent in the series of global scientific assessments issued by the Intergovernmental Panel on Climate Change (IPCC) entitled 'Climate Change 2022: Impacts, Adaptation and Vulnerability' leaves little room for equivocation that the 'extent and magnitude of climate change impacts are larger than estimated in previous assessments' (2022, p.8). Some of the report's findings regarding the inequitable impacts of climate change, air pollution and vulnerability are sobering: 'Hot extremes including heatwaves have intensified in cities, where they have also aggravated air pollution events, and limited functioning of key infrastructure'. The 'observed impacts' were found 'to be concentrated amongst the economically and socially marginalized urban residents'. The report goes on to point out that: 'Global hotspots of high human vulnerability are found particularly in West-, Central- and East Africa, South Asia, Central and South America, Small Island Developing States and the Arctic. Vulnerability is higher in locations with poverty, governance challenges and limited access to basic services and resources, violent conflict and high levels of climate-sensitive livelihoods' (2022, p 11-12)*.

Now more than ever before, there is a global urgency in responding to the needs of those who are doubly threatened by exposure to toxic levels of fossil fuel pollution and vulnerabilities to climatic adversities such extreme heat waves, droughts, flash floods and coastal zone inundation The toll of disease and morbidity burdens accruing at the toxic intersection of air pollution and climatic adversity presents a global imperative that requires looking beyond the textual parsing of three decades of intergovernmental negotiations. The global trend towards urbanization requires ensuring inclusive, community and city-based actions to reduce fossil fuel air pollution and curb short lived climate pollutants (SLCPs). From the perspective of decades of scientific consensus generated by numerous globally relevant institutions including the IPCC, it is important to be absolutely clear that emissions reductions of SLCPs cannot substitute for energy sector related GHG emissions mitigation. But, ignoring the grave impacts of SLCPs, and discounting the regional and national benefits for health, agriculture and food security that result from SLCPs emissions reductions is both ineffective and inequitable.

Reducing particulate matter (PM) pollution) including $PM_{2.5}$ emissions that emanate from the incomplete combustion of fossil fuels, wood and other biomass is critically important from a human health perspective. What has not yet been adequately addressed within the context of global climate change negotiations is that one of the principal components of $PM_{2.5}$ – black carbon (BC) –is known to be a SLCP. BC emissions have also been found to be directly linked to serious, adverse regional and in some case more localized climate change impacts including regional rainfall and weather patterns, and also most importantly in the loss of annual production levels of rice, wheat and maize. What this book argues is that increasing access to clean air and sustainable energy for all is integral to climate responsive action and to reducing the grave human health impacts of energy related air pollution. Curbing $PM_{2.5}$ emissions offers a demonstrable win-win on multiple fronts-climate change, clean air and clean energy. The book builds upon the argument that the interwoven crises of fossil fuel air pollution and climate change are well documented and extract the harshest tolls in the poorest households and communities. It discusses how responses to these interlinked crises cannot be relegated to being addressed via segregated UN global goals and policy silos. It finds that scaled up global partnerships that can jointly address energy-related pollution, mitigate climate change and address the needs of the poorest communities across the world are long over-due. The public health risks and costs that loom for some of the most populous cities in the world as a result of PM pollution should not be ignored any longer. It is now or never for broadening and deepening responses on clean air and clean energy for all.

In writing this book, it is also necessary to recognize that the ties of community and friendships are what sustain in the face of grief and loss. This book could not have been written without those who showed our small family countless acts of kindness. To the five families- the Baron, the DiRusso, the Argyros, the Hinksmon and the Steinscheinder families-who stood up to support our family in the most spontaneous and heartfelt

---

*IPCC (2022) Climate Change 2022: Impacts, Adaptation and Vulnerability. Summary for Policy Makers. Geneva: IPCC. https://report.ipcc.ch/ar6wg2/pdf/IPCC_AR6_WGII_SummaryForPolicymakers.pdf

manner, please know we are so grateful. I owe a debt of gratitude for the wisdom and generous counsel provided by Jeremy Temkin and Priya Raghavan whose assistance meant the world. To my friends, some of whom are the god parents of my boys who encouraged me when I was at my saddest, there are simply no words that can convey what your friendship meant so I will list your names in the hope you can understand how much I value you: Sumant Shrivastava, Veronique Lambert, Ambassador Elizabeth Thompson Ambassador Kwabena Osei Danquah, Mahenau Agha, Dr J. John and Laurie John, Nina Aebi, Ram Manikkalingam and Elaine Ashworth. To my dear circle of 'mom' friends, too many to list and the best kind of friends for a widowed mother to have, I want to say that I feel blessed to have you all in my corner. Deepest thanks are owed to Prof. Peter Haas who has played a crucial role in guiding my work on climate change. I am also grateful for friends and colleagues – Dr. Leena Srivastava, Judith Enck, Nina Orville, Dave Klassen and Dr Karl Hausker – whose ideas and insights have enlivened my work. Special thanks are owed to my editors at Wiley- Andrew Harrison, Rosie Hayden and Frank Weinreich- for greenlighting my proposal and shepherding the publishing process. I would be remiss if I did not also thank my kind and patient copy-editing team at Wiley.

In the end though this book would simply not have been written without the constant love and support from my parents – Elizabeth and Abraham – who have shown me what sacrifice and hard work looks like, and who have raised me to live without prejudice or judgment. This book has been written in memory of my husband and for our sons who have shown me what grace, courage, resilience and grit in the face of adversity and grief looks like. And, finally to the father of my two wonderful boys, I say thank you for giving me the gift of this unscripted, heart-breaking/heart-expanding journey. This book honors the legacy of his work at the UN in representing the smallest and most vulnerable countries in the struggle against climate change – a legacy that is owed to his sons.

Their father/my deceased husband – Ambassador John William Ashe – was one of the original negotiators of the first UN resolution calling for the historic 1992 Earth Summit, where the United Nations Framework Convention on Climate Change was open for signature, and at which historic global agreement on sustainable development and environmental issues was first reached. The first of his family to get a college education, John completed a Master of Science in Bioengineering, and a PhD in Biomedical Engineering from the University of Pennsylvania and had dual undergraduate degrees in Mathematics and Engineering. In an article entitled, "*UN Diplomat Seeks Miracle: Bring Together Rich and Poor*", the New York Times referenced him as '*one of the most influential diplomats*' at the 2002 World Summit on Sustainable Development stating that: '*He is chairman of the meeting's trade and finance committee- the man charged with bringing together the rich and poor nations, which are feuding over how to reduce poverty while preserving the environment.*' When he died, I got hand-written letters of condolence from many of his friends who had worked shoulder to shoulder with him negotiating diverse environmental challenges. One such letter from one of John's dearest friends summarized him best: '*He was indeed a great man, so exceptional that in a profession where titles are the norm, almost everyone called him John, not 'Excellency", not "Ambassador" – just John. A special and unique human being.*'

John's relevant achievements in promoting the sustainable development concerns of developing countries that comprise the majority voice in the United Nations (UN) are a matter of public record. He served in a leadership position, often as Chairman or Co-Chairman in more than 40 committees and organizations of the UN. He worked to secure global consensus on a wide range of international negotiations ranging from climate change to persistent organic pollutants. He was the first chair of the Clean Development Mechanism of the Kyoto Protocol and also chaired the Climate Convention's Subsidiary Body on Implementation. He played a leading role as Co-Chair of the Rio+20 Conference on Sustainable Development in 2012 and its historic global agreement, "*The Future We Want.*" As the President of the 68th Session of the UN General Assembly, his efforts were critical in preparing the UN for reform, the introduction of the High-Level Political Forum, and the preliminary work in reaching consensus on global agreements on Financing for Development as well as the 2015 Paris Climate Agreement. For more than two decades at the UN, he made crucial and unprecedented contributions to sustainable development, and to almost every major multilateral environmental process and agreement within the UN system. On June 30, 2016, the UN General Assembly paid tribute to his memory and his accomplishments. And yet, despite his considerable work in the global arena, he was an inherently quiet and gentle man whose greatest gift and accomplishment was being a loving father to his sons. Always remembered and never forgotten are his acts of kindness to many in need, his calm spirit and infectious smile. May his soul, and the souls of countless others who have fought valiantly to make our shared planet a more inclusive place rest in peace.

The views and opinions expressed in this book are solely those of the author and should not be attributed to any organization or entity.

1

Destroying Lives and Evidenced in Plain Sight

The Intertwined Crises of Climate Change, Lack of Access to Clean Energy and Air Pollution

1.1 Now or Never: The Urgency of Linked Action on Clean Air and Clean Energy in the Struggle Against Climate Change

There is no dearth of scientific and global consensus that anthropogenic or human-induced climate change poses an existential threat to human life. In 1824, Fourier first discussed why the Earth was warmer than could be explained by solar radiation and raised the issue of heat being trapped in the atmosphere. Tyndall then offered an answer by experimentally demonstrating that greenhouse gases (GHGs) such as carbon dioxide (CO_2) can effectively absorb infrared radiation – the greenhouse effect. Building on Tyndall's results, in 1896, Swedish scientist and Nobel Prize winner, Svante Arrhenius produced the first estimate of the sensitivity of global temperatures to increases in CO_2. By 1938, Guy Callendar demonstrated that the production of carbon dioxide by the combustion of fossil fuels was responsible for increasing the average temperature on Earth (Weart 2008; Hawkins and Jones 2013; Seidenkrantz 2018; NASA Earth Observatory website 2000). More recently, NASA, which has conducted a historic program of breakthrough research on climate science, has categorically warned that the Earth is trapping an unprecedented amount of heat, resulting in drastically warmer oceans and land temperature, with most of the warming occurring in the past 40 years and the seven most recent years being the warmest, with 2016 and 2020 tied for the warmest year on record (NASA website 2021).

Global scientific and policy consensus around climate change as a definitive and existential challenge is not a recent phenomenon (Haas 1990; Bernard and Semmler 2015). More than 16 years ago, Bill Allen, editor in chief of *National Geographic*, wrote that he was publishing the first of a three-part series of stories focused on Antarctica, Alaska and Bangladesh on a topic – global climate change – that he was 'willing to bet' would make 'people angry enough to stop subscribing', but was doing so because these stories 'cover subjects that are too important to ignore' and show 'the hard truth as scientists see it'. He added that he 'can live with some cancelled memberships' but '. . . would have a harder time looking myself in the mirror if I didn't bring you the biggest story in geography today' (Allen 2004). Today, that 'biggest story in geography' has already devastated, and is anticipated to destroy vulnerable lives spanning the world from Dhal Char, Bangladesh to New Orleans, Louisiana. Deadly forest fires, flash floods and heat waves span the globe, coral reefs are bleached, marine ecosystems are dying and low-lying coastal cities face the

Air Pollution, Clean Energy and Climate Change, First Edition. Anilla Cherian.

escalating costs of inundation and erosion as both ocean temperatures and sea-levels rise. But the capacity to adapt to and rebuild after calamitous climatic impacts, retreat to safer environs, and find new livelihoods is a luxury that millions who are exposed to endemic levels of fossil fuel related air pollution cannot afford.

The discomfiting truths are that the global community has long known that the morbidity and ill-health burdens associated with climatic adversities will be borne by those who have done the least to contribute to per capita emissions of GHGs; and that the nexus between poverty, exposure to toxic levels of air pollution and inexorable climatic impacts will extract the harshest toll on the least resilient and most vulnerable among us. Back in 2007, the United Nation's (UN) principal development agency, the UN Development Programme (UNDP) issued its *Human Development Report* that warned of five drivers or 'tipping points' by which climate change could stall and actually reverse human development: reduced agricultural productivity and increased food insecurity; heightened water stress and insecurity; rising sea levels and increased exposure to climate disasters; loss of ecosystems and biodiversity and amplified health risks, with the greatest health impacts felt in developing countries. Its warning remains prescient in this current time as the world reels from the combined effects of a global pandemic and the increasing trend of extreme climatic events. The 2007 Report was categorical about the global failure to act conclusively and decisively on climate change: 'Failure will consign the poorest 40 percent of the world's population—some 2.6 billion people—to a future of diminished opportunity. It will exacerbate deep inequalities within countries. . . In today's world, it is the poor who are bearing the brunt of climate change. Tomorrow, it will be humanity as a whole that faces the risks that come with global warming' (UNDP 2007, p. 2). A few paragraphs later, the Report went on to highlight that the world lacked 'neither the financial resources nor the technological capabilities to act' and consequently failure to act, cooperatively on climate change would 'represent not just a failure of political imagination and leadership, but a moral failure on a scale unparalleled in history'. Here, it specifically called attention to the fact that future generations would look harshly upon those who were provided with evidence, and '. . . understood the consequences and then continued on a path that consigned millions of the world's most vulnerable people to poverty and exposed future generations to the risk of ecological disaster' (2007, p. 2).

It is time to acknowledge the fact that the consequences of the global failure to act conclusively on climate change have been known to, and will continue inexorably to be borne by millions living in the poorest households, communities and countries, even as global negotiations to address climate change have been occurring for decades. Within the UN context, climate change was identified more than 30 years ago as a global challenge when the UN General Assembly (UNGA) adopted a resolution sponsored by the Government of Malta, recognizing climate change as a 'common concern of mankind' (UNGA 1988). Intergovernmental climate negotiations have been going on for more than three decades broadly centred around two distinct but interrelated issues, both of which are associated with major technology, financing and capacity related constraints particularly for the smallest and poorest of UN member states:

- Mitigation or the reduction of GHG emissions that are seen as principally responsible for the rise in global surface temperatures.

- Adaptation or the human and/or ecosystem related responses to a range of adverse climatic impacts, such as sea-level rise (SLR), increase in the frequency and intensity of extreme weather related events, effects on fragile marine ecosystems and coastal zone inundation, that accompany a rise in global surface temperatures.

The issue of climate change has galvanized the public more than any other global environmental problem. And yet, a comprehensive and effective global resolution to the climate change crisis that expressly addresses the health and morbidity costs associated with unclean air and polluting forms of energy has proven elusive over the years. 27 years after the first UNGA climate resolution was adopted, the UN Paris Agreement (PA) on climate change was gavelled into history after a marathon final day of negotiations on 12 December 2015. All UN member states universally pledged to undertake ambitious action, and agreed with the PA's serious concern about 'the urgent need to address the significant gap between the aggregate effect of Parties' mitigation pledges in terms of global annual emissions of greenhouse gases by 2020 and aggregate emission pathways consistent with holding the increase in the global average temperature to well below 2 °C above preindustrial levels and pursuing efforts to limit the temperature increase to 1.5 °C above preindustrial levels' (UNFCCC 2015, p. 2). What made this 2015 New Year's planetary resolution different from all prior UN climate resolutions and agreements was that it was the first inclusive, yet completely voluntary global climate change accord that covered all member states. The voluntary rather than legally binding aspect of the PA is in contrast to the overarching UN climate treaty, the 1992 United Nations Framework Convention on Climate Change (UNFCCC), as well as the 1997 Kyoto Protocol (KP) to the UNFCCC. But, it is precisely the PA's inclusion of the widest possible cooperation by all countries, and the entirely voluntary scaling up national climate pledges that serves as the global litmus test for distinguishing between climate hype versus verifiable climate action. On October 21, 2021, the United States (US) Office of the Director of National Intelligence (ODNI) released its first 'National Intelligence Estimate on Climate Change' and offered a stark '*takeaway*': '*Global momentum is growing for more ambitious greenhouse gas emissions reductions, but current policies and pledges are insufficient to meet the Paris Agreement goals. . . Intensifying physical effects will exacerbate geopolitical flashpoints, particularly after 2030, and key countries and regions will face increasing risks of instability and need for humanitarian assistance*' (US ODNI 2021, p. i).

The Intergovernmental Panel on Climate Change (IPCC), the world's largest compilation of scientific expertise ever convened on any global environmental topic, has persistently warned about climate change since 1988 via a series of comprehensive assessment reports (ARs). The IPCC's Fifth Assessment Report (AR5) 'Summary for Policy Makers' (SPM) cautioned that: 'Human influence on the climate system is clear, and recent anthropogenic emissions of greenhouse gases are the highest in history. . . . Warming of the climate system is unequivocal, and since the 1950s, many of the observed changes are unprecedented over decades to millennia. The atmosphere and ocean have warmed, the amounts of snow and ice have diminished, and sea level has risen' (IPCC 2014a, p. 2). But, the warming of the climate system could actually be much worse than anticipated with CO_2 being added to the atmosphere 100 times faster than at any point in pre-industrial human history and more damage being done in the three decades since the IPCC was established than in the whole of human history (Wallace-Wells 2019). The SPM's stark reminder of irreversibility of climate change except in the case of a comprehensive and timely net removal of atmospheric CO_2 emissions is worrisome precisely because even after a complete halt of net CO_2 emissions, surface temperatures will remain elevated for several centuries: 'A large fraction of anthropogenic climate change resulting from CO_2 emissions is irreversible on a multi-century to millennial time scale, except in the case of large net removal of CO_2 from the atmosphere over a sustained period. Surface temperatures will remain approximately constant at elevated levels for many centuries after a complete cessation of net anthropogenic CO_2emissions. Due to the long time scales of heat transfer from the ocean surface to depth, ocean warming will continue for centuries' (IPCC 2013, p. 28).

But now, the most recent IPCC Sixth Assessment Report (AR6) Working Group 1 SPM report has issued a grim warning: '*It is unequivocal that human influence has warmed the atmosphere, ocean and land. Widespread and rapid changes in the atmosphere, ocean, cryosphere and biosphere have occurred.* . . . Since 2011 (measurements reported in AR5), concentrations have continued to increase in the atmosphere, reaching annual averages of 410 ppm for carbon dioxide (CO_2), 1866 ppb for methane (CH_4), and 332 ppb for nitrous oxide (N_2O) in 2019. Land and ocean have taken up a near-constant proportion (globally about 56% per year) of CO_2 emissions from human activities over the past six decades, with regional differences (high confidence). Each of the last four decades has been successively warmer than any decade that preceded it since 1850' (emphasis added, 2021, p. 5).

Figure 1.1 excerpted from AR6 SPM provides a grim schematic view of human-induced global warming. There has been ample global policy recognition of the costs of climate change being inequitably borne by the global poor. In 2009, a UN Report entitled 'The Impact of Climate Change on Development Prospects of the Least Developed Countries and Small Island Developing States' noted that: 'Climate change affects all, but it does not affect us equally. Nor do we possess the same capacity to respond to its challenges. As is often the case, the most vulnerable countries - particularly the Least Developed Countries and Small Island Developing States - find themselves in the worst situation again'. The report went on to call for the 2009 UN Climate Conference in Copenhagen – billed as the Conference to 'seal the deal' – to 'produce tangible commitments for the benefit of the most vulnerable to climate change' (UN 2009, p. 4). But more than a decade later, as the dust settles from the 2021 UN Glasgow Climate Conference, a comprehensive and effective climate deal focused on addressing the needs of the poorest communities and countries is yet to be sealed. Timperly has provided compelling evidence of the broken $100 billion promise of climate finance that was pledged by richer countries in 2009 to help poorer countries adapt to climate change (2021). Also in 2009, the International Energy Agency (IEA) estimated that for each year that passes the window for action on emissions reductions over a given period becomes narrower. It calculated that each year of delay before moving onto the emissions path consistent with a 2 °C temperature threshold would add approximately $500 billion to the global incremental investment cost of $10.4 trillion for the period 2010–2030 and, more significantly, that a delay of just a few years would likely render that goal completely out of reach (IEA 2009, p. 52).

Human influence has warmed the climate at a rate that is unprecedented in at least the last 2000 years

Changes in global surface temperature relative to 1850–1900

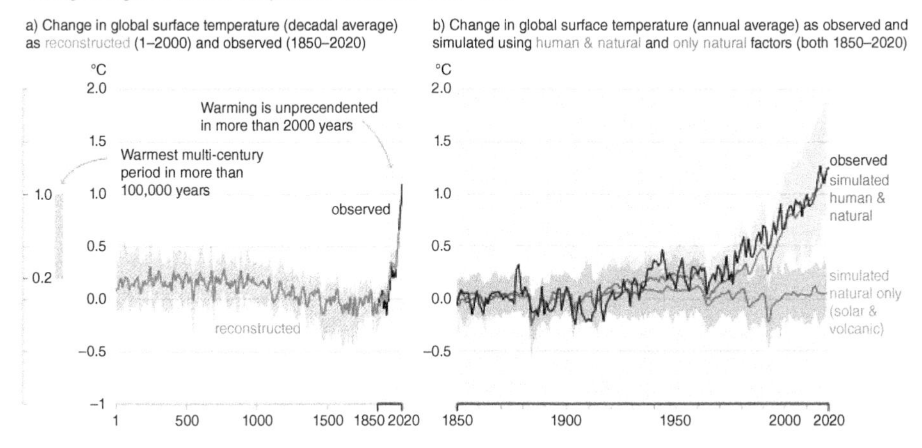

Figure SPM.1: History of global temperature change and causes of recent warming.

Panel a): Changes in gobal surface temperature reconstructed from paleoclimate archives (solid grey line, 1–2000) **and from direct observations** (solid black line, 1850–2020), both relative to 1850–1900 and decadally averages. The vertical bar on the left shows the estimated temperature (*very likely* range) during the warmest multi-century period in at least the last 100,000 years, which occurred around 6500 years ago during the current interglacial period (Holocene). The Last Interglacial, around 125,000 years ago, is the next ago during the current for a period of higher temperature. These past warm periods were caused by slow (multi-millennial) orbital variations. The grey shading with white diagonal lines shows the *very likely* ranges for the temperature reconstructions.

Panel b): Changes in global surface temperature over the past 170 years (black line) relative to 1850–1900 and annually averaged, compared to CMIP6 climate model simulations (scc Box SPM.1) of the temperature response to both human and natural drivers (brown), and to only natural drivers (solar and volcanic activity, green). Solid coloured lines show the multi-model average, and coloured shades show the very likely range of simulations. (see Figure SPM.2 for the assessed contributions to warming).

{2.3.1, 3.3, Corss-Chapter Box 2.3, Cross-Section Box TS.1, Figure la, TS.2.2}

Figure 1.1 History of global temperature change (IPCC/AR6 SPM). *Source:* IPCC (2021) AR6 WGI SPM, p. 7.

In 2016, the World Bank's report 'Shock Waves: Managing the Impacts of Climate Change and Poverty' found that poor people are disproportionately affected by climate impacts and that climate change could push an additional 100 million people into extreme poverty by 2030 (World Bank 2016). Meanwhile, the social upheaval and dislocation of millions living at the cross-roads of extreme poverty and climate vulnerability was documented in 'Groundswell - Preparing for Internal Climate Migration', which focused on Sub-Saharan Africa, South Asia and Latin America (representing 55% of the developing world's population). The report found that 'over 143 million people' could be forced to migrate as a result of SLR, water and food insecurity. The report's key message that '*poorest and most climate vulnerable areas are the hardest hit*, and that vulnerable lives have '*the fewest opportunities to adapt locally or to move away from risk and, when moving, often do so as a last resort*' while those who are '*even more vulnerable, will be unable to move, trapped in increasingly unviable areas*' remains haunting (emphasis added, Rigaud et al. 2018, p. xxi). The grave impacts of climate change on vulnerable households, cities and countries has also been clearly signalled by a 2018 IPCC Special Report, 'Global Warming of 1.5°C':

- '*Populations at disproportionately higher risk of adverse consequences* with global warming of 1.5°C and beyond include *disadvantaged and vulnerable populations, some indigenous peoples, and local communities dependent on agricultural or coastal livelihoods.*
- *Regions at disproportionately higher risk* include Arctic ecosystems, dryland regions, small island developing states, and Least Developed Countries.
- *Poverty and disadvantage are expected to increase* in some populations as global warming increases; limiting global warming to 1.5°C, compared with 2°C, could reduce the number of people both exposed to climate-related risks and susceptible to poverty by up to several hundred million by 2050.
- Exposure to multiple and compound climate-related risks increases between 1.5°C and 2°C of global warming, with *greater proportions of people both so exposed and susceptible to poverty in Africa and Asia*' (emphasis added, IPCC/SPM 2018, pp. 9–10).

Close to 20 years after the adoption of the first UNGA resolution, at a 2007 UN High Level Climate Summit, the Maltese Prime Minister Lawrence Gonzi warned that the UNGA needed 'new mechanisms' for tackling the issue of global warming and its repercussions 'in a more cohesive and concerted manner', or 'future generations would pay the price' (UN News Centre 2007). Today, it is hard to duck around the evidence that current and future generations are indeed paying the price with much heavier morbidity and disease burdens exerted on those who are less able to withstand extreme climatic adversities, and have contributed the least in terms of per capita GHG emissions. Recognition that poorer and more marginalized households, communities, cities and countries will pay the harshest price as a result of their inabilities to withstand climatic impacts has been well documented (African Development Bank 2003; Roberts and Parks 2007; Bullard and Wright 2009).

In dealing with the adversities associated with climate change including dislocation and loss of life, the global policy community has long known that poorer, smaller and more vulnerable communities and countries will be left reeling as they lack resiliency and safety nets that allow for recovery and remediation. The existential threat posed by SLR was forcefully articulated by Prime Minister Lee Hsien Loong of Singapore – a country lying no more than 4 metres above the mean sea level. At the 2019 National Day Rally Lee stated: 'We should treat climate change defences like we treat the Singapore Armed Forces – with utmost seriousness. Work steadily at it, maintain a stable budget year after year . . . do it over many years and several generations. Both the Singapore Armed Forces and climate change defences are existential for Singapore. These are life and death matters. Everything else must bend at the knee to safeguard the existence of our island nation'. In adapting to climate change, Lee went on to point out that Singapore would borrow inspiration from Dutch 'Polders' – reclaimed land enclosed by dykes – that require pumps to remove excess water and are protected by sea-walls (Chin 2019). In stark contrast to wealthier Singapore, which has marshalled resources to address the problem, coping with extreme climatic events will harshly impact poorer countries and communities especially small island developing states (SIDS) and Least Developed Countries (LDCs). But, layered upon the challenge of coping with climatic adversities such as SLR, and evidenced in plain sight yet remaining largely unaddressed for millions of lives is the world's single largest environmental health risk – air pollution.

On 25 March 2014, the World Health Organization (WHO) – the world's primary global organization mandated to respond to public health challenges – reported in a press release for the first time that 7 million people died – one in eight of total global deaths – as a result of air pollution exposure (based on 2012 WHO data). As the WHO put it: 'This finding more than doubles previous estimates and confirms that air pollution is now the world's largest single environmental health risk. Reducing air pollution could save millions of lives' (WHO Media Centre 2014). But the grim reality is that WHO's guidance on the related risks of climate change and air pollution predated its 2014 warning about air pollution. In 1997, just five years after the adoption of the historic UNFCCC, a WHO report entitled 'Health and Environment in Sustainable Development' referenced key environmental threats to human health which included: 'Water pollution from populated areas, industry and intensive agriculture; urban air pollution from motor cars, coal power stations and industry; climate change; stratospheric ozone depletion and transboundary pollution' (1997, p. 2). In 2015, the 68th session of the WHO Assembly adopted a resolution entitled 'Health and Environment: Addressing the health impacts of air pollution': 'Noting with deep concern that indoor and outdoor air pollution are both among the leading avoidable causes of disease and death globally, and the world's largest single environmental health risk. Acknowledging that 4.3 million deaths occur each year from exposure to household (indoor) air pollution and that 3.7 million deaths each year are attributable to ambient (outdoor) air pollution, at a high cost to societies; Aware that exposure to air pollutants, including fine particulate matter, is a leading risk factor for non-communicable diseases in adults, including ischaemic heart disease, stroke, chronic obstructive pulmonary disease, asthma and cancer, and poses a considerable health threat to current and future generations; *Concerned that half the deaths due to acute lower respiratory infections, including pneumonia in children aged less than five years, may be attributed to household air pollution, making it a leading risk factor for childhood mortality. Further concerned that air pollution, including fine particulate matter, is classified as a cause of lung cancer by WHO's International Agency on Research for Cancer'* (emphasis added, World Health Assembly 2015, p. 20). By 2016, the WHO found 80% of outdoor air pollution–related premature deaths were associated with ischaemic heart disease and strokes, 14% with chronic obstructive pulmonary disease (COPD) and acute lower respiratory infections and 6% with lung cancer (WHO 2016a). A landmark 2018 report by the WHO highlighted that 'climate change is the greatest health challenge of the 21st century and threatens all aspects of human society', and expressly highlighted climate change *as a* 'poverty multiplier' (WHO 2018, p. 10).

Finding answers to the interlinked challenges of climatic impacts, energy poverty and toxic levels of energy related air pollution is now a global imperative from a public health perspective, and especially so within cities given the global trend towards urbanization. In the wake of the respiratory-borne COVID-19 pandemic that extracted a huge toll on those already suffering endemic disease burdens and unable to access health services, there is an urgency to understanding how exposure to air pollution and the lack of access to clean energy and health services intersect. This urgency for linked action on clean air and clean energy access is especially relevant to the future of the world's most populous and polluted cities. In his 2019 speech to the C-40 World Mayors Summit, the UN Secretary General highlighted the tremendous potential for cities as the loci for action on clean air, clean energy and climate resilience. He specifically referenced that cities consume more than two-thirds of the world's energy and account for more than 70% of global carbon dioxide emissions. His call for city-based action on climate and clean air is exactly what is needed and yet is glaringly absent in long-standing UN global goal silos on climate and clean energy: 'Friends, *cities are where the climate battle will largely be won or lost.* With more than half the world's population, cities are on the frontlines of sustainable . . . and inclusive development. *With air pollution a grave and growing issue, people look to you to champion better urban air quality. With environmental degradation driving migration to urban areas, people rely on you to make your cities havens for diversity, social cohesion and job creation. You are the world's first responders to the climate emergency*' (emphasis added, UN Press Release 2019).

It is indeed time to more effectively factor in the loci of cities as the frontline actors for climate and clean air responsive action, particularly in Asia and Africa. The global community has run out of excuses for delaying integrated action on polluting forms of energy and toxic levels of air pollution that worsens the lives of those least responsible for causing the problem of historical GHGs. Climate change has been explicitly and consistently highlighted as a 'threat multiplier' by numerous global entities, including the US Department of Defense which issued a publicly available 2014 warning: 'Rising global temperatures, changing precipitation patterns, climbing sea levels and more extreme weather events will intensify the challenges of global instability, hunger, poverty and conflict' and 'will likely lead to food and water shortages, pandemic diseases, disputes over refugees and resources, and destruction by natural disasters in regions across the globe' (US Department of Defense 2014, foreword, WEF 2014). The World Economic Forum's (WEF) 2019 Global Risks Report placed climate change as a primary risk with compounding/multiplier effects on human ill health, food insecurity, biodiversity loss bluntly highlighted the 'climate catastrophe' ahead: 'Of all risks, it is in relation to the environment that the world is most clearly sleepwalking into catastrophe' (2019a, p. 15). It is time to see that the propensity for extended sleepwalking into the entwined climate and air pollution crises is based on the illogic of having UN global silos that segregate increasing access to clean energy, curbing air pollution and addressing climate change.

By examining existing UN-negotiated goal silos on climate change and sustainable energy, the aim is to evidence the need for integrated and localized action on the inherently linked climate and air pollution crises. Linkages between climate vulnerability, poverty and exposure to fossil fuel air pollution have been well documented by numerous UN and global entities, but persistent global goals/negotiations silos that segregate energy for sustainable development and poverty reduction goals from climate change goals have impeded the practice of integrated partnerships (Cherian 2015). Non-nation-state actors (NNSAs) such as local/municipal actors and the clean energy sector are principal responders to the integrated frontline on clean air, access to clean energy for all and climate change in urban areas that cope with toxic levels of air pollution. The aim of this chapter and this book is to focus on new frameworks for action by NNSAs on curbing short-lived climate pollutants (SLCPs) which are associated with particulate matter (PM) pollution and offer public health and environmental benefits, but are not factored into

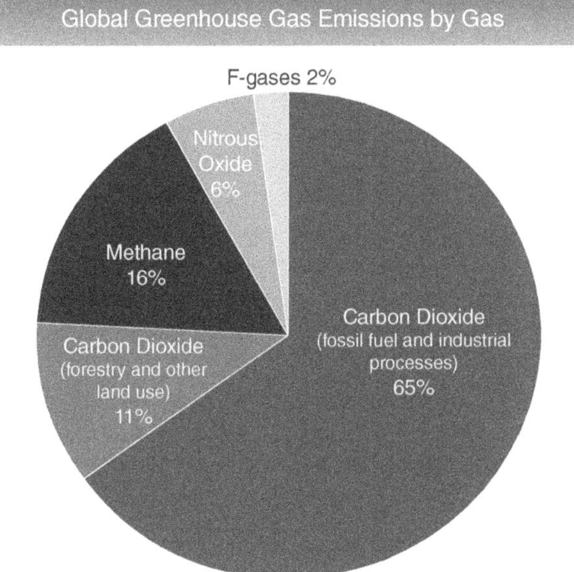

Figure 1.2 Global GHG emissions by gas. *Source:* US EPA website (2021). (Note: Details about the sources included in these estimates can be found in the *Contribution of Working Group III to the Fifth Assessment Report of the Intergovernmental Panel on Climate Change*).

the historic PA. To be clear from the outset, the need for linked action on clean air, clean energy access for the poor and climate change via NNSA-driven partnerships and modalities cannot be viewed as a means to replace comprehensive global GHG reductions. New forms of city-based measures should be seen as essential for addressing the more immediate imperatives of public health and reduced morbidity burdens associated with PM air pollution.

According to the US Environmental Protection Agency (EPA), Figure 1.2 (derived from IPCC's AR5), the key GHGs emitted as a result of human activities at the global level are as follows:

- **Carbon Dioxide (CO_2)** Fossil fuel use is the primary source of CO_2. CO_2 can also be emitted from direct human-induced impacts on forestry and other land use, such as through deforestation, land clearing for agriculture and degradation of soils. Likewise, land can also remove CO_2 from the atmosphere through reforestation, improvement of soils and other activities.
- **Methane (CH_4)** Agricultural activities, waste management, energy use and biomass burning all contribute to CH_4 emissions.
- **Nitrous Oxide (N_2O)** Agricultural activities, such as fertilizer use, are the primary source of N_2O emissions. Fossil fuel combustion also generates N_2O.
- **Fluorinated gases (F-gases)** Industrial processes, refrigeration and the use of a variety of consumer products contribute to emissions of F-gases, which include hydrofluorocarbons (HFCs), perfluorocarbons (PFCs) and sulphur hexafluoride (SF_6).

Mitigating GHGs should be viewed as critical to the future of human development. On 11 May 2019, sensors at the Mauna Loa Observatory (the premier atmospheric baseline station of the US National Oceanic and Atmospheric Administration [NOAA]) confirmed, for the first time in recorded history, that monthly concentrations of CO_2 breached the 400 parts per million (ppm) threshold (NOAA/ESRL website 2019). Two of the world's leading scientific organizations, the Royal Society (UK) and the National Academy of Sciences (US) provided a sobering assessment that even if emissions of CO_2 stopped altogether, '. . . surface temperatures would stay elevated for at least a thousand years, implying a long-term commitment to a warmer planet due to past and current emissions. . . . *The current CO_2-induced warming of Earth is therefore essentially irreversible on human timescales.* The amount and rate of further warming will depend almost entirely on how much more CO_2 humankind emits' (emphasis added, 2020, p. 22). But, here it is also important to point out that there remains an imbalance between aggregate CO_2 emissions by countries and CO_2 emissions based on the population of each country (i.e. per capita emissions). Interestingly, the IEA over a decade ago also pointed out that GHG emissions from developing countries are likely to exceed those of developed countries within the first half of this century (IEA 2009) Although aggregate GHG emissions have increased dramatically over time, the major countries responsible for the largest aggregate shares of emissions have not changed significantly. Seven countries have consistently been amongst the top emitters on an annual basis and have driven emissions growth since 1850, namely, the United States, the United Kingdom, Germany, France and Russia and more recently India and China. By way of comparison, three-quarters of the 50 lowest emitting countries in 2014 are the same countries as in 1850 (Lebling et al. 2019). It is the stark distinction between per capita emissions seen in conjunction with the burden of disease accruing fossil fuel related air pollution that merits attention.

The inequitable morbidity and disease costs borne by those who rely on polluting solid fuels and who are exposed to toxic levels of fossil fuel related air pollution occurs within a global context where GHG emissions are unmistakably on the rise. The UN Environment Programme (UNEP) has produced an annual Emissions Gap Report for 10 years detailing where GHG missions are headed in comparison to where they should be to avoid the worst impacts of climate change. The 2019 Emissions Gap Report provided a stark reminder that GHG emissions continue to escalate, despite numerous scientific warnings and political commitments: 'There is no sign of GHG emissions peaking in the next few years; every year of postponed peaking means that deeper and faster cuts will be required. By 2030, emissions would need to be 25 per cent and 55 per cent lower than in 2018 to put the world on the least-cost pathway to limiting global warming to below 2°C and 1.5°C respectively' (2019, p. xiv). But the 2020 Emissions Gap report summary is even more sombre in its assessment: 'Are we on track to bridging the gap? Absolutely not. Although 2020 emissions will be lower than in 2019 due to the COVID-19 crisis and associated responses, GHG concentrations in the atmosphere continue to rise, with the immediate reduction in emissions expected to have a negligible long-term impact on climate change. However, the unprecedented scale of COVID-19 economic recovery measures presents the opening for a low-carbon transition that creates the structural changes required for sustained emissions reductions. Seizing this opening will be critical to bridging the emissions gap' (2020, p. iv). Whether this opportunity to build back better, cleaner and greener is actually seized is quite literally up in the air, because on 20 April 2021, the IEA announced that in spite of COVID lockdowns, global energy related CO_2 emissions are on course to surge by 1.5 billion tonnes in 2021 – the second-largest increase in history – reversing most of last year's decline caused by the COVID-19 pandemic. But the real question is what is being done and what will happen to those who are both climate vulnerable and lack access to clean energy in the near future? IEA's *Global Energy Review 2021* has estimated that CO_2 emissions will increase by almost 5% in 2021 to 33 billion tonnes – biggest annual rise in emissions since 2010, during the carbon-intensive recovery from the global financial crisis. The key driver is coal demand, which is set to grow by 4.5%, surpassing its 2019 level and approaching its all-time peak from 2014, with the electricity sector accounting for three-quarters of this increase (IEA press release 2021) Figures 1.3 and 1.4 excerpted from the UNEP (2020a) Emissions Gap Report outline the growth in GHGs as well as the differences between absolute versus per capita emissions of the world's six top emitters.

To better understand the linkages between clean energy, air pollution and climate change, it is useful to point out that access to energy (sources, services and technologies) has widely viewed as essential to human development in all parts of the globe. Smil in his detailed history of how energy has shaped all aspects of human society from pre-agricultural foraging to fossil-fuel driven civilization argued that energy is the only universal currency that enables all things to get done (2017). Conversely, the lack of access to cost-effective, reliable energy as well as reliance on polluting solid fuels has been shown to impact negatively on income poverty, nutrition, gender and health inequalities, access to livelihoods and educational opportunities (Goldemberg et al. 1988; Sokona et al. 2004; Modi et al. 2006). The topic of 'energy poverty' was outlined very early on in the climate and energy global debate as a

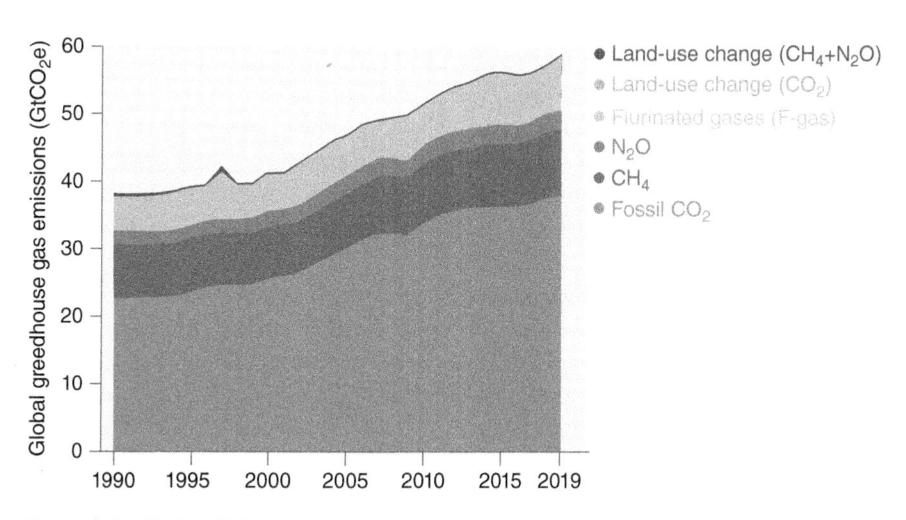

Figure 1.3 Global GHG emissions from all sources. *Source:* UNEP (2020b, p. v).

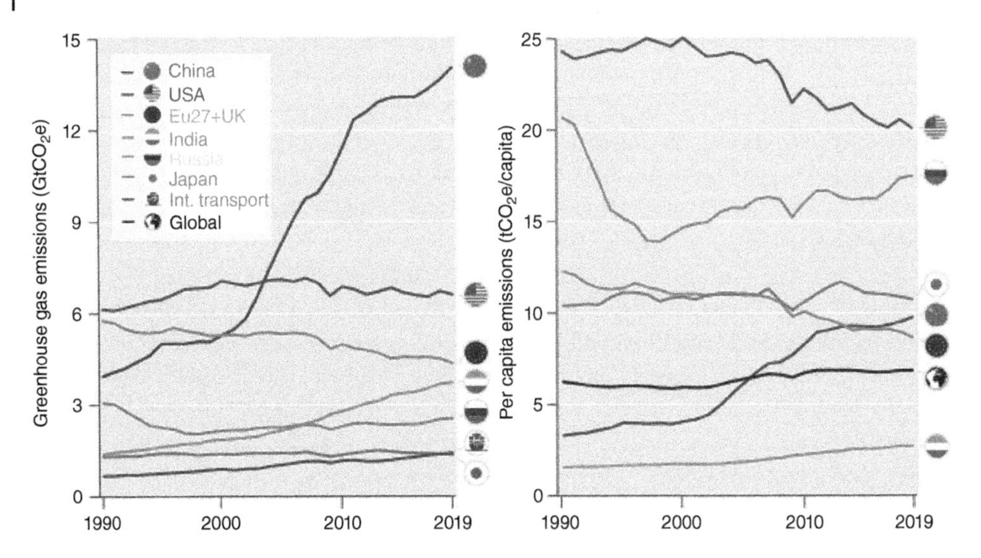

Figure 1.4 Absolute GHG emissions of the top six emitters (excluding Land Use Change emissions) and international transport (left) and per capita emissions of the top six emitters and the global average (right). *Source:* UNEP (2020b, p. vi).

causal link to income poverty, health and gender inequities by Goldemberg et al. in *Energy for a Sustainable World* (1988). The multidimensional linkages between energy and inequalities related to poverty, gender and urbanization were documented in an early 2000 joint report (prepared by two of the principal UN development agencies and the World Energy Congress), which called for energy issues to be 'brought to centre stage and given the same importance as other major global issues' (UNDP/UNDESA/WEC 2000, p. 40). Access to modern energy services has been deemed essential for socio-economic development and poverty reduction across countries, communities and households (Bazilian et al. 2010; Sovacool 2012). Srivastava et al. in their literature review of 'energy access' pointed to the fact that the terms 'energy poverty' and 'energy access' have been used interchangeably, and highlighted 'an important distinction' in that 'energy poverty is more amenable to be defined as a benchmark', while 'energy access can be presented as a continuum linked to different income levels reflecting different stages of development' (2012, p. 12). The recognized linkage between access to modern, cost-effective, energy and environmental objectives related to climate change are what make the concept of increasing access to sustainable energy for all a central element of the debate on sustainable development (Rehman et al. 2012, p. 27).

The WHO's previously referenced 1997 report express focus on poverty and inequity 'as two of the most important contributory factors to poor environmental conditions and poor health'; and its reference as to how integrated environmental and health policy interventions matter for air pollution abatement are worth recalling over 24 years later (1997, p. 6). The WHO emphasized that: '*Indoor air pollution can be particularly hazardous to health because it is released in close proximity to people. The most prominent source of indoor air pollution in developing countries is household use of biomass and coal for heating and cooking, usually involving open fires or stoves without proper chimneys.* A large number of studies in recent years have shown remarkable consistency in the relationship observed between changes in daily ambient suspended particulate levels and changes in daily mortality. Two different methods for estimating the total global mortality from suspended particulate air pollution exposures arrive at very similar total numbers (i.e. 3 million and 2.7 million), with indoor air pollution accounting for the vast majority of total deaths' (emphasis added, 1997, p. 15).

The linkages between the lack of access to clean energy and air pollution were referenced in UNDP's 2002 report entitled 'Energy for Sustainable Development: A Policy Agenda' which outlined the socio-economic costs of the energy and air pollution imbalance experienced by poorer households: 'Worldwide, 2 billion people are without access to electricity, and the same number use traditional fuels - fuelwood, agricultural residues, dung - for cooking and heating. Over 100 million women spend hours each day gathering and carrying fuelwood and water, and then spend additional hours cooking in poorly ventilated spaces. *The stoves used often lead to significant health impacts, through the generation of pollutants that expose women and children to air pollution corresponding to smoking two packs of cigarettes a day*' (emphasis added, UNDP 2002, p. 30). In 2009, a joint UNDP & WHO study pointed out that the number of people estimated to die every year due to household air pollution (HAP) from poorly combusted biomass fuels was anticipated to rise by 2030 to around 1.5 billion (WHO/UNDP 2009). The health consequences of using biomass for cooking, lighting and heating in poor households were found to be staggering. In reviewing the lack of access to clean energy in developing countries, the joint report noted that 44% of those who die each year from HAP are children, while women account for 60% of all adult deaths (2009).

The health and morbidity dangers of small PM emissions from the ineffective combustion of fossil and solid fuels have been well-documented and yet remain as unprecedented environmental health risks. As early as 2005, the WHO issued air quality guideline limits on PM pollutants that measured 2.5 µm or less – $PM_{2.5}$- and 10 µm or less – PM_{10},- based 'on the close, quantitative relationship between exposure to high concentrations of small particulates (PM_{10} and $PM_{2.5}$) and increased mortality or morbidity, both daily and over time'

Box 1.1 WHO air quality guideline values issued in 2005.

Particulate matter (PM): WHO air quality guideline values*

Fine particulate matter ($PM_{2.5}$)

- 10 µg/m^3 annual mean
- 25 µg/m^3 24-hour mean

Coarse particulate matter (PM_{10})

- 20 µg/m^3 annual mean
- 50 µg/m^3 24-hour mean

*According to the WHO, PM pollutants such as nitrates and black carbon penetrate deep into the lungs and into the cardiovascular system, posing the greatest risks to human health. $PM_{2.5}$ was identified as one of the principal air pollutants directly linked with causing strokes, ischaemic heart disease; chronic obstructive pulmonary disease and lung cancer.

Source: WHO (2005) Air Quality Guidelines.

(WHO (2005)) Box 1.1 specifies the WHO guidelines. In doing so, WHO made clear that: 'Small particulate pollution has health impacts *even at very low concentrations – indeed no threshold has been identified below which no damage to health is observed*. Therefore, the WHO 2005 guideline limits aimed to achieve the lowest concentrations of PM possible' (emphasis added, WHO 2005).

A decade later, in 2015, in response to air pollution being identified as a major global public health threat, the 194 WHO member states adopted the first World Health Assembly resolution to 'address the adverse health effects of air pollution'. Member states agreed on a Road Map aimed at providing an enhanced global response to the adverse health effects of air pollution. Amongst the main elements of this road map were the monitoring and reporting of air pollution and enhanced systems, structures and processes for monitoring and reporting health trends associated with air pollution (WHO 2015). The WHO's Secretariat resolution was grounded in a report which was stark and succinct as to the extensive and yet inequitable morbidity burden of air pollution:

- '*Air pollution is one of the main avoidable causes of disease and death globally*. About 4.3 million deaths each year, most in developing countries, are associated with exposure to household (indoor) air pollution. A further 3.7 million deaths a year are attributed to ambient (outdoor) air pollution.
- Even at relatively low levels air pollution poses risks to health, and because of the large number of people exposed it causes significant morbidity and mortality in all countries. However, although all populations are affected by air pollution, *the distribution and burden of consequent ill-health are inequitable*. The *poor and disempowered*, including slum dwellers and those living near busy roads or industrial sites, *are often exposed to high levels of ambient air pollution, levels that appear to be worsening in many cities. Women and children in households* that have to use polluting fuels and *technologies for basic cooking, heating and lighting bear* the brunt of exposure to indoor air pollution.
- *Most air pollutants are emitted as by-products of human activity*, including heat and electricity production, energy-inefficient transport systems and poor urban development, industry, and burning waste and brush or forests' (emphasis added, WHO 2015, pp. 1–2).

Figure 1.5 excerpted from the WHO's 2018 special report on climate and health provides global contributions of different sectors to the GHG emissions and shows that the sources of climate change and air pollution are broadly the same. It outlines why measures to curb the grave public health impacts of urban PM ambient pollution matter for millions of lives.

PM air pollution has been identified as a risk factor for many of the leading causes of death including heart disease, stroke, lower respiratory illnesses, lung cancer, diabetes and COPD (WHO 2016a; Cohen et al. 2017). WHO's 2018 report included a critical finding as to why addressing air pollution matters so much for the lives of so many across the globe: 'The most direct link between climate change and ill health is air pollution' (p. 16). A significant finding by the WHO was that if climate change is not mitigated, global income inequality could increase grossly especially since the health impacts of climate change are unevenly distributed (WHO 2018, p. 25). But more recently, a cross-national collaborative study spanning the world's leading medical research and scientific institutions has found that exposure to ambient PM pollution is 'several fold larger' than previously estimated, 'suggesting that outdoor particulate air pollution is an even more important population

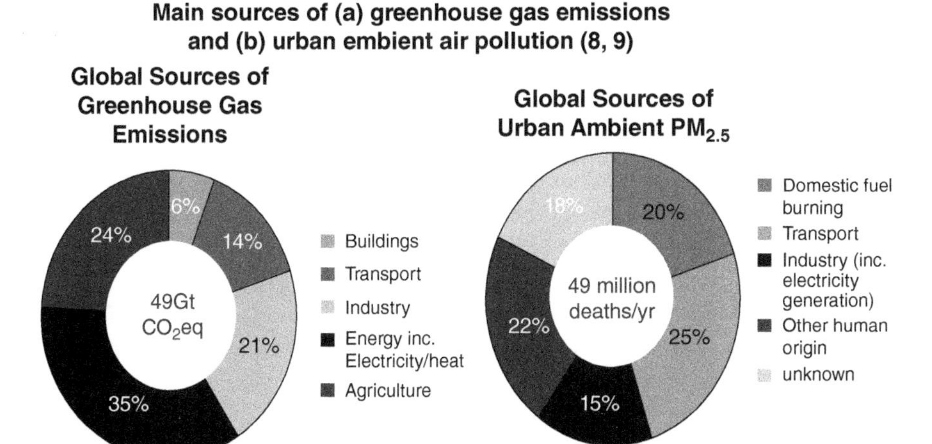

Figure 1.5 Main sources of GHG emissions and urban ambient air pollution.
Source: WHO (2018, p. 17).

health risk factor'. The study based on data from 41 cohort studies of outdoor air pollution from 16 countries – the Global Exposure Mortality Model (GEMM) estimated 8.9 million deaths in 2015, a figure 120% larger than the risk function used in the more established Global Burden of Disease (GBD) metrics. The researchers conclude that: 'The results suggest that $PM_{2.5}$ exposure may be related to additional causes of death than the five considered by the GBD and that incorporation of risk information from other, non-outdoor, particle sources leads to underestimation of disease burden, especially at higher concentrations' (Burnett et al. 2018, abstract).

The disconcerting human development reality is that the layering of climatic adversities, poverty, and marginalization, combined with the health impacts of exposure to air pollution, and the lack of access to non-polluting energy sources puts millions of lives at stake. There is irrefutable evidence that climate change cannot be ameliorated without massive reductions of CO_2 and other GHGs. But, a central argument advanced is that the global community as a whole has not done enough to address the linkages between climate change and air pollution and, in particular, that global policy silos on clean energy access and climate change have relegated the curbing on SLCPs associated with toxic levels of PM pollution to the margins. As evidenced by their name, and unlike CO_2 emissions which are longer lived in the atmosphere, SLCPs are pollutants that persist for a short duration/time span in the atmosphere. But, SLCPs have been found to be extremely potent in terms of their global warming potential compared to longer lasting GHGs such as CO_2. Reducing SLCPs like black carbon (BC), methane (CH_4 – which is also a recognized GHG) and tropospheric ozone (O_3) offers multiple benefits – short-term climate mitigation and public health.

A 2011 scientific assessment released by UNEP and the World Meteorological Organization (WMO) found that measures targeting SLCPs could achieve 'win–win' results for the climate, air quality and human well-being over a relatively short timeframe (UNEP/WMO 2011). This 2011 joint report entitled 'Integrated Assessment of Black Carbon and Tropospheric Ozone' found that reducing SCLPs 'now will slow the rate of climate change within the first half of the century' and that a small number of focused emission reduction strategies targeting BC and O_3 could immediately 'slow the rate of climate change within the first half of the century' (UNEP/WMO 2011, p. 1). BC is a major component of soot and exists as particles in the atmosphere. Emissions that result from the incomplete/inefficient combustion of solid fuels and traditional biomass are released as a mixture of health-damaging indoor air pollutants such as BC that have short atmospheric life spans and result in significant negative impacts on human health and climate change particularly at the national and regional levels. Ozone, at the ground level, is an air pollutant that is harmful to health and ecosystems and is a major component of urban smog. Throughout the troposphere, or lower atmosphere, ozone is also a significant GHG. Ozone is not directly emitted but is produced from emissions of precursors, of which CH_4 and carbon monoxide are of particular relevance in this study. By contrast, ozone in the stratosphere is considered to be beneficial in protecting life from the sun's harmful ultraviolet (UV) radiation (UNEP/WMO 2011, p. 7).

In 2012, the governments of Bangladesh, Canada, Ghana, Mexico, Sweden and the United States, along with UNEP, came together to initiate efforts to treat SLCPs as an urgent and collective challenge. Together, they formed the Climate & Clean Air Coalition (CCAC), which now comprises a voluntary partnership of governments and non-governmental organizations (NGOs), and aimed at including cities and other NNSAs, to support linked action on several challenges: climate, public health, energy access and food security. Box 1.2 excerpted from the CCAC website provides a succinct overview of the importance of curbing SLCPs.

Box 1.2 What are short-lived climate pollutants? (Excerpted from Climate & Clean Air Coalition)

What are short-lived climate pollutants?

Short-lived climate pollutants are powerful climate forcers that remain in the atmosphere for a much shorter period of time than carbon dioxide (CO_2), yet their potential to warm the atmosphere can be many times greater. Certain short-lived climate pollutants are also dangerous air pollutants that have harmful effects for people, ecosystems and agricultural productivity.

The short-lived climate pollutants black carbon, methane, tropospheric ozone, and hydrofluorocarbons are the most important contributors to the man-made global greenhouse effect after carbon dioxide, responsible for up to 45% of current global warming. If no action to reduce emissions of these pollutants is taken in the coming decades, they are expected to account for as much as half of warming caused by human activity.

Source: CCAC website 2021.

So, the first question is what is the UN-led global community doing about curbing the four principal SLCPs: BC, methane (CH_4), tropospheric ozone (O_3) and HFCs? A joint paper published by World Resources Institute (WRI) and Oxfam highlighted several key issues including that: 'early and ambitious action' to reduce SLCPs are 'essential to achieving the goals of' the PA and the Sustainable Development Goals (SDGs); actions to curb 'highly potent' SLCPs are critical in ensuring that the threshold of 1.5°C is not crossed 'which will affect poor and vulnerable communities first and worst'; and actions aimed at reducing SLCPs also 'delivers multiple benefits for development and human well-being, supporting efforts to improve health, enhance food security, and alleviate poverty'. Equally significantly, the paper also pointed out that: 'actions to mitigate these potent pollutants were often underrepresented in the first nationally determined contributions (NDCs) submitted by Parties to the Paris Agreement' (Ross et al. 2018, p. 1). But, the main challenge to date is that there is no global or regionally relevant pollution mitigation strategy or protocol that covers emissions of SLCPs, particularly BC that covers the needs of those countries and communities where the problem of BC-related PM pollution is most severely experienced. More recently, there has been an increasing push towards addressing the mitigation of HFCs within the context of the Kigali Amendment to the Montreal Protocol, and CH_4 is a recognized GHG that is technically covered within the context of the voluntary national reporting requirements of the PA. To be clear, the focus of this book is primarily on the linkages between PM pollution, health and climate inequities and more circumscribed to understanding the relevance of curbing BC.

From the immediate perspective of this book, the decades-old intergovernmental negotiations on climate change have consistently not addressed the issue of SLCPs like BC and O_3, both of which have major public health impacts that negatively constrain the lives of poorest communities within developing countries and cities, particularly those in Asia and Africa. BC is a solid particle or aerosol, not a gas, and results from emissions from gas and diesel engines, coal-fired power plants and other sources including solid biomass (solid fuels). Atmospheric BC concentrations have been related to anthropogenic activities, and BC emission reductions represent a potential mitigation strategy that could reduce global climate forcing from anthropogenic activities in the short term and slow the associated rate of climate change (UNEP/WMO 2011; Bond et al. 2013). Curbing SLCPs and thereby mitigating toxic levels of air pollution can offer short-term climate mitigation benefits, but what is often ignored and urgently needs to be highlighted is that increasing access to clean energy for all and transitioning to low carbon energy future provides valuable cost savings from human health perspectives in individual countries.

Reducing PM 2.5 emissions are critically important from a human health perspective, but what is often not reflected is that one of the principal components of PM 2.5 – BC emitted as a result of incomplete combustion of solid fuels has also been identified as an SLCP. While emission reductions of CO_2 are absolutely integral to addressing anthropogenic climate change, SLCPs like BC, a component of PM pollution, have been found to contribute directly to adverse impacts on human health, leading to premature deaths worldwide, and also negatively impact on agriculture and rainfall patterns. An extensive, landmark cross-national research assessment of the role of BC emissions specified that the predominant sources of BC are combustion related, namely, fossil fuels for transportation, solid fuels for industrial and residential uses and open burning of biomass. The assessment estimated BC to be: '. . . the second most important human emission in terms of its climate forcing in the present-day atmosphere; only carbon dioxide is estimated to have a greater forcing. Sources that emit black carbon also emit other short-lived species that may either cool or warm climate' (Bond et al. 2013, p. 5381).

Arguably, the nexus between increasing access to clean air, curbing fossil fuel related PM pollution and addressing climate change matters now more than ever for millions of lives that are amongst the poorest and most vulnerable in cities and communities. Highlighting the findings of new collaborative research on clean air and climate change conducted by Duke University, NASA and other entities on 48 contiguous US states and major cities before the US Congress, Drew Shindell (who heads CCAC Scientific Advisory Panel and is a lead author for the IPCC) provided a forceful case to prioritize climate change related health benefits of clean air in the US: '*Over the next 50 years, keeping to the 2°C pathway would prevent roughly 4.5 million premature deaths, about 3.5 million hospitalizations and emergency room visits, and approximately 300 million lost workdays in the US*. These large impacts reflect our updated understanding of the severe toxicity of air pollution and the dangers of heat exposure. Although it does not appear on death certificates it is indirectly responsible for a substantial fraction of heart diseases, including strokes, and respiratory diseases, including lower respiratory infections and chronic obstructive pulmonary disease. The economic value of these health and labor benefits is enormous. *The avoided deaths are valued at more than $37 trillion*. The avoided health care spending due to reduced hospitalizations and emergency room visits exceeds $37 billion, and the increased labor productivity is valued at more than $75 billion. *On average, this amounts to over $700 billion per year in benefits to the US from improved health and labor alone, far more than the cost of the energy transition*' (emphasis included, Shindell 2020).

The significance of the finding that reducing the severe toxicity of air pollution reduces morbidity, ill-health burdens and costs that not just offset but are actually far greater than the costs of transitioning to clean energy has paramount importance not just for the US, but for those households, cities and countries faced with exposure to hazardous levels of air pollution. It is now or never for addressing the layering of the double threats – energy related air pollution and climate vulnerabilities – both of which are pressing challenges for the broader UN-led global sustainable development agenda (SDA).

1.2 Time to Look Beyond UN SILOS on Sustainable Energy and Climate Change to Curb Toxic Air Pollution: Why Non-Nation-State Actors (NNSAs) Matter in the Fight for Clean Air, Clean Energy and Climate

Responding to the nexus between energy related air pollution, public health and climate mitigation becomes an unmistakably urgent policy imperative for all relevant stakeholders when seen in conjunction with UN's broader goal of poverty eradication by 2030. According to the World Bank, 10% of the world's population or 734 million people lived on less than $1.90 a day and experienced extreme poverty in 2015, compared to nearly 36% or 1.9 billion people in 1990. Extreme poverty was found to have declined to 8.6% in 2018. But the World Bank has recently estimated that the COVID-19 pandemic will have a disproportionate impact on the global poor and will 'push an additional 88 million to 115 million people into extreme poverty', with 'the total rising to as many as 150 million by 2021, depending on the severity of the economic contraction': 'Had the pandemic not convulsed the globe, the poverty rate was expected to drop to 7.9% in 2020' (World Bank 2020). It is within this development context of poverty reduction that the climate related health impacts and morbidity burdens of energy related air pollution needs to be understood. But there is now more than ever a pressing urgency to move beyond UN global goal silos on climate and sustainable energy towards more localized and integrated measures that are responsive to the needs of cities and communities on curbing fossil fuel related air pollution.

Since the adoption of the historic 1992 UNFCCC, which took eleven (11) intergovernmental negotiating sessions to be adopted, more than 25 annual cycles of intergovernmental negotiating meetings have occurred under the aegis of Conference of Parties (COPs) to the UNFCCC (Gupta 2014; Cherian 2012). Notwithstanding the rapid growth in global climate negotiations fora, securing a legally binding global climate change agreement still remains a quixotic goal. There has also been no shortage of policy prognostications and research related to climate negotiations ranging from game theory, regime, institutional governance analysis and climate justice perspectives (Haas et al. 1993; Luterbacher and Sprinz 2001; Giddens 2011; Stern et al. 2014; Bernard and Semmler 2015; Sjöstedt and Penetrante 2015). Robinson and Herbert (2001) outlined the need for integrating climate change early on with sustainable development needs. There is a considerable body of literature on sustainable development negotiations (Sachs 2015; Chasek et al. 2017; Kanie et al. 2017).

Enabling the active participation of NNSAs in global environmental governance to reduce institutional failure has been viewed as one of the most important tasks for policymakers seeking to improve the effectiveness of global governance (Hemmati 2001; Esty and Ivanova 2002). Gemmill and Bamidele-Izu focused on the role of NGOs and civil society actors in global environmental governance and identified five major roles that civil society can play in global environmental governance: (i) information collection and dissemination; (ii) policy development consultation; (iii) policy implementation; (iv) assessment and monitoring and (v) advocating environmental justice (2002, p. 78). While climate change has galvanized civil society stakeholders' actions ranging from the world's most powerful CEOs to student activists, what needs to be reflected upon is that the implementation of the UN segregated goals on climate change, clean energy and curbing air pollution neither yield results nor allows for dynamic partnerships between NNSAs such as local governments and the energy sector. The COVID pandemic with its grave public health precautions of social distancing and global travel shut-downs caused the necessary postponement of the 26th annual COP (COP-26). But, the reality is that the decades-old process of convening annual climate COPs which are now massive global juggernauts resulting in an alphabet soup of newly formed technical groups, ad hoc committees and escalating global air-travel related emissions associated with transporting participants to diverse cities where these massive global climate conferences/summits are held. Despite the push towards ensuring a carbon neutral COP-26, according to preliminary estimates by the host government's official carbon accounting firm-Arup- COP-26 was 'responsible for 102,500 tons of carbon dioxide equivalents, which was twice as much CO_2 equivalent associated with previous COPs held in Madrid in 2019 and Paris in 2015 and 'four times as much as the earlier climate summits in Copenhagen and Durban, South Africa, according to figures compiled by The Washington Post' (Booth and Stevens 2021).

UNEP's Executive Director, Inger Andersen in her foreword to the UNEP Emissions Gap Report 2020 pointed to the urgency of 'changes in consumption behaviour by individuals and the private sector' which included the need to redesign cities and make housing more efficient. But she was categorical about assigning blame for GHG emissions: 'The wealthy bear the greatest responsibility in this area. The combined emissions of the richest 1 per cent of the global population account for more than twice the combined emissions of the poorest 50 per cent. This elite will need to reduce their footprint by a factor of 30 to stay in line with the Paris Agreement targets' (UNEP 2020a, p. xiii). Ironically, evidence as to the rich versus poor emission imbalance was apparent when a slew of 400 private jets were used by uber-wealthy climate celebrities to attend COP-26 (Parsons 2021).

Feigning ignorance regarding carbon inequality is hard to justify. Over five years ago, climate change was inextricably linked inequality in Oxfam's Report, 'Extreme Carbon Inequality'. The report estimated 'carbon inequality' to be such that '*the poorest half of the global population* – around 3.5 billion people – are responsible for only around *10% of total global emissions* attributed to individual consumption, yet *live overwhelmingly in countries most vulnerable to climate change*. Around 50% of these emissions meanwhile can be attributed to the richest 10% of people around the world, who have average carbon footprints 11 times as high as the poorest half of the population, and 60 times as high as the poorest 10%. *The average footprint of the richest 1% of people globally could be 175 times that of the poorest 10%*' (emphasis added, Oxfam 2015, p. 1). Calling attention to the massive scale of loss, devastation and dislocation imposed by climate change, the UN Special Rapporteur on extreme poverty and human rights pointed out: 'Perversely, while people in poverty are responsible for just a fraction of global emissions, they will bear the brunt of climate change, and have the least capacity to protect themselves . . . We risk a "*climate apartheid*" scenario where the wealthy pay to escape overheating, hunger and conflict while the rest of the world is left to suffer'(emphasis added, UN News 2019a).

Another recent example of this 'climate apartheid' scenario albeit by the well-intentioned wealthy was the 2019 convening of a by-invitation only, celebrity-focused, three-day Google camp on climate change. According to an article in Ecowatch, this event which cost upwards of $20 million meant that the Palermo airport had to be readied 'for the expected arrival of 114 private jets not to mention private helicopters, yachts and limousines used for the transportation of the various guests'. As the article notes, the modalities used to convene this event were in stark contrast to the event's promised mission, 'as a flight from New York to Palermo, Sicily, generates around 4.24 metric tons of CO2', which is 'a lot of carbon for just a few people. And, that doesn't include the greenhouse gasses emitted by the 2,300 horsepower diesel-engine private yachts' that several attendees used (Davidson 2019). The imbalance in per capita GHG emissions is unambiguous across and within countries and cities and makes the dissonance between those who lack access to non-polluting energy sources yet also contribute the least in term of per capita emissions versus those who proclaim the need for climate activism by jetting around the world harder to ignore.

IPCC's AR6 Working Group 1 SPM leaves little room for equivocation or doubt as to the global urgency of climate action. The climate response to a broader range of GHGs, land use and air pollutant futures than assessed in AR5 are considered consistently via five new emissions scenarios and include:

'Scenarios with high and very high GHG emissions (SSP3-7.0 and SSP5-8.5) and CO2 emissions that roughly double from current levels by 2100 and 2050, respectively, Scenarios with intermediate GHG emissions (SSP2-4.5) and CO2 emissions remaining around current levels until the middle of the century, and

Scenarios with very low and low GHG emissions and CO2 emissions declining to net zero around or after 2050, followed by varying levels of net negative CO2 emissions23 (SSP1-1.9 and SSP1-2.6)' (2021, p. 15).

The SPM highlighted that: 'Based on the assessment of multiple lines of evidence, global warming of 2°C, relative to 1850–1900, would be exceeded during the 21st century under the high and very high GHG emissions scenarios considered in this report (SSP3-7.0 and SSP5-8.5, respectively). Global warming of 2°C would extremely likely be exceeded in the intermediate scenario (SSP2-4.5). Under the very low and low GHG emissions scenarios, global warming of 2°C is extremely unlikely to be exceeded (SSP1-1.9), or unlikely to be exceeded (SSP1-2.6)' (2021, p. 18). Making assumptions about the global political will to move towards a very low and zero GHG emissions scenario currently strain credulity, but what needs to be publicly acknowledged is the scale of devastation that awaits millions across the world who lack the basic resources to adapt towards compounded climate, health and humanitarian crises.

Close to seven years after UN member states universally pledged to undertake ambitious action to implement the PA on climate change, it is long overdue time to ask what exactly is being done by the UN global community (including the global elite) to address the health and morbidity burdens that millions face at the harsh intersection between climate adversities and fossil fuel related air pollution. What set the 2015 PA apart from all prior UN climate resolutions and agreements was that it was the first inclusive, yet completely voluntary global climate change accord that covers all member states and therefore was markedly different from its parent treaty the 1992 UNFCCC and the 1997 KP to the UNFCCC. But, what this voluntary and non-legally binding implementation of the PA means is that the scaling up of the PA's commitments hinges entirely on the level of 'ambition' of the voluntary national pledges of climate action referred to as 'independent Nationally Determined Contributions' (NDCs). In other words, the future implementation record of global climate action within the PA depends entirely on the individual national commitments of the world's largest aggregate GHG emitting countries. What is unambiguous, however, is that the implementation of the PA is directly linked to global efforts to reduce poverty, food insecurity and human rights inequities, as reflected in the introductory or 'chapeau' section of the PA: 'Emphasizing the intrinsic relationship that climate change actions, responses and impacts have with equitable access to sustainable development and eradication of poverty. . . Acknowledging that climate change is a common concern of humankind, Parties should, when taking action to address climate change, respect, promote and consider their respective obligations on human rights, the right to health, the rights of indigenous peoples, local communities, migrants, children, persons with disabilities and people in vulnerable situations and the right to development, as well as gender equality, empowerment of women and intergenerational equity' (UNFCCC 2015, p. 21).

The PA's explicit recognition of the eradication of poverty, the right to health, along with the human rights and the rights of historically marginalized and vulnerable communities is critical in locating the argument that heavy reliance on polluting forms of solid fuels (biomass), and access to clean air should be considered as integral to and not separated from climate responsive actions. The PA entered into force on 4 November 2016 after ratification by three of the world's largest aggregate GHG emitters – China,

US and India. On 1 June 2017, the world's second-largest aggregate GHG emitter announced its decision to unilaterally withdraw from the PA and called for new negotiations. This call was rejected by several UN donor countries, including France and Germany who said that the PA 'cannot be renegotiated' (Stokols 2017). Meanwhile, the Foreign Minister of India (the world's third-largest aggregate GHG emitter) pushed back against the US claim that India's ratification of the PA was 'contingent upon receiving billions and billions and billions of dollars of foreign aid from developed countries' by categorically stating: 'India signed the Paris Agreement not because of any pressure or out of greed. We are committed to the environment and this commitment is 5,000 years old . . . I clearly dismiss both accusations' (Indian Express 2017), But, India and China pushed back on the phrase 'phase out' in relation to coal and moved to replace it with 'phase down' in the recently concluded COP-6 Glasgow deal. This move by India and China resulted in COP-26 Chair, Alok Sharma stating that the deal struck in Glasgow was a 'fragile win' and that China and India would need to 'justify their actions to nations that are more vulnerable to the effects of global warming' (Cursino and Faulkner, 2021). The decades-old reality that veterans of global climate change negotiations will attest to are that consensus-based intergovernmental outcomes secured after two weeks of late-night, back-room discussions, are not governed by the need to protect the smallest and most vulnerable countries. Even before, the 2017 withdrawal of the US from the PA, the WEF highlighted the collective failure of the intergovernmental climate negotiations: 'The risk of global governance failure which lies at the heart of the risk map, is linked to the risk of climate change. Negotiations on climate change mitigation and adaptation are progressing by fits and starts, perpetually challenged to deliver a global legal framework' (WEF 2014, pp. 21–22).

Ciplet et al. (2015) demonstrated the international climate change policymaking arena has failed to act conclusively to fully address climate change. Harlan et al. (2015) provide compelling reasons for climate change being seen as a 'justice issue' given that marginalized communities and countries use considerably less fossil fuel energy while adverse climatic impacts are experienced vastly differently by the rich and poor as are climate responsive measures. As it turns out, the 2013 COP held in Warsaw, Poland, was sponsored by a range of energy polluting industries and ran alongside the International Coal Summit (which featured a keynote address by the then head of the UNFCCC) (Goodman 2013); and the 2018 COP was held in Katowice, also located in Poland and home to the largest coal mine in the European Union (EU) (Mathiesen 2017). Speaking to the 2018 Katowice COP assembly, the young Swedish climate change activist, Greta Thunberg provided a bleak assessment of UN climate negotiations: 'We cannot solve a crisis without treating it as a crisis. We need to keep the fossil fuels in the ground, and we need to focus on equity . . . We have not come here to beg world leaders to care. You have ignored us in the past and you will ignore us again. We have run out of excuses and we are running out of time' (Sutter and Davidson 2018). Speaking on 29 January 2019, Thunberg upped the urgency when she told yet another global assemblage gathered at WEF in Davos: 'We are facing a disaster of unspoken sufferings for enormous amounts of people. And now is not the time for speaking politely or focusing on what we can or cannot say . . . I want you to panic. I want you to feel the fear I feel every day. I want you to act. I want you to act as you would in a crisis. I want you to act as if the house is on fire, because it is' (WEF 2019b).

Intergovernmental climate negotiations have not yielded conclusive agreement on key elements related to the PA related to climate adaptation needs of the smallest and poorest countries. The return to the PA negotiations by the second-largest aggregate emitter sent a clear message, but the question that persists is whether the textual confines of the two-week-long cycle of intergovernmental negotiations will suffice in terms of integrated action clean air and clean energy for all? Tense and protracted climate negotiations have been the norm for decades. The final, plenary session of the 2018 24th UN Climate Conference in Poland had to be postponed no less than six times. In the waning days of 2019, climate activists across the world were left disappointed that the 25th UN annual climate negotiations meeting ended in utter disarray. Despite extending the two-week Madrid conference for an additional two days, and after 25 consecutive annual climate conferences convened in diverse cities, countries failed to deliver essential outcomes such as setting a rulebook for the PA and designing a global carbon market. In spite of extending the two-week 2019 25th COP in Madrid, Parties to the PA could only agree to a 'watered-down text', which 'reflects a failure to agree on the key outcomes that were needed at the summit: setting a rulebook for the Paris Agreement and designing a global carbon market' (Keating 2019). In its summary of COP-25, the UNFCCC Secretariat sounded hopeful that: 'Despite Parties falling short of agreeing on issues related to Article 6 of the Paris Agreement and on the launch of cooperative instruments – essential tools for enhancing the efficiency of mitigation efforts and increasing finance for adaptation – most technical issues relating to the market-based and non-market approaches under Article 6 were resolved in 2019. COP 26 will be tasked with sealing the deal on Article 6' (UNFCCC 2020, p. 24).

The idea that comprehensive action on climate change and clean energy for the poorest and most vulnerable can be catalyzed by well-intentioned celebrities who fly around imploring the world, or by the glacial pace of nation-state-driven textually based negotiations, while exposure to toxic levels of fossil fuel related air pollution for those cannot afford air filters/purifiers continue unabated is hard to justify. Decades ago, the UN sustainable development community universally agreed on the primacy of poverty eradication as a global goal and reaffirmed the same in both the 2030 SDA and the PA. The UN-led global community has to face up to the facts that action on climate change needs to first and foremost address the needs of the poor and vulnerable, and that avoiding linked action to curb fossil fuel related air pollution contravenes the logic of its ambitious SDA. Today, it is hard to deny that our shared human propensity to pollute extends into the air we breathe, the food and water we consume, and is especially pernicious for millions who live at the intersection between energy related air pollution and adverse climatic impacts. It is precisely this linkage between inefficient and polluting energy sources and air pollution that merits urgent attention not just by the UN member states and organizations focused on the SDA and PA but also by NNSAs including the local and municipal governments, civil society groups, the energy private sector, as well as city and regionally based entities.

The UN's SDA is anchored by 17 SDGs, including two entirely separate goals on affordable and clean energy (SDG 7) and climate change (SDG 13) (see Figure 1.6). What is unambiguous is that the SDGs and the PA both make poverty eradication by 2030 a central priority.

Figure 1.6 SDGS in the UN's 2030 sustainable development agenda. *Source:* UN website, About the Sustainable Development Goals (2021).

The SDA's 'pledge that no one will be left behind' is expressly focused on poverty eradication, but it includes two separate SDGs on sustainable energy and climate change: *SDG 7:* 'Ensure access to affordable, reliable, sustainable and modern energy for all' (SDG 7); and *SDG 13:* 'Take urgent action to combat climate change and its impacts'. What makes the SDA and the PA historically unique is that they share a priority on poverty eradication and are intended to be implemented in an integrated manner. Together, these two global agreements were envisaged to dramatically improve human well-being and sustainable development in all countries. But the problem is that a detailed examination of two separate intergovernmental negotiating tracks on energy for sustainable development and climate change that were used to arrive at the SDGs and the PA revealed that the existence of segregated policy goals and targets on sustainable energy and climate change and an overall lack of integrated action on energy related air pollution (Cherian 2015).

With the 2030 full implementation deadline looming for the SDGs and the PA, it is time to ask what is being done in terms of integrated action to curb SLCPs and reduce public health and morbidity burdens associated with energy related air pollution exposure? Both the PA and the SDGs are voluntary agreements with no legal penalties or obligations for individual countries to meet commitments. It is time to ask whether UN member states should be the sole arbiters of global climate goals based on tense intergovernmental textually driven negotiations that have to date not considered actions to curb SLCPs and toxic levels of air pollution. The role of NNSAs – including local/

municipal governments and the private sector – are arguably important in addressing the enormity of the challenges associated with intersecting challenges of climate change, lack of access to clean energy and air pollution. The gravity of the linked crises of climate change and air pollution crises necessitates looking beyond segregated UN policy and goal silos on clean energy and climate change and asking:

- How exactly has air pollution been addressed at the global and regional level within the UN's broad SDA?
- Is there an integrated global policy nexus on clean air, clean energy and climate action that is responsive to the needs of those most impacted by energy related air pollution, lack of access to clean energy and climate vulnerabilities?
- What is the role for NNSA partnerships including regional and city-based modalities that integrate action on clean air, clean energy and climate challenges in countries like India where the scale and scope of increasing access to clean energy and curbing air pollution intersect for millions of lives?

The identification of air pollution as the world's single largest environmental health risk by the WHO merits a careful examination as to how globally responsive action on air pollution has been integrated with SDG 7 (increasing access to clean energy for all) and SDG 13 (climate). Are segregated policy goals on sustainable energy for all (SDG 7), sustainable cities and communities (SDG 11) and climate change (SDG 13) adequately positioned to address the inequitable health burdens imposed on those who have contributed the least in terms of per capita GHG emissions but who rely heavily on polluting solid fuels for their basic needs? The year 2020 was billed as the year for conclusive UN agreement on the PA's modalities. But, the COVID-19 global pandemic revealed a less than palatable global truth, namely, a collective global failure or, more euphemistically put, a global disconnect between the early identification of climate change within the context of poverty eradication and the growing public health crisis of air pollution, both of which disproportionately and negatively impact on poorer households, communities and cities.

The historical role of NNSAs – fossil fuel and cement producers – in terms of anthropogenic GHG emissions has been far more significant than previously discussed. As highlighted in 'Tracing anthropogenic carbon dioxide and methane emissions to fossil fuel and cement producers, 1854–2010' (Heede 2014). Heede's analysis focused on commercial and state-owned entities responsible for producing fossil fuels and cement as the primary sources of GHGs that have driven and continue to drive climate change, and found that '*nearly two-thirds of historic carbon dioxide and methane emissions can be attributed to 90 entities*' (emphasis added, 2014, p. 241). The powerful role of fossil fuel-driven industry groups and countries in stymieing action at the national and global level is hardly surprising, but still worth highlighting in terms of the tenacious duration and hold of such lobbying efforts.

Baumgartner et al.(2009) demonstrated the impact of lobbying on policy changes – essentially in determining who wins and loses. Efforts to mitigate climate change unleashed enormous lobbying interests both within countries like the US and at the global level from the very start of global negotiations on climate. Over two decades ago, Ross Gelbspan highlighted what he termed 'a most impressive campaign', which he characterized as an 'assault on our sense of reality' within the US Congress which allowed for a successful war waged on science and which has been a 'potent weapon on the international stage, permitting the corporate coal and oil interests - in tandem with the Organization of Petroleum Exporting Countries (OPEC) and other coal and oil exporting nations - to frustrate diplomatic attempt to address the crisis meaningfully' (Gelbspan 1997, p. 9). More recently, Brulle (2018) estimated that the lobbying expenditures related to climate change legislation in the US Congress from 2000 to 2016 was over $2 billion, constituting 3.9% of total lobbying expenditures. Led by fossil fuel and transportation corporations, utilities and affiliated trade associations, these lobbying efforts were found to dwarf those of environmental organizations and renewable energy corporations.

It is time to question whether the UN-led global SDA can respond effectively to the needs of socio-economically marginalized households, communities and countries lacking access to clean energy and clean air whilst suffering from adverse climatic impacts? What exactly does an examination of global or regional partnerships or protocols on climate, energy and air pollution reveal in terms of integrated and responsive action that can transform lives for millions across the globe? The COVID pandemic has exposed the tragic costs borne by those who are consigned to live at the nexus of public health neglect and existing chronic respiratory illnesses worsened by exposure to PM pollution that emanates from fossil fuel combustion and industrial pollution. According to a 2020 study conducted by researchers at Harvard University's T.H. Chan School of Public Health, 'Air pollution and COVID-19 mortality in the United States: Strengths and limitations of an ecological regression analysis', individuals with long-term exposure to air pollution are more likely to die of COVID-19. This first of its kind study examined the link between long-term exposure to $PM_{2.5}$ – generated largely from fuel combustion from cars, refineries and power plants – and the risk of death from COVID-19 in the US. It looked at 3000 counties across the US compared levels of fine particulate air pollution with coronavirus death counts for each area. After adjusting for population size, hospital beds, number of people tested for COVID-19, weather and socio-economic and behavioural variables such as obesity and smoking, the researchers found that a small increase in long-term exposure to $PM_{2.5}$ leads to a large increase in the COVID-19 death rate. More specifically, the study found that someone who lives for decades in a US county with high levels of fine particulate pollution is 8% more likely to die from COVID-19 than someone who lives in a region that has just one unit (1 $\mu g/m^3$) less of such pollution (Wu et al. 2020). This happens to be a terrifying finding for millions who have lived with chronic exposure to $PM_{2.5}$ in the most polluted cities in the world. Additionally, what is troubling about COVID-19 morbidity burdens on poorer, black and brown lives is what Gibbons outlined, which is that detailed studies of 'past epidemics show the same tragic pattern repeating again and again', infectious diseases more easily take hold in groups with pre-existing illnesses, who live and work in crowded conditions and who also lack access to adequate health care.

Another 2020 study by Cambridge University researchers entitled 'Links Between Air Pollution and COVID-19 in England' also found a link between the severity of COVID-19 infection and long-term exposure to air pollutants, including nitrogen oxides and ground-level ozone from car exhaust fumes or burning of fossil fuels. Researchers explored the links between major fossil fuel related air pollutants and SARS-CoV-2 mortality in England. The study compared real-time SARS-CoV-2 cases and morbidity from public databases to air pollution data monitored across over 120 sites in different regions in the UK. It found that $PM_{2.5}$ was a major contributor to COVID-19 cases in England, and an increase of 1 m^3 in the long-term average of $PM_{2.5}$ was associated with a 12% increase in COVID-19 cases in England. The study concluded that 'a small increase in air pollution leads to a large increase in the COVID-19 infectivity and mortality rate in England'; and that the study itself could be used as 'a framework to guide both health and emissions policies in countries affected by this pandemic' (Travaglio et al. 2021, abstract). Meanwhile, 'Air Pollution and Risk of Death due to COVID-19 in Italy' also confirmed the 'existence of a link between pollution and the risk of death due to the disease', and more specifically reiterated 'the need to act in favour of policies aimed at reducing pollutants in the atmosphere, by means of speeding up the already existing plans and policies, targeting all sources of atmospheric pollution: industries, home heating and traffic' (Dettori et al. 2021, abstract). The conclusions and recommendations of all three studies are of particular relevance for cities like Delhi, India, which ranks amongst the most air polluted in the world. According to a 20 October 2020 BBC report, $PM_{2.5}$ levels in Delhi averaged around 180–300 $\mu g/m^3$ – 12 times higher than the WHO's safe limits – a depressing reversal as Delhi residents were able to breathe relatively clean air because a stringent lockdown brought industries and traffic to a grinding halt (Pandey 2020). However, a 26 March 2021 Reuters article referenced the fact that Delhi had been ranked as the most polluted city in the world for the third year in a row by researchers at IQAir, an organization that measures air quality levels of $PM_{2.5}$ across the world's capital cities (Arora 2021). The harsh reality is that there is a paucity of data on the full extent of PM air pollution on human morbidity in cities in the developing world, let alone those that examine the impacts on long-term exposure to PM air pollution on COVID-19 given the current burden of disease and death experienced in countries like India, Brazil and South Africa where COVID variants are decimating lives.

There is now an increased urgency in focusing on the linkage between PM pollution related to the curbing of SLCPs within the context of sustainable cities in countries like India because cities are the loci where two inherently linked crises combine – climate vulnerabilities and toxic levels of energy related air pollution. The magnitude of the public health crisis that air pollution poses is impossible to escape in many cities. Cities are undeniably on the frontline for linked action on clean energy, air pollution reduction and climate resiliency solutions. Winning the interconnected battle necessitates integrated action that is directed via innovative and responsive city-driven measures that do not need to depend on stalemated intergovernmental climate outcomes. The role of cities and the linkages between air pollution, clean energy and climate responsive action are discussed in greater detail in the chapters that follow. From the outset, it is important to underscore that climate and clean energy action falls clearly within the purview of individual nation-states, but the urgency of the intertwined climate, clean energy and clean air access crises is such that there is a need to broaden the arena of participants and critically examine the track record of UN outcomes related to air pollution, climate change and clean energy access. Have intergovernmental consensus-driven declarations of willingness to address climate change focused on specific energy related air pollution targets and goals that are directly responsive to those cities and communities where PM air pollution urgently needs to be abated? Put simply, it is necessary to confront the protracted pace of intergovernmental negotiations and look towards new forms of partnerships and modalities that can more effectively integrate access to clean air and access to clean energy for the poor via the reduction of SLCPs.

The main argument advanced is that addressing the nexus between curbing SLCPs and increasing access to clean energy for the poor does, in fact, require integrated and innovative partnerships that look beyond nation-state-driven intergovernmental outcomes and existing policy silos on clean energy, pollution and poverty reduction and climate action. While it is undeniable that national governments – sovereign UN member states – have typically set the rules and frameworks for climate and clean energy, it is also clear that innovative action has not waited for the fractured pace of intergovernmental negotiations. The annual pilgrimages of climate cognoscenti have not resulted in verifiable improvements in the lives of those who lack access to clean energy and are exposed to toxic levels of PM air pollution. Here, it is also useful to ask why the UN's one and only successfully implemented regional air pollution protocol – the Convention on Long-range Transboundary Air Pollution (CLRTAP) – focuses only on Europe?

Regulatory efforts to curb air pollution have been and remain highly effective within the context of advanced industrialized countries, particularly in Europe. But there is a glaring absence of regional regulatory agreements aimed at reducing emissions of air pollutants within developing countries where the problem of energy related air pollution and lack of access to clean energy happens to be most pervasive. According to UNEP/WMO (2011), BC, which exists as particles in the atmosphere and is a major component of soot, has been demonstrated to be an SLCP. Both BC and O_3 are air pollutants harmful to human health, ecosystems and agriculture/food security. BC emissions result from the incomplete combustion of fossil fuels, wood and other biomass, and its negative impacts are felt by poor households that lack access to clean energy while O_3 is the third most important GHG contributor after carbon dioxide and methane. The UN's 1979 CLRTAP, which includes the Gothenburg Protocol (established in 1999), sets legally binding emission reductions commitments for 2020 and beyond for all major air pollutants shown to damage human health including sulphur dioxide (SO_2), nitrogen oxides (NOx), ammonia (NH_3), volatile organic compounds (VOCs) and fine particulate matter ($PM_{2.5}$). Administered by the UN's European Commission on Europe (UNECE), the 7 October 2019 entry into force of the amended Protocol makes it the first ever binding agreement to target emission reduction of $PM_{2.5}$, which happens to be both an SLCP and a predominant public health concern for polluted cities and megacities across the world.

The 40-year history of implementing CLRTAP and its protocols has been hailed as achievements that are 'unparalleled', and include the decoupling emissions and economic growth, cutting back certain air pollutants by 40–80%, recovering forest soils from acidification and avoiding some 600,000 premature deaths per year (UN News 2019b). But, there is a major global and regional disconnect in terms of addressing the linkages between air pollution and urban ill health for the rest of the world. The problem of PM air pollution impacts current and future generations living in developing countries of Asia and Africa at a scale vastly different from that experienced in Europe which successfully implemented one of the UN's only regional air pollution protocols – the CLRTAP. The problem that needs to be addressed is that there are no similar regional protocols on air pollution that cover other regions of the world including South Asia where the scope and scale of the problem of air pollution is urgent, massive and brought to the fore by COVID related respiratory burdens of disease and death. The intrinsically linked PM air pollution and climate crises in countries like India – the third largest aggregate CO_2 emitter but the 21st per capita CO_2 emitter – is such that depending on the tenuous, incremental intergovernmental consensus-driven outcomes can no longer suffice. The immediate sections provide context as to why the world's largest environmental health risk – air pollution – intersects with access to clean energy for the poor and focuses on the need for integrated, localized NNSA inclusive action on air pollution, clean energy and climate actions.

1.3 Mapping the Scope of the World's Largest Environmental Health Risk: Why Curbing Particulate Matter Air Pollution Matters for Millions of Lives

Poverty and socio-economic marginalization have been evidenced as putting women, children and the elderly in a seriously disadvantaged position in terms of coping with adverse impacts of climate change (UNDP 2007; World Bank 2016). Back in 2014, the Fifth Assessment Report (AR5) of the IPCC identified three pathways by which climate change impacts on human health:

'(1) Direct impacts, which relate primarily to changes in the frequency of extreme weather including heat, drought, and heavy rain;
(2) Effects mediated through natural systems, for example, disease vectors, water-borne diseases, and air pollution; and
(3) Effects heavily mediated by human systems, for example, occupational impacts, undernutrition, and mental stress' (IPCC 2014a, p. 716).

The IPCC finding that the health effects of climate change impact differentially and negatively on the global poor, including in the case of health risks associated with air pollution is categorically clear: 'Climate change is an impediment to continued health improvements in many parts of the world. If economic growth does not benefit the poor, the health effects of climate change will be exacerbated. In addition to their implications for climate change, essentially all the important climate-altering pollutants other than carbon dioxide (CO2) have near-term health implications (very high confidence). *In 2010, more than 7% of the global burden of disease was due to inhalation of these air pollutants* (high confidence)' (emphasis added, IPCC 2014a, p. 713). As referenced by AR5: 'Put into terms of disability-adjusted life years (DALYs), particle air pollution was responsible for about 190 million lost DALYs in 2010, or about 7.6% of all DALYs lost. This *burden* puts *particle air pollution among the largest risk factors globally*, far *higher* than *any other environmental risk* and *rivaling or exceeding all of the five dozen risk factors examined*, including malnutrition, smoking, high blood pressure, and alcohol' (emphasis added, IPCC 2014a, p. 728).

The largest concentrations of the *'energy poor'* (that is, people who are both poor and lack access to sustainable modern forms of energy) are currently in Sub-Saharan Africa and South Asia where the direct use of solid biomass has been well-documented to be widespread (Energy Policy: Srivastava et al. 2012). HAP resulting from the burning of solid fuels (wood, crop wastes, charcoal, coal and dung) imposes natural resource constraints and destroys the lives of women and children who spend more time in front of polluted hearths (Gordon et al. 2014). Heavy reliance on solid fuel use has been associated with acute lower respiratory infections, COPD, lung cancer and other illnesses at the household level, and burning biomass also impacts on local environments by contributing to deforestation and outdoor air pollution. Quaderi and Hurst's summary findings and conclusions in 'The Unmet Burden of COPD' (2018) regarding COPD's 'under-recognition and inequities' being particularly grave for low and middle-income countries is highlighted in Box 1.3. The authors pointed out that: 'Those who have never smoked tobacco *can* still get COPD – think 'biomass COPD' and find that peak levels of PM_{10} in biomass-using homes can be as high as 10,000 $\mu g/m^3$, 200 times more than the standard in high-income countries. $PM_{2.5}$ are finer particles which penetrate deep into the lung and have the greatest health-damaging potential" (Quaderi and Hurst 2018, p.2).

Box 1.3 Chronic obstructive pulmonary disease: under-recognition and inequitable impacts on the poor.

'*COPD remains a growing but neglected global epidemic. It is under-recognised, under-diagnosed and under-treated resulting in millions of people continuing to suffer from this preventable and treatable condition.* The lower an individual's socio-economic position, the higher their risk of poor health: women and children living in severe poverty have the greatest exposures to HAP.

In the poorest countries, cooking with solid fuels can be the equivalent of smoking two packs of cigarettes a day. A 1-year old would have accumulated a two pack year smoking history having never seen tobacco. Inaction to mitigate COPD therefore exacerbates health inequalities.

Climbing the "energy ladder" occurs gradually as most LMIC households use a combination of fuels. The poorest, at the bottom of the ladder, use crop waste or dung which is the most harmful when undergoing incomplete combustion. Those at the top of the ladder use electricity or natural gas. Increasing prosperity and development has a direct positive correlation with increasing use of cleaner and more efficient fuels for cooking.

The unmet global burden of COPD is a silent killer in LMICs. In conclusion, we suggest that given the high and rising global burden of COPD, a revolution in the diagnosis and management of COPD and exacerbations of COPD in LMICs must be an urgent priority'.

Source: emphasis added, Quaderi and Hurst (2018).

Box 1.4 excerpted from the WHO's special report on climate change and human health lists key linkages between energy related air pollution, climate change and public health impacts.

Box 1.4 Strong linkages between climate change, air pollution and health.

- The human activities that are destabilizing the Earth's climate also contribute directly to ill health.
- Burning fossil fuels for power, transport and industry is the main source of the carbon emissions that are driving climate change and a major contributor to health-damaging air pollution, which every year kills over seven million people due to exposure inside and outside their homes.
- Over 90% of the urban population of the world breathes air containing levels of outdoor air pollutants that exceed WHO's guidelines.
- Air pollution inside and outside the home is the second leading cause of deaths from non-communicable diseases (NCDs) worldwide; it is responsible for 26% of deaths from ischaemic heart disease, 24% of those from strokes, 43% from chronic obstructive pulmonary disease and 29% from lung cancer.
- The sectors that produce most GHGs – energy, transport, industry, agriculture, waste management and land use – are also the main sources of fine particulate matter and other important air pollutants. These include short-lived climate pollutants such as black carbon, methane and ground-level ozone, which also threaten human health.
- Approximately 25% of urban ambient air pollution from fine particulate matter is contributed by traffic, 15% by industrial activities including electricity generation, 20% by domestic fuel burning, 22% from unspecified sources of human origin and 18% from natural sources.
- Effectively all exposure to indoor air pollution, which causes almost four million deaths a year, is from use of solid fuels for cooking in poor households.

Source: WHO (2018, p. 16).

What is also significant to underscore is that the health impacts of air pollution are felt disproportionately within poorer communities and particularly amongst women and children from poorer households. The role of gender in determining activities that expose women and young girls to ill health and lack of access to education and employment for instance is important not just in terms of responding to HAP but also in terms of gender-assigned household responsibilities and hours spent in terms of searching for the very forms of solid fuels that contribute to increased health risks. More than 25 years ago, Indian climate change and energy expert, Jyoti Parikh highlighted the need for greater attention and data collection on gender and energy policy: 'The fact that there are no gender specific data available in the energy sector- which is one of the most quantified sectors is largely due to lack of concern and understanding about gender issues by energy policy makers and analysts' (1995, p. 7). Parikh pointed out gender-specific disparities lessen as economic development increases. However, growing health and economic disparities resulting from the COVID pandemic are a major future development concern in terms of gender and racial inequities. Key findings from a special report 'Burning Opportunity: Clean Household Energy for Health, Sustainable Development, and Well Being of Women and Children' show that little room for equivocation as to who bears the negative health impacts of exposure to HAP and include the following facts:

- 'Household air pollution is the single most important environmental health risk factor worldwide.
- Women and children are at a particularly high risk of disease from exposure to HAP. Sixty percent of all premature deaths attributed to household air pollution occur in women and children. Women experience higher personal exposure levels than men, owing to their greater involvement in daily cooking and other domestic activities. The single biggest killer of children aged under five years worldwide is pneumonia. This disease cuts short almost a million young lives each year. More than 50% of those pneumonia deaths are caused by exposure to HAP.
- Improving health in urban environments depends in part on addressing pollution from household fuel burning. In India, for example, new research estimates that almost 30% of outdoor air pollution is from household sources.
- Women and girls are the primary procurers and users of household energy services, and bear the largest share of the health and other burdens associated with reliance on polluting and inefficient energy systems. Owing to the considerable amount of time spent in proximity to polluting combustion sources, women and children are at particularly high risk of disease from exposure to HAP.
- New analyses find that reliance on polluting fuels and technologies is associated with significant drudgery and time loss for children – especially girls. Data on wood and water gathering from 30 countries show that both boys and girls in clean fuel-using households spent less time gathering wood or water than those from homes cooking mainly with polluting fuels. Girls living in households that cook mainly with polluting fuels bear the greatest time-loss burden collecting wood or water' (WHO 2016c, p. ix).

Exposure to air pollution was found to vary greatly by socio-economic status according to a 2015 review of existing literature on air pollution funded by US National Institute of Environmental Health Sciences. 'Socioeconomic Disparities and Air Pollution Exposures: A Global Review' found that poorer communities tend to be exposed to higher concentrations of air pollution, compared to richer communities (Hajat et al. 2015). But there is an additional generational inequity that is being passed on to children from poorer households that was highlighted by the UN Children's Fund (UNICEF) 2016 report entitled *'Clear the Air for Children'*. Based on satellite imagery of outdoor air pollution, this report provided for the first time the staggering scale and scope of the air pollution crisis faced by children. South Asia was found to have the largest number of children living with toxic air pollution – 620 million, followed by East Asia and Pacific region – 450 million children; West and Central Africa – 240 million; Eastern and Southern Africa – 200 million, and when taken together all the regions comprised a total of 2 billion children living in areas that exceed WHO guideline limits of $10\,\mu g/m^3$ (2016, p. 63). Some of the report's key findings regarding air pollution provide evidence of the costs and disproportionate burdens borne by children in low- and middle-income countries and across regions (see Box 1.5).

Box 1.5 Burdens of air pollution on children: key findings related to disproportionate burdens.

Around 300 million children currently live in areas where the air is toxic – exceeding international limits by at least six times

In total, around 2 billion children live in areas that exceed the World Health Organization annual limit of 10 μg/m³ (the amount of micrograms of ultra-fine particulate matter per cubic metre of air that constitutes a long term hazard)

Globally, air pollution affects children in low- and middle income countries more. Up to 88 per cent of all deaths from illnesses associated with outdoor air pollution16 and over 99 per cent of all deaths from illnesses associated with indoor air pollution occur in low- and middle-income countries.

Asia currently accounts for the vast bulk of total deaths attributable to air pollution. The proportions, however, are changing. In Africa, increasing industrial production, urbanization and traffic is causing the rapid rise of outdoor air pollution. *As this happens, the number of African children exposed to outdoor air pollution is likely to increase, especially as the continent's share of the global child population is set to increase markedly.* By mid-century, more than one in three children globally is projected to be African.

Outdoor air pollution tends to be worse in lower-income, urban communities. Lower-income areas are often highly exposed to environmental pollutants such as waste and air pollution. Factories and industrial activity are also more common near lower-income areas, and there is often less capacity to manage waste. This can result in burning, including of plastics, rubber and electronics, creating highly toxic airborne chemicals which are highly detrimental to children. Poorer families are also less likely to have resources for good quality ventilation, filtration and air conditioning to protect themselves from harmful air.

Indoor air pollution is most common in lower-income, rural areas. Over 1 billion children live in homes where solid fuels are used in cooking and heating. While outdoor air pollution tends to be worse in poor urban communities, indoor air pollution tends to be worse in rural communities where biomass fuels are more frequently used in cooking and heating due to lack of access to other forms of energy.

Source: UNICEF (2016, pp. 8–9).

There is an urgent need to focus on policy measures that derived from UNEP/WMO for HAP as well as, those that may be gleaned from the experiences of LRTAP and its Gothenburg Protocol for their relevance and feasibility for cities in Asia and Africa with a particular focus on India where both HAP and outdoor/ambient air pollution have reached crippling toxic levels. In their analysis of HAP in LMICs, for instance, Ochieng et al. noted that improved biomass cookstoves have for a long time been considered as most immediate policy intervention, but the ability of improved biomass cookstoves to reduce exposure to HAP that meet health standards remains questionable and there is limited evidence as to adoption and use barriers. Ochieng et al. call, therefore, for additional research on policy interventions that can reduce exposure including focusing on poverty eradication as the means to advance towards cleaner energy (Ochieng et al. 2018).

It is also necessary to highlight the massive gaps in knowledge and data on air pollution – indoor and outdoor – directly relevant to developing countries. While studies have examined the relationship between air pollution and cardiorespiratory diseases, there is a shortage of data and assessments of the health risks across regions and within vulnerable populations. Hajat et al. (2015) pointed out that most North American studies have shown that areas where socio-economically marginalized communities dwell experience higher concentrations of air pollutants. Research from Asia, Africa and other parts of the world has shown a general trend similar to that of North America, but research in these parts of the world is limited (Hajat et al. 2015, p. 440). Lelieveld et al. (2015) pointed out that it has proven difficult to quantify premature mortality related to air pollution, in regions where air quality is not monitored, and also because the toxicity of particles from various sources varies. Using a global atmospheric chemistry model to investigate the link between premature mortality and seven emission source categories in urban and rural environments, outdoor air pollution, mostly by $PM_{2.5}$, was estimated to lead to 3.3 million premature deaths per year worldwide, predominantly in Asia. Under a business-as-usual emission scenario, the contribution of outdoor air pollution to premature mortality was estimated to double by 2050 (Lelieveld et al. 2015, p. 367). Curbing fossil fuel related air pollution in an urgent global imperative. Access to clean air is directly linked to access to clean energy for households that rely on polluting forms of energy, as well as growing levels of urban outdoor air pollution in many of the most congested and populous cities of the world. The lack of access to clean air and energy is an essential element in not just carbon inequality as defined by Oxfam but also by extension pollution inequality. From the immediate perspective of this book, access to clean air is viewed as fundamental to improving human health and well-being and poverty reduction. The following section outlines the scope of work undertaken in the remaining chapters.

1.4 Outlining Scope of Work: Brief Overview and Caveats as to Limitations

In outlining the scope of work in the chapters that follow, it is useful to begin with issuing caveats that circumscribe the limitations of the chapters and also flag the realization that synergies on clean air, clean energy and climate change requires exponentially increased responsive and innovative actions by NNSAs and governments. IPCC's AR6 SPM has outlined just how pervasive climatic impacts are likely to be: 'With further global warming, every region is projected to increasingly experience concurrent and multiple changes in climatic impact-drivers. Changes in several climatic impact-drivers would be more widespread at 2°C compared to 1.5°C global warming and even more widespread and/or pronounced for higher warming levels. All regions are projected to experience further increases in hot climatic impact-drivers (CIDs) and decreases in cold CIDs (high confidence). Further decreases are projected in permafrost, snow, glaciers and ice sheets, lake and Arctic sea ice (medium to high confidence) 40. These changes would be larger at 2°C global warming or above than at 1.5°C (high confidence). For example, extreme heat thresholds relevant to agriculture and health are projected to be exceeded more frequently at higher global warming levels (high confidence)' (2021, p. 32). In spite of the consistent build-up in warnings about global warming emanating from the IPCC and other global entities, there is evidence of global negotiations and policy silos that have consistently separated out energy for sustainable development from climate change objectives, which is a concern that is explored further in this book.

There have been a growing number of powerful voices calling climate change the biggest existential crisis to face our shared planet. Within the UN's broad umbrella of sustainable development, the topic of climate change as an institutional/governance challenge has been widely researched for years. But, responding to global climate change has proven to be a vastly complicated task, and one that is made even more complex when contextualized within the SDA and PA's universally agreed upon priority of poverty eradication. Then, there is additional challenge of providing a means of understanding three different yet linked topics – air pollution, access to clean energy and climate change a which have been consistently addressed by within segregated UN negotiations and policy tracks. Having separated out policy tracks and negotiated outcomes – silos – on energy for sustainable development and climate action are hard to rationalize within the broader SDA framework. Simply put, having long-standing global silos on clean energy access, climate change and air pollution reduction is illogical since fossil fuel energy is a key driver in climate change and PM pollution. And yet, it is precisely these goal silos on clean energy, climate and air pollution that need to be broken down and transformed into an integrated action-oriented agenda.

It is important to emphasize the fact that there is a tremendous amount of policy and analytical work being done by a wide range of entities and researchers on all three of the topics – clean/sustainable energy, clean air and climate change which, the remaining chapters in the book cannot remotely attempt to cover in any complete manner. Consequently, the aim of the remaining chapters are to attempt to build upon and tie together diverse threads of excellent research that has been done by so many others. The discussion contained also needs to be understood as being subjective and cannot be construed as being anywhere near as comprehensive in scope as needed to fully address the scale of PM pollution and climate change experienced by those most vulnerable within developing countries. Additionally, it is important to underscore that the broad topics of climate change, sustainable development and clean energy are some of the most heavily researched today. At the outset, it should be noted that the chapters of this book are narrowly circumscribed in their scope, and also do not purport to provide an in-depth or historical view of UN climate change or energy for sustainable development negotiations. Key topics such as climate finance, adaption and resilience building as well as, energy and climate justice are recognized as crucial but nevertheless fall outside of the immediate purview of the discussion undertaken. Equally importantly, the overall focus on providing broad perspectives based on a categorization of countries as 'developing' should be seen as nuanced, and by no means can be considered as definitive and are reflective only as globally understood within the context of the UN SDA and PA.

At the core what is being argued is that it is time to look beyond the confines of intergovernmental negotiations, and to ask what can be done if access to clean air and clean energy is considered integral to responding to climate change by NNSAs including cities/local communities. The entwined climate change and fossil fuel air pollution crises pose double burdens for millions of lives and need to be addressed in an integrated and inclusive manner. The central aim therefore is to provide evidence that PM pollution extracts the heaviest tolls on the poorest and most vulnerable people and communities and that the global community has done little to address the world's single largest environmental health risk. The heavy reliance on inefficient energy devices (open fires and traditional cook stoves) as well as inefficient sources of energy such as solid fuels and traditional biomass in poor households in developing countries, particularly those in sub-Saharan Africa and South Asia has been documented to result in incomplete combustion of and release of SLCPs including BC which has serious health and short-term climatic impacts (UNEP 2013). Emissions that result from the incomplete/inefficient combustion of solid fuels and traditional biomass are released as a mixture of health-damaging indoor air pollutants such as BC, that have short atmospheric life spans but result in significant negative impacts to human health and climate change at the national and regional level (UNEP/WMO 2011). A key premise advanced is that the nexus between climate resilience, clean energy, poverty reduction and urbanization urgently needs to be transformed by the increasingly dynamic role of NNSAs including municipal and energy sector stakeholders.

Some of the key issues addressed and questions raised in the chapters are as follows:

- What is the scope of agreed global guidance on NSA partnerships related to the proposed global partnership mechanism for SDGs (SDG 17) that are directly relevant to NNSAs including private sector-driven partnerships focused on meeting clean energy for all (SDG 7) and climate change objectives (SDG 13)?
- Given the central role of the energy sector in climate change and the linkages between the lack of access to clean energy and SLCPs, what exactly is the existing scope of global guidance that allows for distinguishing between partnership mechanisms/modalities related to SDGs 7 (Sustainable Energy) and SDG 13 (Climate Change)?
- What does an examination of UN-related partnership databases and portal reveal in terms of integrated action on clean energy access, curbing air pollution and addressing climate change?
- How can integrated action that mitigates SLCPs and increases access to clean air and clean energy be scaled up for regions where these challenges are most urgent? What broad lessons can be learned from the CLRTAP and its Gothenburg Protocol, as well as the landmark assessment done by UNEP/WMO (2011) that are relevant for the most polluted cities in world?What are the broad parameters for scaling up linked action on clean air, access to clean energy and climate change for countries like India and Indian cities which are impacted by some of the highest levels of PM pollution?

The constant thread in all the chapters is that partnerships which integrate action on clean air, clean energy and climate action are crucial. The UN-led global community has long focused on partnerships for sustainable development (PSD) as a means of delivering on an overarching pledge to eradicate poverty for the past decades so what for example is the track record of such partnerships in delivering on poverty eradication and access to clean energy? Chapters 2 and 3 examine the record of key UN global goals and partnership efforts focused on climate, increasing access to clean energy and air pollution. Chapter 2 begins by providing context as to the global neglect of climate related health impacts that have particular relevance for poorer and more vulnerable communities and countries. It then focuses on the agreed global guidance and the implementation record of the global partnership mechanism for the precursor to the SDGs – the Millennium Development Goals (MDGs) to see whether increasing access to sustainable energy and curbing air pollution have been dealt with in the context of the SDGs. Chapter 2 provides a historical analysis of the global record on partnerships mechanisms/modalities related to sustainable energy and climate action within the context of UN's 2030 SDA. It provides evidence as to whether definitional clarity or confusion exists in terms of PSDs in particular those related to sustainable energy, climate change and clean air.

Chapter 3 focuses on examining the agreed UN record on global partnerships mechanisms related to climate change including the Clean Development Mechanism (CDM) of the 1997 KP and the PA's proposed Article 6 mechanism. Globally agreed climate change outcomes within the UNFCCC framework are examined in terms of integrated partnerships aimed at addressing poverty eradication, reducing energy related air pollution and climate change goals. The chapter concludes by reviewing the online partnership portal – the Non-state Actor Zone for Climate Action (NAZCA) hosted by the UNFCCC. Given that the NAZCA portal is expressly aimed at partnership actions that cities, companies, investors and regions – i.e. NNSAs – are taking to address climate change, the aim is to critically examine this portal and the extent to which NNSAs actions within it are aimed at addressing SDGs on clean energy and climate change including most importantly SDG 1 – poverty eradication.

The need to focus on integrated action to curb SLCPs and air pollution that are not nation-state driven but sectoral and city/community-driven is the focus of Chapters 4–6. The most recent projection by the UN's Population Division is that, 'Africa will have 22 per cent of the world's urban population- 1.5 billion urban dwellers, while Asia, with 3.5 billion persons residing in urban areas, will have 52 per cent of the world's urban population by 2050' (UNDESA 2018, pp. 24–25). Integrated partnership actions focused on addressing the linkages between urbanization, air pollution and poverty are critical in the most populous cities in Africa and Asia. In a world where more than half the global population currently lives in cities, civil society/NSAs based in the largest cities of the world arguably holds the key to global action on clean air, clean energy and climate change. Cities are the loci where climate, clean air and clean energy needs and related health/morbidity burdens are the greatest, and where integrative change can occur.

Innovative forms of NNSA partnerships on curbing air pollution, mitigating SLCPs and increasing access to clean energy for the poor are essential in averting the air pollution catastrophe. The role of NNSAs including cities in driving climate responsive and clean energy actions especially in regions of the world where air pollution threatens lives is undeniable. Chapter 4 examines the urgency of urban air pollution by focusing on the linkages between urbanization and PM air pollution. It also discusses the role of cities as the new loci for integrated climate and clean air responsive action and then moves on highlighting the significance of PM pollution for India and Indian cities. India is the third-largest aggregate GHG emitter, but it has the lowest per capita emissions amongst the top ten leading national GHG emitters, and it has long argued that poverty eradication and equity considerations need to be factored in its climate change response.

But now the toxic problem of air pollution threatens future development in India at many levels. The chapter's conclusion urges the need for cleaner, inclusive and greener future for cities.

Chapter 5 delves into the issue of curbing SLCPs as a means to address not just climate change but also health, food and water insecurities experienced primarily in developing countries. Global environmental problems such as climate change are broadly defined as global challenges which cannot be contained within the confines of a state, or a set of adjoining states, and are therefore viewed as requiring multilateral or global responses within the UN. In contrast, air pollution within the UN context has been defined as a transboundary regional problem for the most part. Chapter 5 focuses on the importance of curbing SLCPs in particular BC emissions. It also focuses on providing a brief overview of the UN's only regional air pollution treaty – the CLRTAP a and its Gothenburg Protocol, which is the world's first protocol aimed at $PM_{2.5}$ emissions reductions. The idea is to highlight a few key lessons learned and implications for other regions in the world that are faced with the toxic levels of PM pollution. This chapter also focuses on specific measures for curbing SLCPs that have been highlighted by CCAC which endorsed the 2011 recommendations made by UNEP/WMO in their joint integrated assessment. The role of the CCAC as the only globally relevant yet voluntary partnership network that could serve as a model for improving air quality and mitigating SLCPs is also briefly touched upon.

Chapter 6 focuses on summarizing that integrated action on the nexus between climate change, air pollution reduction and access to clean energy matters for millions whose voices are not heard in intergovernmental negotiating fora. It further highlights the role of NNSAs such as the private sector by looking at shifts towards low and zero carbon strategies proposed by the CEOs of the two of world's biggest corporations and asks what if anything fossil-free divestments mean for clean energy and clean air for the poor? The chapter then provides an overview of the importance of innovative partnership initiatives, modalities and measures that are expressly focused on the reduction of SLCPs. It concludes by providing a brief update on the outcomes of the 2021 COP-26 (Glasgow) and by arguing that segregated goals/silos on sustainable energy and climate do not enable SLCPs and PM pollution reduction. The absence of integrated and inclusive clean air, clean energy and climate action only worsens the lives of those who already suffer the costs of carbon and air pollution inequities.

References

African Development Bank (2003). *Poverty and Climate Change: Reducing the Vulnerabilities of the Poor through Adaptation*. Paris: OECD.

Allen, B. (2004). Global warming-bulletins from a warmer world: from the editor. *National Geographic* (September 2004).

Arora, N. (2021). *New Delhi is World's Most Polluted Capital for Third Straight Year – IQAir Study*. Reuters. https://www.reuters.com/article/us-india-pollution/new-delhi-is-worlds-most-polluted-capital-for-third-straight-year-iqair-study-idUSKBN2B817F. (accessed 20 April 2021).

Baumgartner, F. et al. (2009). *Lobbying and Policy Change: Who Wins, Who Loses, and Why*. Chicago: University of Chicago Press.

Bazilian, M. et al. (2010). Understanding the scale of investment for universal energy access. *Geopolitics of Energy (Special Issue: Energy Poverty and Development)* 32 (10): 21–42.

Bernard, L. and Semmler, W. (ed.) (2015). *The Oxford Handbook of the Macroeconomics of Global Warming*. London: Oxford University Press.

Booth, W. and Stevens, H. (2021), COP26 in Glasgow may have a record carbon footprint, despite low flush loos and veggie haggis. *Washington Post* (12 November 2021). https://www.washingtonpost.com/world/2021/11/12/cop26-carbon-emissions-offsets/ (accessed 21 November 2021).

Bond, T.C. et al. (2013). Bounding the role of black carbon in the climate system: a scientific assessment. *Journal of Geophysical Research: Atmospheres* 118: 5380–5552. https://doi.org/10.1002/jgrd.50171 (accessed 20 April 2021).

Brulle, R. (2018). The climate lobby: a sectoral analysis of lobbying spending on climate change in the USA, 2000–2016. *Climatic Change* 149: 289–303. https://link.springer.com/article/10.1007/s10584-018-2241-z (accessed 27 February 2021).

Bullard, R.D. and Wright, B. (2009). *Race, Place and Environmental Justice after Hurricane Katrina*. Philadelphia: Westview Press.

Burnett, R. et al. (2018). Global estimates of mortality associated with long-term exposure to outdoor fine particulate matter. *Proceedings of the National Academy of Sciences of the United States of America (PNAS)* 115 (38): 9592–9597. https://www.pnas.org/content/115/38/9592 (accessed 27 February 2021).

CCAC (2021). *Who we are* and What are Short –Lived Climate Pollutants. http://ccacoalition.org/en/content/who-we-are; https://www.ccacoalition.org/en/content/short-lived-climate-pollutants-slcps (accessed 20 April 2021).

Chasek, P. et al. (2017). *Global Environmental Politics*. New York: Routledge.

Cherian, A. (2012). Grapping with the Global Climate Challenge. In: *Global Environmental Issues* (ed. F. Harris). Chichester: Wiley-Blackwell.

Cherian, A. (2015). *Energy and Global Climate Change: Bridging the Sustainable Development Divide*. Oxford: Wiley Blackwell.

Chin, N.C. (2019). A 'life and death matter': Singapore could spend US$72b on climate adaptation over next century. *Eco-Business* (20 August 2019). https://www.eco-business.com/news/a-life-and-death-matter-singapore-could-spend-us72b-on-climate-adaptation-over-next-century (accessed 1 March 2021).

Ciplet et al. (2015). *Power in Warming World: The New Global Politics of Climate Change and the Remaking of Environmental Inequality*. Cambridge (MA): MIT Press.

Cohen, A.J. et al. (2017). Estimates and 25-year trends of the global burden of disease attributable to ambient air pollution: an analysis of data from the Global Burden of Diseases Study 2015. *Lancet* 389: 1907–1918.

Cursino, M. and Faulkner, D. (2021). COP26: China and India must explain themselves, says Sharma. *BBC News* (14 November 2021). https://www.bbc.com/news/uk-59280241 (accessed 22 November 2021).

Davidson, J. (2019). Google camp on climate crisis attended by rich and famous in private jets and mega yachts. *Ecowatch* (2 August 2019). https://www.ecowatch.com/google-camp-climate-crisis-2639622926.html (accessed 24 January 2021).

Dettori, M. et al. (2021). Air pollution and the risk of death due to COVID-19 in Italy. *Environmental Research* 192: https://www.sciencedirect.com/science/article/pii/S0013935120313566 (accessed 2 April 2021).

Esty, D.C. and Ivanova, M. (2002). *Global Environmental Governance: Options & Opportunities*, Forestry & Environmental Studies Publications Series. 8. New Haven: Yale University Press.

Gelbspan, R. (1997). *The Heat is On*. New York: Addison-Wesley Publishing.

Gemmil, B. and Bamidele-Izu, A. (2002). The role of NGOs and civil society in global environmental governance. In: *Global Environmental Governance* (ed. D. Esty et al.), 77–100. New Haven: Yale University Press.

Giddens, A. (2011). *The Politics of Climate Change*. Cambridge, UK: Polity Press.

Goldemberg, J. et al. (1988). *Energy for a Sustainable World*. New Delhi: Wiley Eastern Limited.

Goodman, A. (2013). Warsaw climate conference: polluting corporations welcome. *The Guardian* (21 November 2013). https://www.theguardian.com/commentisfree/2013/nov/21/cop-19-warsaw-climate-change-corporations-conference. (accessed on 17 July 2021)

Gordon, S.B. et al. (2014). Respiratory risks from household air pollution in low and middle income countries. *The Lancet Respiratory Medicine* 2: 823–860.

Gupta, J. (2014). *The History of Global Climate Governance*. Cambridge: Cambridge University Press.

Haas, P. (1990). Obtaining International Environmental Protection through Epistemic Consensus. *Journal of International Studies*. Sage Publications 19: 347–363.

Haas, P. et al. (1993). *Institutions for the Earth: Sources of Effective International Environmental Protection*. Cambridge, MA: MIT Press.

Hajat, A. et al. (2015). Socioeconomic disparities and air pollution exposure: a global review. *Current Environmental Health Reports* 2: 440–450 (2015). https://doi.org/10.1007/s40572-015-0069-5 (accessed 20 April 2021).

Harlan, S. et al. (2015). Climate Justice and Inequality. In: *Climate Chagne and Society* (ed. R. Dunlap and R. Brulle), 127–163. Oxford: Oxford University Press.

Hawkins, E. and Jones, P. (2013). On increasing global temperatures: 75 years after Callendar. *Quarterly Journal of the Royal Meteorological Society*. https://rmets.onlinelibrary.wiley.com/doi/pdf/10.1002/qj.2178 (accessed 22 June 2021).

Heede, R. (2014). Tracing anthropogenic carbon dioxide and methane emissions to fossil fuel and cement producers, 1854–2010. *Climatic Change* 122 (1–2): 229–241. https://link.springer.com/article/10.1007/s10584-013-0986-y#citeas (accessed 20 April 2021).

Hemmati, M. (2001). *Multi-Stakeholder Processes for Governance and Sustainability: Beyond Deadlock and Conflict*. London: Earthscan Publications.

IEA (2009). *World Energy Outlook: 2009*. Paris: OECD/IEA.

IEA (2015). *World Energy Outlook: 2015*. Paris: OECD/IEA.

IEA (2021). *Global Energy Review 2021*. Paris: IEA. https://www.iea.org/reports/global-energy-review-2021# (accessed 22 April 2021).

IEA/UNDP/UNIDO (2010). *Energy Poverty: How to Make Energy Access Universal*. Paris: OECD/IEA.

Indian Express (2017). Sushma Swaraj rubbishes Donald Trump's claim, says India didn't sign Paris climate deal for money. *Indian Express* (5 June 2017). http://indianexpress.com/article/india/sushma-swaraj-external-affairs-minister-mea-annual-press-conference-donald-trump-qatar-dubai-paris-climate-agreement-4689877 (accessed 7 January 2021).

IPCC (2013). *Summary for Policymakers in Climate Change 2013: The Physical Basis. Contribution of Working Group 1 to AR5 of the IPCC*. Cambridge: Cambridge University Press.

IPCC (2014a). Synthesis Report. Summary for Policymakers. https://www.ipcc.ch/site/assets/uploads/2018/02/AR5_SYR_FINAL_SPM.pdf (accessed 28 February 2021).

IPCC (2014b). Fifth Assessment Report (AR-5) Synthesis Report. https://www.ipcc.ch/site/assets/uploads/2018/02/SYR_AR5_FINAL_full.pdf (accessed 27 February 2021).

IPCC (2019). *IPCC Special Report on Climate Change, Desertification, Land Degradation, Sustainable Land Management, Food Security, and Greenhouse Gas Fluxes in Terrestrial Ecosystems*; Chapter 5. Geneva: IPCC/WMO. https://www.ipcc.ch/site/assets/uploads/2019/08/2f.-Chapter-5_FINAL.pdf (accessed 22 January 2021).

IPCC (2021). *Summary for Policymakers-Climate Change 2021: The Physical Science Basis. Contribution of Working Group I to the Sixth Assessment Report of the Intergovernmental Panel on Climate Change*. Cambridge: Cambridge University Press. https://www.ipcc.ch/report/ar6/wg1/downloads/report/IPCC_AR6_WGI_SPM.pdf (accessed 9 August 2021).

IPCC Summary for Policy Makers (SPM) (2018). Global Warming of 1.5°C – an IPCC special report on the impacts of global warming of 1.5°C above pre-industrial levels and related global greenhouse gas emission pathways, in the context of strengthening the global response to the threat of climate change, sustainable development, and efforts to eradicate poverty. https://www.ipcc.ch/site/assets/uploads/sites/2/2018/07/SR15_SPM_version_stand_alone_LR.pdf (accessed 20 January 2021).

IUCC (1992). *United Nations Framework Convention on Climate Change, UNEP/WMO Informational Unit on Climate Change*. Geneva, Switzerland: IUCC.

Keating, D. (2019). Failure in Madrid as COP25 ends in disarray. *Forbes* (15 December 2019). https://www.forbes.com/sites/davekeating/2019/12/15/failure-in-madrid-as-cop25-climate-summit-ends-in-disarray/#328fa2953d1f (accessed 2 April 2021).

Kanie, N. et al. (2017). *Governing through Goals: Sustainable Development Goals as Governance Innovation*. Cambridge: MIT Press.

Lebling, K. et al. (2019). Five charts show how global emissions have changed since 1850. *WRI Blog* (2 April 2018), https://www.wri.org/blog/2018/04/5-charts-show-how-global-emissions-have-changed-1850 (accessed 27 February 2021).

Lelieveld, J. et al. (2015). The contribution of outdoor air pollution sources to premature mortality on a global scale. *Nature* 17 September 2015 525 (7569): 367–371. https://pubmed.ncbi.nlm.nih.gov/26381985 (accessed 2 March 2021).

Lustgarten, A. (2020) Climate change will force a new American migration. *Propublica* (15 September 2020). https://www.propublica.org/article/climate-change-will-force-a-new-american-migration (accessed 22 January 2021).

Luterbacher, U. and Sprinz, D. (ed.) (2001). *International Relations and Global Climate Change*. Cambridge, MA: MIT Press.

Mathiesen, K. (2017). UN climate conference 2018 heads to heartland of Polish coal. *Climate Home News*. https://www.climatechangenews.com/2017/06/01/un-climate-conference-2018-heads-heartland-polish-coal/ (accessed on 17 July 2021).

Modi, V. et al. (2006). *Energy and the Millennium Development Goals*. New York: UNDP/UN Millennium Project/World Bank.

NASA Earth observatory website (2000). Svante Arrhenius. https://earthobservatory.nasa.gov/features/Arrhenius (accessed 7 April 2021).

NASA website (2021). Climate Change: How do we know? https://climate.nasa.gov/evidence (accessed 22 April 2021).

NOAA/ESRL website (2019). About Mauna Loa Observatory. https://www.esrl.noaa.gov/gmd/obop/mlo/aboutus/aboutus.html (accessed 7 April 2021).

Ochieng, C. et al. (2018). Household air pollution in low and middle income countries. *Oxford Research Encyclopedia of Environmental Science*. https://oxfordre.com/environmentalscience/view/10.1093/acrefore/9780199389414.001.0001/acrefore-9780199389414-e-25 (accessed 23 February 2021).

Oxfam (2015). Extreme Carbon Inequality. https://www-cdn.oxfam.org/s3fs-public/file_attachments/mb-extreme-carbon-inequality-021215-en.pdf (accessed 12 March 2021).

Parikh, J. (1995). *Gender Issues in Energy Policy*, vol. 160-1995. Bombay: Indira Gandhi Institute of Development Research.

Parsons, J. (2021). Billionaires fly in to COP26 climate conference in parade of polluting private jets. *Metro News* (1 November 2021). https://metro.co.uk/2021/11/01/billionaires-fly-to-cop26-climate-conference-in-parade-of-private-jets-15521675/ (accessed on 21 November 2021).

Pandey, V. (2020). COVID 19 and pollution: Delhi starting at coronavirus disaster. *BBC News*. https://www.bbc.com/news/world-asia-india-54596245 (accessed 20 April 2021).

Pellow, D.N. and Brulle, R.J. (ed.) (2005). *Power, Justice and the Environment: A Critical Appraisal of the Environmental Justice Movement*. Cambridge: MIT Press.

Quaderi, S.A. and Hurst, J.R. (2018). The unmet global burden of COPD. *Global Health, Epidemiology and Genomics, 3*, e4. doi: https://doi.org/10.1017/gheg.2018.1 https://www.ncbi.nlm.nih.gov/pmc/articles/PMC5921960/#ref3 (accessed 22 January 2021).

Rehman, I.H. et al. (2012). Understanding the political economy and key drivers of energy access in addressing national energy access priorities and policies. *Energy Policy* 47 (1): 27–37.

Requia, W. et al. (2018). Global association of air pollution and cardiorespiratory diseases: a systematic review, meta-analysis, and investigation of modifier variables. *American Journal of Public Health* 108 (Suppl. 2): 123–130. https://www.ncbi.nlm.nih.gov/pmc/articles/PMC5922189 (accessed 23 January 2021).

Rigaud, K. et al. (2018). *Groundswell: Preparing for Internal Climate Migration*. Washington, DC: World Bank http://documents.worldbank.org/curated/en/846391522306665751/pdf/124719-v2-PUB-PUBLIC-docdate-3-18-18WBG-ClimateChange-Final.pdf (accessed 12 March 2021).

Roberts, T. and Parks, B. (2007). *A Climate of Injustice: Global Inequality, North-South Politics, and Climate Policy*. Cambridge: Cambridge University Press.

Robinson, J. and Herbert, D. (2001). Integrating climate change and sustainable development. *International Journal of Global Environmental Issues* 1 (2): 2001.

Ross, K. et al. (2018). *Strengthening Nationally Determined Contributions to Catalyze Actions that Reduce Short Lived Climate Pollutants. WRI/Oxfam Working Paper*. Washington DC: WRI. https://files.wri.org/d8/s3fs-public/18_WP_SLCPs_toprint2.pdf (accessed 22 June 2020).

Royal Society and US National Academy of Sciences (2014). *Climate Change: Evidence and Causes*. Washington DC: National Academies Press.

Royal Society and US National Academy of Sciences (2020). Climate Change: Evidence and Causes. https://royalsociety.org/-/media/Royal_Society_Content/policy/projects/climate-evidence-causes/climate-change-evidence-causes.pdf (accessed 27 February 2021).

Sachs, J. (2015). *The Age of Sustainable Development*. New York: Columbia University Press.

Seidenkrantz, M.S. (2018) Eighty years since the first calculations showed that the Earth was warming due to rising greenhouse gas emissions. *Phy.org*. https://phys.org/news/2018-06-years-earth-due-greenhouse-gas.html (accessed 7 April 2021).

Shindell, D (2020). Health and Economic Benefits of a 2°C Climate Policy: Testimony to the House Committee on Oversight and Reform Hearing on "The Devastating Impacts of Climate Change on Health" 5 August 2020. https://nicholas.duke.edu/sites/default/files/documents/Shindell_Testimony_July2020_final.pdf (accessed on 12 March 2021).

Sjöstedt, G. and Penetrante, A.M. (ed.) (2015). *Climate Change Negotiations: A Guide to Resolving Disputes and Facilitating Multilateral Cooperation*. London: Routledge.

Smil, V. (2017). *Energy and Civilization: A History*. Cambridge: MIT Press.

Sokona, Y. et al. (2004). *Energy Services for the Poor in West Africa Sub-Regional 'Energy Access' Study of West Africa. Prepared for 'Energy Access' Working Group, Environnement et D é veloppement du Tiers Monde & Global Network on Energy for Sustainable Development*. New York: UNEP.

Sovacool, B. (2012). The political economy of energy poverty: a review of key challenges. *Energy for Sustainable Development* 16 (3): 272–282.

Srivastava, L. et al. (2012). Energy access: revelations from energy consumptions patterns in rural India. *Energy Policy* 47 (1): 11–20.

Stern, N. et al. (ed.) (2014). *The Global Development of Policy Regimes to Combat Climate Change*. London: World Scientific.

Stockholm Environmental Institute et al. (2020). The Production Gap Report: 2020 Special Report. https://productiongap.org/wp-content/uploads/2020/12/PGR2020_FullRprt_web.pdf (accessed 5 April 2021).

Stokols, E. (2017). Trump quits global climate change accord. *Wall Street Journal* (2 June 2017).

Sutter, J. and Davidson, L. (2018) Teen tells climate negotiators they aren't mature enough. *CNN news* (17 December 2018). https://www.cnn.com/2018/12/16/world/greta-thunberg-cop24/index.html (accessed 7 January 2021).

Timperly, J. (2021). The broken $100 billion promise in climate finance- and how to fix it. *Nature* https://www.nature.com/articles/d41586-021-02846-3 (accessed on November 1, 2021).

Travaglio, M. et al. (2021). Links between air pollution and COVID-19 in England. *Environmental Pollution* 268. Part A. https://www.sciencedirect.com/science/article/pii/S0269749120365489?via%3Dihub (accessed 20 April 2021).

UN (1987). *Our Common Future: Report of the World Commission on Environment and Development*. New York: UN.

UN (2008). *Climate Change: Impacts, Vulnerabilities and Adaptation in Developing Countries*. Bonn: UNFCCC Secretariat.

UN (2009). *The Impact of Climate Change on the Development Prospects of the Least Developed Countries and Small Island Developing States*. New York: UN.

UN (2015). Transforming our World: The 2030 Sustainable Development Agenda. https://sustainabledevelopment.un.org/post2015/transformingourworld (accessed 22 June 2020).

UN News (2019a). World faces 'climate apartheid', 120 more million in poverty: UN expert. *UN News*. https://news.un.org/en/story/2019/06/1041261 (accessed 30 June 2021).

UN News (2019b) Ground breaking clean air protocol to guard human health and planet enters into force. *UN News*. https://news.un.org/en/story/2019/10/1048612 (accessed 1 July 2021).

UN News Centre (2007). General Assembly mechanism needs to devise climate change response – Malta. https://news.un.org/en/story/2007/09/233152-general-assembly-mechanism-needed-devise-climate-change-response-malta (accessed 2 April 2021).

UN Press Release (2019). Climate Battle 'Will Largely Be Won or Lost' in Cities, Secretary-General Tells World Mayors Summit. https://www.un.org/press/en/2019/sgsm19808.doc.htm (accessed 12 March 2021).

UN website (2021). Sustainable Development Goals. https://www.un.org/sustainabledevelopment/sustainable-development-goals (accessed 30 July 2021).

UNDP (2002). *Energy for Sustainable Development: A Policy Agenda*. New York: UNDP.

UNDP (2007). *Human Development Report 2007/2008: Fighting Climate Change*. New York: UNDP.

UNDP/UNEP/WEC (2000). *World Energy Assessment*. New York: UNDP.

UNDESA (2018). *World Urbanization Prospects: 2018 Revision*. New York: UNDP.

UNEP (2013). *Year Book: Emerging Issues in Our Global Environment*. Nairobi: UNEP.

UNEP (2019). *Emissions Gap Report 2019*. Nairobi: UNEP. https://wedocs.unep.org/bitstream/handle/20.500.11822/30797/EGR2019.pdf?sequence=1&isAllowed=y (accessed 14 April 2021).

UNEP (2020a). *Emissions Gap Report 2020: Full Report*. Nairobi: UNEP. https://www.unep.org/emissions-gap-report-2020 (accessed 14 April 2021).

UNEP (2020b). *Emissions Gap Report: Executive Summary*. Nairobi: UNEP. https://wedocs.unep.org/bitstream/handle/20.500.11822/34438/EGR20ESE.pdf?sequence=25 (accessed 20 April 2021).

UNEP/WMO (2011). *Integrated Assessment of Black Carbon and Tropospheric Ozone: Summary for Policy Makers*. Nairobi, Kenya: UNEP/WMO.

UNFCC (2015). Paris Agreement. Report of the Conference of the Parties on its twenty-first session. FCCC/CP/2015/10/Add.1. https://unfccc.int/sites/default/files/english_paris_agreement.pdf (accessed 7 April 2021).

UNFCCC (2015). Report of the Conference of the Parties on its twenty first session. Actions taken by the Conference of the Parties and Adoption of the Paris Agreement FCCC/CP/2015/10/Add.1. https://unfccc.int/sites/default/files/resource/docs/2015/cop21/eng/10a01.pdf (accessed 23 April 2021).

UNFCCC (2020). *Climate Change: Annual Report 2019*. Bonn: UNFCC Secretariat. https://unfccc.int/sites/default/files/resource/unfccc_annual_report_2019.pdf (accessed 2 April 2021).

UNGA (1988). *Protection of Global Climate for Present and Future Generations of Mankind*, A/Res/43/53. New York: UN.

UNGA (2019) Climate Change and Poverty. A/HRC/41/39 (17 July 2019). https://undocs.org/en/A/HRC/41/39 (accessed 3 January 2021).

UNICEF (2016). *Clear the Air for Children*. New York: UNICEF https://www.unicef.org/publications/files/UNICEF_Clear_the_Air_for_Children_30_Oct_2016.pdf (accessed 22 April 2021).

Union of Concerned Scientists (2020). Each Country's share of CO_2 Emissions Report. https://www.ucsusa.org/resources/each-countrys-share-co2-emissions (accessed 19 November 2020).

US Department of Defense (2014). *2014 Climate Change Adaptation Road Map*. Alexandria, VA: Office of the Deputy Under Secretary of Defense for Installations and Environment.

https://www.acq.osd.mil/eie/downloads/CCARprint_wForward_e.pdf (accessed 1 June 2021).

US EPA Website (2021). Global Greenhouse Gas Emissions Data. https://www.epa.gov/ghgemissions/global-greenhouse-gas-emissions-data (accessed 20 April 2021).

US ODNI (2021). National Intelligence Estimate: Climate Change and International Responses Increasing Challenges to US National Security Through 2040" 21 October 2021. https://www.dni.gov/files/ODNI/documents/assessments/NIE_Climate_Change_and_National_Security.pdf (accessed 30 October 2021).

Wallace-Wells, D. (2019). *The Uninhabitable Earth*. London: Allen Lane.

Weart, S. (2008). *The Discovery of Global Warming*. Cambridge, MA: Harvard University Press.

World Health Assembly (2015). Sixth-Eigth World Health Assembly: Resolutions and Decisions. https://apps.who.int/gb/ebwha/pdf_files/WHA68-REC1/A68_R1_REC1-en.pd (accessed 24 April 2021).

WEF (2014). *Global Risks 2014*. Geneva: World Economic Forum.

WEF (2019a). *Global Risks Report 2019*. Geneva: WEF http://www3.weforum.org/docs/WEF_Global_Risks_Report_2019.pdf (accessed 10 January 2021).

WEF (2019b). Greta Thunberg's World Economic Forum 2019 Special Address. http://opentranscripts.org/transcript/greta-thunberg-world-economic-forum-2019 (accessed 12 March 2021).

WHO (1997). *Health and Environment in Sustainable Development*. Geneva: WHO. https://ise.unige.ch/isdd/IMG/pdf/WHO_EHG_97.12_eng.pdf (accessed 7 January 2021).

WHO (2005) Air Quality Guidelines. https://apps.who.int/iris/bitstream/handle/10665/69477/WHO_SDE_PHE_OEH_06.02_eng.pdf;sequence=1 (accessed on 19 December 2021)

WHO (2006). *WHO Air Quality Guidelines for Particulate Matter, Ozone, Nitrogen Dioxide and Sulfur Dioxide Global Update 2005*. Geneva: WHO. https://apps.who.int/iris/bitstream/handle/10665/69477/WHO_SDE_PHE_OEH_06.02_eng.pdf;jsessionid=354F2E4677FA9D4CD1A15F12EB6321BE?sequence=1 (accessed 2 April 2021).

WHO (2007). *Indoor Air Pollution: National Burden of Disease Estimates*. Geneva: WHO.

WHO (2015). *Health and the Environment: Addressing the Health Impact of Air Pollution: Report by the WHO Secretariat for 68th World Health Assembly*. A68/18. Geneva: WHO. https://apps.who.int/iris/bitstream/handle/10665/253237/A68_R8-en.pdf?sequence=1&isAllowed=y (accessed 24 January 2021).

WHO (2016a). Ambient air pollution: A global assessment of exposure and burden of disease. https://apps.who.int/iris/bitstream/handle/10665/250141/9789241511353-eng.pdf?sequence=1 (accessed 24 January 2021).

WHO (2016b). Air pollution level rising in many of the world's poorest cities. WHO Press release. http://www.who.int/en/news-room/detail/12-05-2016-air-pollution-levels-rising-in-many-of-the-world-s-poorest-cities 12 May 2016 (accessed 27 February 2021).

WHO (2016c). *Burning Opportunity: Clean Household Energy for Health, Sustainable Development and Wellbeing of Women and Children*. Geneva: WHO. https://apps.who.int/iris/bitstream/handle/10665/204717/9789241565233_eng.pdf;jsessionid=08AD6DE7F290AE96DC998826D0647C81?sequence=1 (accessed 27 February 2021).

WHO (2017). *Don't Pollute My Future*. Geneva: WHO. https://apps.who.int/iris/bitstream/handle/10665/254678/WHO-FWC-IHE-17.01-eng.pdf;jsessionid=9AF0B0BC2669EB1419CDB9CDCBE9D312?sequence=1 (accessed 12 March 2021).

WHO (2018). *COP-24: Special Report – Health and Climate Change*. Geneva: WHO. https://apps.who.int/iris/handle/10665/276405 (accessed 22 April 2021).

WHO Media Centre (2014). *House Hold Air Pollution and Health. Fact Sheet 292*. Geneva: WHO. https://www.who.int/mediacentre/news/releases/2014/air-pollution/en (accessed 2 April 2021).

WHO Website (2021). Key Facts: Ambient – Outdoor Air Pollution in 2018. https://www.who.int/news-room/fact-sheets/detail/ambient-(outdoor)-air-quality-and-health (accessed 7 April 2021).

WHO/UNDP (2009). *The Energy Access Situation in Developing Countries*. New York: UNDP.

World Bank (2016). *Shock Waves: Managing the Impacts of Climate Change and Poverty*. Washington DC: World Bank.

World Bank (2020). COVID-19 to add as many as 150 million extreme poor by 2021. World Bank Press Release. https://www.worldbank.org/en/news/press-release/2020/10/07/covid-19-to-add-as-many-as-150-million-extreme-poor-by-2021 (accessed 12 February 2021).

Wu, X., Nethery, R.C., Sabath, M.B., Braun, D., and Dominici, F., 2020. Air pollution and COVID-19 mortality in the United States: strengths and limitations of an ecological regression analysis. *Science Advances*, 6(45), p. eabd4049. https://advances.sciencemag.org/content/6/45/eabd4049 and https://projects.iq.harvard.edu/covid-pm (accessed 20 April 2021).

2

Identifying the Locus for Global Action on Clean Energy and Climate Change within the UN

Confronting Segregated Global Goals and Partnership Silos

2.1 Background: Confronting Global Neglect of Climate Related Health Risks

The global prognosis for severe climatic impacts has always been and remains bleak for the poorest communities across the world. On 8 August 2013, super-typhoon Haiyan ripped through the middle of the Philippines, killing 6300 and displacing 4 million more. The after-shocks of livelihoods lost meant that one of the most powerful tropical cyclones ever recorded pushed more than a million more Filipinos further into poverty (World Vision 2019). According to the 2019 Asia Pacific Disaster Report by the UN Economic and Social Commission for Asia and the Pacific (UNESCAP), an unceasing barrage of extreme climatic disasters ranging from raging floods to crippling droughts in the Asia–Pacific region causes the greatest damage in poor and marginalized communities. The report highlights the vicious cycle of exposure to extreme climatic adversities and disasters worsening existing poverty and inequality. 'Poor populations typically lose more because they are overexposed to disasters and have less ability to cope and recover, especially if they have little social protection or post disaster support. Moreover, disasters often have permanent impacts on their education and health thereby locking people into intergenerational poverty traps' (UNESCAP 2019, p. 11). The 2019 State of Climate in Africa Report outlined that climate variability, extreme weather events are amongst the key drivers of the recent increase in global hunger: 'After decades of decline, food insecurity and undernourishment are on the rise in almost all sub-regions of sub-Saharan Africa. In drought-prone sub-Saharan African countries, the number of undernourished people has increased by 45.6% since 2012 according to UN's Food and Agriculture Organization. The year 2019 recorded a deteriorating food security situation in sub-Saharan Africa, as well as increased population displacement and the increased food insecurity of those displaced people. Refugee populations often reside in climate "hot spots", where they are exposed to and affected by slow and sudden-onset hazards, resulting in some cases in secondary displacements' (WMO 2020, p. 18).

Highlighting the 'tone deaf' outcome of COP-26 to the challenges faced by millions of the poorest communities who have done the least to contribute to per capita GHG emissions, the Executive Director of Oxfam issued an excoriating statement: 'Clearly some world leaders think they aren't living on the same planet as the rest of us. It seems no amount of fires, rising sea levels or droughts will bring them to their senses to stop

Air Pollution, Clean Energy and Climate Change, First Edition. Anilla Cherian.

increasing emissions at the expense of humanity. Punishing extreme weather is already wrecking the lives of the most vulnerable. People are barely clinging on, having little resources to cope with the constant threat of losing all they own. The world's poorest have done the least to cause the climate emergency, yet are the ones left struggling to survive while also footing the bill' (Oxfam Press Release 2021). What happens when existing health and disease burdens accruing from exposure to solid fuel/fossil fuel combustion are layered upon adverse socio-economic and climate vulnerabilities? The UN's SDA enshrines poverty reduction as the priority goal but have the health impacts of exposure to fossil fuels been linked to intergovernmental action on climate and clean energy access with the context of the UN's SDA? How have UN SDG partnerships addressed the intersecting challenges of fossil fuel energy related air pollution, lack of access to clean energy and climate change?

PM air pollution needs to be factored in as one of the world's largest environmental health risks which is inherently linked to climate change, but there is an even broader context to grapple with, namely, global neglect of climate related health impacts. Confronting the policy nexus between energy related air pollution, climate change and health inequities requires recognizing that concepts such as equality/inequality, fairness/unfairness and justice/injustice terms have been varyingly defined, analyzed and responded to in terms of policy implementation. While this chapter does not focus on providing analytical inputs on the concept of environmental or climate injustice/inequity, it is important to reference the complexities of analyzing environmental injustices that do not translate into treating marginalized communities as being homogenous groups. What this means is recognizing for example as Detraz does that 'there are multiple conceptualizations of justice at play' which is 'not unusual, as there are different perceptions about which aspects of justice should take precedence over others' (2017, p. 52). But what is significant to note is that the disproportionate impacts of environmental risks on poor communities, communities of colour and indigenous people has been amply evidenced within the realm of 'environmental justice' studies (United Church of Christ 1987; Bullard 1993, 2005; Pellow and Brulle 2005; Banerjee 2014). In 1987, the United Church of Christ (UCC's) Commission on Racial Justice Report 'Toxic Wastes and Race in the US' coined the term 'environmental racism' and issued a finding that still resonates today: 'Race proved to be most significant among variables tested in association with the location of commercial hazardous waste facilities. This represented a consistent national pattern' (1987, p. xiii). Troublingly, a follow-up report 20 years later found that '. . . racial disparities in the distribution of hazardous wastes are greater than previously reported' and that 'people of color make up the majority of those living in host neighborhoods within 3 kilometers (1.8 miles) of the nation's hazardous waste facilities. Racial and ethnic disparities are prevalent throughout the country' (Bullard et al. 2007, p. x).

The US EPA definition builds on the idea of environmental justice as an equality-driven principle: 'Environmental justice is the fair treatment and meaningful involvement of all people regardless of race, color, national origin, or income, with respect to the development, implementation, and enforcement of environmental laws, regulations, and policies. EPA has this goal for all communities and persons across this Nation. It will be achieved when everyone enjoys the same degree of protection from environmental and health hazards and equal access to the decision-making process to have a healthy environment in which to live, learn, and work' (US EPA 2013). But, it is precisely the systematic and endemic lack of access to a healthy environment faced by millions of poorer and mostly black and brown lives in the context of layered threats of pollution and climate vulnerabilities that is key to understanding the imperative for linked community-driven action. Here, it is also important to keep in mind the 'environmentalism of the poor' as referenced by Martinez-Alier (2002, 2014). Bullard and Wright (2012), for instance, contrast how official government responses to disasters ranging from environmental and public health emergencies, toxic contamination, industrial accidents, bioterrorism threats show that African Americans are disproportionately affected and argue that differentiated disaster responsiveness impacts most negatively on vulnerable lives.

Within the sustainable development arena, issues of intergenerational fairness and equity were raised early on by Weiss (1989, 1992), while the role of gender equality and sustainable development was discussed by Leach (2015). The globalization of environmental injustice including climate injustice was examined by Shiva (2008) in her book *Soil not Oil: Environmental Justice in an Age of Climate Crisis*, and a range of research has been done on intersectionalities between gender, environment and climate change justice (Hartmann 2010; Alston 2014; Detraz 2014). Live, learn, work has been defined as the arena for environmental justice, but it is important to keep in mind that within the UN context, climate justice was not explicitly referenced in any specific globally agreed SDG in terms of the unjust burdens of ill health and morbidity borne by those who are least responsible for per capita GHG emissions. The triple intersection of poverty, ill health and climate adversities is, however, real and was highlighted by WHO (2018) as well as in 'Mental Health and Our Changing Climate' (Clayton et al. 2017). The latter report noted that the impacts of climate change on human physical, mental and community health arise both 'directly' as a result of coping with extreme events such as floods, storm wildfires and heat waves exacerbated by climate change, and 'indirectly' as a result of weakened infrastructure and food insecurity which can also impact on physical and mental health. It found that: 'Some communities and populations are more vulnerable to the health related impacts of climate change. Factors that increase sensitivities to mental health impacts include geographic location, presence of pre-existing disabilities or chronic illnesses, and socioeconomic and demographic inequalities such as education level, income and age' (2017, p. 6). Highlighting this triple threat concern, the influential *Countdown on Climate Change and Health* pointed out that human impacts of climate change are irreversible – affecting the health of populations around the world – and called attention to the disproportionate impact climate change has on the world's most marginalized populations if social and environment justice concerns are not addressed (Watts et al. 2018). So how have the UN's ambitious SDA and its SDGs responded to the linked health and human development concerns resulting from the lack of access to clean energy, exposure to PM pollution and climate action?

In order to understand how air pollution and climate action have been addressed within the UN's SDA, it is necessary to first focus briefly on the issue of climate related health inequities. Climatic impacts such as increasing global temperatures, extreme changes in precipitation, ecosystem damage and disruption are some of the key causal factors contributing to the modern-day spread of vector-borne diseases (VBDs) like Dengue and Zika with the potential to wreak havoc on those who already lack access to adequate public health care. The transmission of VBDs like Zika, for example, provides a worrisome answer to the question of how climatic impacts worsen existing socio-economic inequalities and also demonstrates global neglect of health inequities. With the world reeling from the COVID-19 pandemic, it is possible that VBDs like Zika might not be occupying public attention, but it is worth remembering that Zika was identified as the 'new global threat for 2016' (Lancet Editorial 2016). This fear was echoed by leaders of two US Health Agencies in a joint news conference on 11 April 2016. Dr. Anne Schuchat, Principal Deputy Director of the Centers for Disease Control (CDC), noted that the range of the Aedes mosquito species responsible for spreading Zika was far larger than previously thought and the disease's effects were more damaging than initial medical studies suggested. Dr. Fauci, then head of the National Institute of Allergy and Infectious Diseases at the National Institute of Health, stressed the importance of funding for research and prevention stating that: 'I don't have what I need right now, what I've done is take money from other areas of non-Zika research to start' (Mohney 2016). A CDC report entitled 'Possible Zika Virus Infection Among Pregnant Women — United States and Territories' issued a clear warning: 'Zika virus is a cause of microcephaly and brain abnormalities, and it is the first known mosquito-borne infection to cause congenital anomalies in humans' (Simeone et al. 2016). Zika, like two other VBDs – Dengue and Chikungunya – spreads among humans via the *Aedes aegypti* mosquito. But, it is what is not known about Zika including the full extent of its impacts on serious birth defects, partial paralysis, its possible connection to those previously infected by Dengue, as well as the first reported case of Zika via sexual transmission in the US, that had international health experts and organizations panicked. Discovered in 1947, what was known was that the impacts of the Zika virus were largely contained within the equatorial belt in Africa and Asia. Until very recently, hardly any attention was paid to it by the global health community. The image (see Figure 2.1) of a small stack of papers on a desk, with the caption 'Entire world literature on Zika. 50 years of neglect', tweeted out by then Director of the CDC, Dr. Frieden, served as a visual shock tool (Frieden 2016).

Global neglect of the intersection between the lack of access to public health services, poverty and climate vulnerability in the case of exposure to both HAP and VBDs like Zika reveals gender inequities. Zika's suspected linkages to cognitive development birth defects

Entire world literature on Zika. 50 years of neglect.

10:11 AM. Feb 12, 2016. Twitter for iPhone

Figure 2.1 Image of entire world literature on Zika: Director of CDC (2016). *Source:* Frieden (2016).

worsened existing reproductive, maternal and child health care dilemmas faced by poor women and their families. As Basu stated: 'Being poor is bad enough in this context. But being poor and female is even worse, because it increases the possibility that the outcome – an infection with Zika – will be more devastating than the rash or fever or conjunctivitis that typically describes the infection' (Basu 2016). Here, it is useful to reference Detraz's point that 'gendered inequality' has 'been much less discussed as a component of environmental (in)justice than has race or class' (2017, p. 53). In light of the health care costs and burdens of the COVID pandemic being inequitably borne by poorer countries and communities, a 2015 Harvard independent panel report on the global response to Ebola – 'Will Ebola Change the Game: Ten Essential Reforms Before the Next Pandemic' was foreboding. The report found that the Ebola epidemic 'exposed deep inadequacies in the national and international institutions responsible for protecting the public from the far-reaching human, social, economic, and political consequences of infectious disease outbreaks'. The report's key recommendations related to preventing major disease outbreaks are revelatory in light of current and future globally infectious pandemics:

'Recommendation 1: The global community must agree on a clear strategy to ensure that governments invest domestically in building such capacities and mobilise adequate external support to supplement efforts in poorer countries. This plan must be supported by a transparent central system for tracking and monitoring the results of these resource flows. Additionally, all governments must agree to regular, independent, external assessment of their core capacities.

Recommendation 2: WHO should promote early reporting of outbreaks by commending countries that rapidly and publicly share information, while publishing lists of countries that delay reporting. Funders should create economic incentives for early reporting by committing to disburse emergency funds rapidly to assist countries when outbreaks strike and compensating for economic losses that might result' (Moon et al. 2015, p. 2204).

So, what happens when adverse climatic impacts are added to the mix of global public health neglect? Increasing global temperatures, extreme changes in precipitation, ecosystem damage and disruption are causal factors contributing to the spread of VBDs with the potential to decimate the poorest and most vulnerable. These are not hypothetical concerns that can be ignored. The 2014 AR5 of the IPCC identified three pathways by which climate change impacts on human health: (i) Direct impacts, which relate primarily to changes in the frequency of extreme weather including heat, drought, and heavy rain; (ii) Effects mediated through natural systems, for example, disease vectors, water-borne diseases, and air pollution; and (iii) Effects heavily mediated by human systems, for example, occupational impacts, undernutrition, and mental stress (IPCC, Synthesis Report 2014).Zika and Chikungunya were not mentioned in IPCC's AR5 which stated that mosquito-borne diseases – Malaria and Dengue – are 'the best-studied diseases associated with climate change', due to sensitivity to climatic factors such as temperature, precipitation and humidity (IPCC 2014, p. 723). What makes the spread of climate-sensitive VBDs like Zika and Dengue more challenging is that poorer communities and countries are already more vulnerable to adverse impacts of climatic change. Dengue is a VBD that merits particular global attention, but the reality is that research on these linked VBDs has been globally neglected. As Gubler, in 'Economic Burden of Dengue', highlighted, dengue has been 'ignored for many years, and only recently has the potential magnitude of the dengue problem been acknowledged by policymakers and funding agencies'. The paper goes on to point out that: 'In 2012, dengue is the most important vector-borne viral disease of humans and likely more important than malaria globally in terms of morbidity and economic impact. The latest studies estimate 3.6 billion people living in areas of risk, over 230 million infections, millions of cases of dengue fever, over 2 million cases of the severe disease, and 21,000 deaths' (Gubler 2012, p. 743). Gubler's argument that the lack of adequate planning, housing, water, sewage and waste management created ideal conditions for the early spread of dengue virus has policy resonance. Emerging pathogen experts have argued that tackling Zika, Dengue and VBDs requires broader development measures aimed at alleviating urban poverty (Vittor 2016). The need for integrated global public health services and research that is responsive to needs of poor urban women and newborns was seen as essential to its 'Transforming Our World: The 2030 Sustainable Development Agenda' that makes poverty eradication a priority and 'pledges to leave no one behind' (UNGA 2015a). But, if funds for research and prevention are currently lacking in countries like the US, it is incomprehensible to assume that minimum core capacities to combat VBDs exist in poorer countries.

In addition to the health inequities associated with spread of climate-sensitive VBDs, one of the most neglected health challenges for developing countries remains the linkages between climate change and mental health. Back in 2007, the Inter-Agency Standing Committee (IASC), supported by the WHO, created a set of guidelines on mental health and psychosocial support in the event of extreme weather or disaster events. The IASC guidelines recommend epidemiological surveys of mental disorders and distress of the general population following extreme weather events that consider indicators of risk (e.g. marginalized status) and protective factors (e.g. access to mental health care) (IASC 2008). But, access to resources (institutional, human and financial) has not yet been reliably extended in the poorest communities and countries of the world hit hard by extreme climatic catastrophes. Hayes and Poland (2018) in their comprehensive review of the global literature on mental health and climate change provided an extensive table summarizing a wide range of mental health impacts of climate change hazards and the populations of concern. But the world's largest climate related health risk –fossil fuel energy related air pollution – is absent in this mapping tool. The 'climate hazard' of polluting fossil fuel emissions including those emanating from the inefficient combustion of solid fuels in households has not been factored in; and the focus on the populations of concern in the case of VBDs does not reflect the scope of diseases like Dengue, Malaria, and Zika which make human health devastation extensive in developing countries. However, a 2021 report sponsored by the American Psychological Association entitled 'Mental Health and Our Changing Climate' acknowledged that: 'The inequities of climate change are vast and significant. While the destructive impacts of climate change will be felt by everyone, the burdens will fall heavily on those oppressed by historic and present day social, economic, and political power dynamics. No group is homogenous, however those who are economically disadvantaged, from communities of color, are indigenous, children, older, or women, have disabilities or pre-existing mental health conditions, or are outdoor workers may be more prone to mental health difficulties as climate change exacerbates preexisting vulnerabilities' (Clayton et al. 2021, p. 6).

The 2014 IPCC Synthesis Report had high confidence in its finding that the health effects of climate change impact differentially and negatively on the global poor, including the health risks associated with air pollutants: 'Climate change is an impediment to continued health improvements in many parts of the world. If economic growth does not benefit the poor, the health effects of climate change will be exacerbated. In addition to their implications for climate change, essentially all the important climate-altering pollutants (CAPs) other than carbon dioxide (CO_2) have near-term health implications (very high confidence). In 2010, more than 7% of the global burden of disease was due to inhalation of these air pollutants' (IPCC 2014, p. 713). As referenced by the IPCC's AR5: 'Put into terms of disability-adjusted life years (DALYs), particle air pollution was responsible for about 190 million lost DALYs in 2010, or about 7.6% of all DALYs lost. *This burden puts particle air pollution among the largest risk factors globally, far higher than any other environmental risk and rivaling or exceeding all of the five dozen risk factors examined, including malnutrition, smoking, high blood pressure, and alcohol*' (emphasis added, IPCC 2014, p. 728). This IPCC finding about the scope of risk associated with PM pollution provides an added urgency in avoiding any delay related to global and regional partnerships and protocols to curb PM pollution. It also lends credence to the global need of leveraging integrated action on clean air, clean energy and climate mitigation. The question is whether the lack of access to clean energy and fossil fuel related air pollution has been addressed within the broad context of the UN SDA? The immediate sections below provide an overview of the UN's evolution towards the SDA and its separate SDGs on energy and climate change. In examining key UN-agreed outcomes on sustainable development, the question raised is whether policy confusion or clarity on clean air, clean energy and climate partnerships exists as a result of over 25 years of intergovernmental negotiated outcomes on energy for sustainable development and climate change.

2.2 Segregated UN Goal Silos on Clean Energy and Climate Change

Within the UN context, intergovernmental negotiated responses to both sustainable energy and climate change are located within the broader framework of sustainable development, which is one the five overall focal areas of the UN. The UN has a massive global mandate to implement and works on behalf its member states in the following five broad areas to:

- Maintain international peace and security;
- Protect human rights;
- Deliver humanitarian aid;
- Promote sustainable development and climate action;
- Uphold international law (UN.org website 2021a).

In the over 75 intervening years since the inception of the UN Charter, the world has witnessed a dramatic increase in global challenges including the current concern over the spread of global pandemics. The concept and implementation of sustainable development within the UN has evolved over time primarily as a result of the agreed global outcomes of a series of intergovernmental negotiations. Sustainable development has emerged as one of the five areas of the UN's mandate as a result of landmark UN negotiations

Conferences/Summits which initially focused on concepts such as 'environment and development' and then honed in on the term 'sustainable development' (Sachs 2015). The key global conferences that were influential in shaping the UN's evolving agenda on this topic include the following:

- UN Conference on the Human Environment (UNCHE) (1972)
- World Commission on Environment and Development (WCED) (1987)
- UN Conference on Environment and Development (1992)
- General Assembly Special Session on the Environment (1997)
- World Summit on Sustainable Development (WSSD) (2002)
- UN Conference on Sustainable Development (2012)

The UN benchmark adoption of the term 'sustainable development' can be traced back to the 1987 World Conference on Environment and Development (WCED) Report, which defined the concept of sustainable development. 'Sustainable development is development that meets the needs of the present without compromising the ability of future generations to meet their own needs. It contains within it two key concepts:

- the concept of needs, in particular the essential needs of the world's poor, to which overriding priority should be given; and
- the idea of limitations imposed by the state of technology and social organization on the environment's ability to meet present and future needs (UN 1987, para 4).

Both these key concepts – needs of the world's poor to which priority is accorded and the environment's ability to meet present and future needs – resonate decades later in the UN's SDA with its pressing 2030 implementation deadline. The agreed premise about sustainable development meeting the basic needs of all, and extending opportunities for human well-being can be seen therefore as essential in terms of the evaluation of efforts made towards responding to air pollution and climate related health risks disproportionately experienced by the world's global poor. To understand how clean energy, climate change and clean air are dealt with in the UN context, it is necessary to briefly highlight the complexity of UN global environmental treaty making and implementation apparatus that comprises a gargantuan global SDA. Here, it is useful to recognize what Haas delineated as the inherent challenges in delivering on the promise of sustainable development: 'Sustainable development requires a reorientation of collective understanding and of formal institutions to focus on the key intersecting and interacting elements of complex problems. Technically, efforts to cope with environmental threats must be comprehensive if they are to address the complex array of causal factors associated with them. Yet comprehensiveness is difficult to achieve, because few governments or international institutions are organized to cope with the multiple dimensions of environmental problems, and many states lack the technical resources to develop and apply such efforts' (2004, p. 570).

Najam focused on the specific challenges related to global environmental governance for developing countries (2005). Dealing simultaneously with poverty eradication while responding to an array of global environmental problems ranging from biodiversity loss to climate change, all while increasing access to public health, clean and cost-effective energy and curbing air pollution are massive challenges for states that lack capacities and resources, and that is exactly why priority must and should be given to linked responsive actions such as the policy nexus between clean air and clean energy for the poor with the UN context. As Kanie in 'Governance with Multilateral Environmental Agreements: A Healthy or Ill-Equipped Fragmentation' noted, the concept of sustainable development 'calls for simultaneous and concerted efforts to deal with pollution, economic development, unequal distribution of economic resources, and poverty reduction' (2007, p. 69). Deriving from this and put simply, linkages matter for implementing sustainable development. An integrated policy agenda on clean energy, clean air, climate action and poverty eradication can and should be seen as imperative to effective global action on multilaterally agreed upon SDGs. But does an examination of the global track record on UN outcomes on energy, air pollution and climate change reveal an integrated policy nexus on these challenges or segregated policy and goal silos?

Evaluating the extent to which clean air/curbing air pollution has been factored into climate change and energy for sustainable development requires confronting the fact that the UN and its member states have been addressing global and regional environmental challenges via a complex web of multilateral agreements that are continually growing in scope. Ever since the historic 1972 UNCHE, also known as the Stockholm Conference, there has been a rapid growth in a series of UN multilateral or global agreements/treaties focused on addressing an array of global environmental challenges. These multilateral environmental agreements (MEAs) have been put into place to address a host of environmental problems ranging from: loss of biological diversity and species, elimination of harmful chemicals on the ozone layer, combating drought and desertification and stabilizing atmospheric GHG concentrations to prevent anthropogenic climate change, control of transboundary movements of hazardous wastes and their disposal, protection of the ozone layer, international trade in endangered species and the protection of wetlands, to name just a few. An in-depth review of the global negotiation processes resulting from the proliferation of MEAs, the roles of governments, NGOs and Convention Secretariats, as well as the implementation challenges of a range of Conventions including the Framework Convention on Climate Change, the Convention to Combat Desertification, the Convention on Biological Diversity, the Commission on Sustainable Development, the UN Forum on Forests, the chemicals conventions (Stockholm, Basel and Rotterdam), the Montreal Protocol on Substances that Deplete the Ozone Layer, the Convention on International Trade in Endangered Species (CITES), the Convention on Migratory Species and the International Treaty on Plant Genetic Resources for Food and Agriculture, over a 20-year period was offered by Chasek and Wagner (2012). In *Global Environmental Politics*, Chasek and Downie (2021) examine an exhaustive range of UN environmental action ranging from climate change, endangered species, ozone depletion, desertification, whaling, hazardous wastes, toxic chemicals and biodiversity and detail the ongoing development of major environmental treaty regimes, mercury and marine issues, SDGs, the 2030 Agenda for Sustainable Development, trade and environment and the impact of the COVID-19 pandemic. The more circumscribed aim of this chapter is not to analyze the negotiations processes for climate change and energy but instead to focus on examining the record of the agreed global UN outcomes and goals on climate, clean energy and clean air for their guidance and responses in terms of partnership-based actions.

According to the UNEP's 'Glossary of Terms for Negotiators of Multilateral Environmental Agreements', an MEA is defined as: 'A generic term for treaties, conventions, protocols, and other binding instruments related to the environment. Usually applied to instruments of a geographic scope wider than that of a bilateral agreement (i.e., between two States)' (UNEP 2007, p. 62). But, there is another category of international environmental agreements which are non-binding agreements also referred to as 'soft law agreements'. In this sense, as per UNEP, the term 'agreements' broadly encompasses diverse instruments, such as treaties, conventions, protocols or oral agreements and yet is also a specific term used to designate international instruments that are not as formalized, correspond to soft law and deal with a narrower range of subject matter than treaties (2007, p. 10). MEAs are standalone agreements, but they can also be 'framework agreements' such as the UNFCCC and the Convention on Biological Diversity in which further agreements (Protocols – the KP or the Cartagena Protocol on Biosafety) provide the necessary standards, procedures and other requirements to further implement the MEA in question. Other forms of MEAs rely heavily on appendices, such as CITES. Unlike other global policy fields such as trade, labour or health, global environmental policymaking hinges on the diverse outcomes of a range of MEAs, with sometimes overlapping issue areas, disparate mandates and conflating institutional arrangements.

UN-affiliated multilateral treaty-making that is legally binding has largely been predicated on a two-step convention–protocol approach (Susskind 1994). But the problem in terms of evaluating MEAs as UNEP itself has pointed is that: 'There are hundreds of Multilateral Environmental Agreements (MEAs) dealing with various environmental issues and they are the main method available under international law for countries to work together on global issues. The assessment of the implementation, compliance and effectiveness of multilateral environmental agreements is in many cases complicated and plagued with gaps in data, conceptual difficulties and methodological problems' (UNEP 2010, p. 2). Adding to the evaluation challenge, MEAs like the UNFCCC and its KP that are legally binding have been combined with non-binding global agreements such as the PA which depend completely on individual countries scaling up their voluntary commitments. Global environmental governance within the UN context can therefore be characterized as closely linked to the institutional structures and status/progress of implementation associated with a diverse set of MEAs. Evaluating environmental governance via the MEAs has proven to be complicated: 'Arguably, the absence of a consistent, globally agreed on set of methods to assess and track MEAs complicates the evaluation of an international environmental governance system which is primarily organised around the development of MEAs, correlating to particular environmental challenges. Different groups of MEAs have been categorised under different thematic areas, depending on the organization that is tracking them and not on the basis of a globally agreed thematic categorization' (Cherian 2012, p. 40). The first ever recent global assessment on the environmental rule of law by UNEP found that, 'a dramatic 38-fold growth in environmental laws and agencies, plus massive investment in environmental agreements by donors, has not led to an equally pronounced improvement in the enforcement of those laws. It still comes down to political will' (UNEP 2019).

While there has been an escalation in MEAs, there has also been a dramatic increase in the number of civil society – NNSAs – involved in implementing them, so in addition to distinguishing between global MEAs, it is also relevant to underline the importance of NNSAs within the context of MEAs. As Kanie and Haas (2004) noted, the role of new and emerging forces in the form of non-state actors, when combined with the multiplicity of MEAs within an interdependent and globalized world, has major impacts on the roles/abilities of states to implement and enforce international environmental law and agreements. But, Kanie and Haas raised a point that remains accurate decades later: 'To date international environmental policy-making has generally been segregated on the basis of topic, sector, or territory. The result is the negotiation of treaties that often overlap and conflict with one another. This engenders unnecessary complications at the national level as signatories struggle to meet their obligations under multiple agreements. At the international level, some coordination efforts exist between environmental institutions through mechanisms such as the Inter-agency Coordination Committee and the Commission for Sustainable Development, but these institutions are far too weak to integrate the three dimensions of sustainable development effectively. They seem to have served more as a pooling regime rather than an effective coordination regime' (2004, pp. 2–3).

Within the UN context, the broadening of NNSAs involved in sustainable development can be seen in reference to nine 'major groups' that emerged in the voluminous 1992 UN global agreement – Agenda 21. The 'major groups' referenced in Agenda 21 include business and industry, children and youth, farmers, indigenous peoples, local authorities, NGOs, the scientific and technological community, women and workers and trade unions. The active involvement of all major groups was viewed as necessary to address global SDGs within the UN (Chasek and Wagner 2012). From 1992 onwards, the role of 'major groups' which comprise NNSAs has been seen as increasingly important to the success of numerous UN sustainable development agreements such as the 2002 Johannesburg Plan of Implementation (JPOI), the 2012 Future We Want (FWW) and the UN's 2030 SDA. However, a detailed review of the above global outcomes revealed a consistent lack of an integrated policy nexus between sustainable energy, poverty eradication and climate mitigation, combined with a lack of definitional clarity on the concept of partnerships related to non-state actors (Cherian 2015). The sections below examine NNSA partnerships aimed at addressing climate change and sustainable energy within one of the UN's most ambitious and universally adopted global agreements on sustainable development, namely, the 2030 SDA (UN 2015a).

Implementing globally agreed goals and partnerships on clean air, clean energy and climate change are crucial to the overall success of the UN's SDA, particularly in light of dwindling funding from official development assistance (ODA). Even before the global constraints resulting from COVID-19 pandemic, an example of the ODA funding shortfall was evidenced by the UN Secretary General Guterres's statement that the UN 'needed $4.4 billion by the end of March 2017 to avert a catastrophe' threatening the lives of 20 million famine victims in Africa because 'just $90 million has actually been received so far – around two cents for every dollar' (UN News Centre 2017). With economic distress apparent across the globe, it is increasingly clear that without the active participation of NNSA partnerships, the UN cannot tackle the scope and scale of the climate, clean energy and clean air challenges. Implementing action on specific MEAs such as the PA, as well as SDGs 7 and 13, hinges on the role of effective partnerships between state and non-state actors/stakeholders at the global, regional, national, city and community level. But, are the SDGs and the PA positioned to scale up linked partnership action?

To better understand how energy and climate goals have or have not been articulated and agreed upon within the UN context, it is necessary to begin with the adoption of the MDGs in 2000. The eight MDGs became the de facto policy tools for major institutions dealing with poverty eradication, human development and sustainability including the International Monetary Fund (IMF, Organization for Economic Cooperation and Development (OECD, the World Bank, regional development banks and developmental agencies, and the deadline for their achievement was set as 2015. Figure 2.2 references the eight MDGs including MDG 8, which called for a different kind of global partnership – 'a global partnership for development' (GPD).

Figure 2.2 The eight UN millennium goals (MDGs). *Source:* UN.org website, Millennium Development Goals (2021b).

The 2015 MDG progress report pointed to significant global progress towards the overarching priority goal of eradicating extreme hunger and poverty:

- 'Extreme poverty has declined significantly over the last two decades. In 1990, nearly half of the population in the developing world lived on less than $1.25 a day; that proportion dropped to 14 per cent in 2015.
- Globally, the number of people living in extreme poverty has declined by more than half, falling from 1.9 billion in 1990 to 836 million in 2015. Most progress has occurred since 2000.
- The proportion of undernourished people in the developing regions has fallen by almost half since 1990, from 23.3 per cent in 1990–1992 to 12.9 per cent in 2014–2016' (UN 2015b, p. 4).

But there was no focus on increasing access to sustainable energy for the poor, or climate change, much less on curbing energy related air pollution in any of the MDGs.

Neither the GPD nor the other MDGs focused on the role of key productive sectors such as agriculture, manufacturing and energy in addressing the overarching objective of poverty eradication. Additionally, MDG 8 – GPD – envisioned a very limited role for the NNSAs, and the role of the private sector was circumscribed to two specific targets. The GPD included only two private sector–specific targets – 8.E and 8.F related to access to affordable essential medicines and new technologies focused on information and communication but did not include any private sector role in increasing access to non-polluting and modern energy services. The UN's assessments of the overall effectiveness of the GPD reveal mixed results as per Box 2.1.

Given the in-built thrust of the MDGs implementation by governments/public-sector commitments, there was an expectation that developed countries would provide significant resource transfers to poorer ones to fight poverty, and private sector participation was limited to the targets listed in MDG 8. In hindsight, non-inclusion of access to sustainable energy which has been shown as crucial to poverty eradication was short-sighted as was limiting NNSA actions related to issues such as public transportation and access to clean energy. Rehman et al. (2017) argued for efficient and urgent action via a pro-poor hybrid model that linked public and private sectors in increasing access to energy and reducing the number of energy poor across the globe. Lack of access to sustainable public transportation remains a key factor for poor households. One of the largest continuing studies of upward mobility based at Harvard found that commuting time emerged as the single strongest factor in the odds of escaping poverty in the US. The longer an average commute in a given county, the worse the chances of low-income families were in moving up the economic mobility ladder (Bouchard 2015). In 'Stranded: How America's Failing Public Transportation Increases Inequalities', the issue of how fare hikes in many US cities places additional constraints on low-income households and how public transportation options such as bike shares are placed in wealthier neighbourhoods and require credit cards for rentals which poorer households lack was highlighted (White 2015).

The lack of foresight in engaging with the energy related private sector can be attributed to the idea that MDG implementation was primarily viewed through the lens of governmental action, which the outcome of the UN's 2002 World Summit on Sustainable Development – the JPOI-referred to as Type 1 initiatives, which were distinct from NNSAs voluntary Type 2 initiatives.

Box 2.1 Assessment of progress towards MDG 8 targets related to private sector.

Target 8.E: In cooperation with pharmaceutical companies, provide access to affordable essential drugs in developing countries

:: From 2007 to 2014, on average, generic medicines were available in 58 per cent of public health facilities in low-income and lower-middle-income countries.

Target 8.F: In cooperation with the private sector, make available benefits of new technologies, especially information and communications

:: Globally, the proportion of the population covered by a 2G mobile-cellular network grew from 58 per cent in 2001 to 95 per cent in 2015.

:: Internet use penetration has grown from just over 6 per cent of the world's population in 2000 to 43 per cent in 2015. 3.2 billion people are linked to a global network of content and applications.

Source: UN.org website, Millennium Development Goals (2021b).

Increasing access to cost-effective and non-polluting energy services, sources and technologies was not agreed upon as a critical cross-cutting driver for reducing poverty and addressing other MDGs. The expunging of clean energy goals from the MDGs is hard to rationalize because the UNGA convened a special session focusing on energy as far back as 1997 which clearly articulated that: 'In developing countries, sharp increases in energy services are required to improve the standard of living of their growing populations. The increase in the level of energy services would have a beneficial impact on poverty eradication by increasing employment opportunities and improving transportation, health and education. Many developing countries, in particular the least developed face the urgent need to provide adequate modern energy services, especially to billions of people in rural areas. This requires significant financial, human and technical resources and a broad-based mix of energy sources' (UNGA 1997, para. 45). In spite of the above recognition by the UNGA, a close examination of two decades of UN global policy records indicated two relevant findings:

- Intergovernmental negotiations on energy for sustainable development have consistently been held on a separate negotiating track from intergovernmental climate change negotiations convened under the aegis of the UNFCCC.
- Within the UN context, from 1992 till 2015 adoption of SDG 7, there was no globally agreed goal on sustainable energy (Cherian 2015).

Within the UN context, the topic of 'energy for sustainable development' fell under the institutional purview of the UN's Commission for Sustainable Development (CSD) until 2013, and was negotiated within this forum. Intergovernmental negotiations and debate on energy for sustainable development have been consistently convened separately from intergovernmental climate change negotiations convened under the aegis of the UNFCCC process. (Chapter 3 traces the evolution of these climate change negotiations and examines the role of clean energy and air pollution reduction partnerships.) The CSD was the forum envisioned by Agenda 21 and formally established in 1992 by UNGA Resolution 47/19. CSD was a subsidiary body of the UN Economic and Social Council composed of 53 rotating members whose election to the CSD was based on geographical allocations: 13 seats from Africa, 11 from Asia, 6 from Easter Europe, 10 from Latin America and the Caribbean and 13 from Western Europe and North America. It held its first substantive session in 1993 and held sessions every year until September 2013 when it was replaced by the High-Level Political Forum (HLPF). The HLPF, like its predecessor, was created as the institutional outcome of the UN's 2012 FWW universally agreed outcome document of the Rio+20 Summit, and it was subsequently adopted as UNGA Resolution 66/288.

Intergovernmental negotiations on energy occurred first within the context of the ninth session of the CSD (CSD-9), which was the first intergovernmental forum in which multi-stakeholder participation occurred on energy as a distinct topic. At CSD-9, the discussion focus on energy was categorized as follows:

- Accessibility of energy
- Energy efficiency

- Renewable energy
- Advanced fossil fuel technologies
- Nuclear energy technologies
- Rural energy
- Energy and transport

CSD-9 also focused on a series of cross-cutting issues: (i) research and development, (ii) capacity building, (iii) technology transfer, (iv) information sharing and dissemination, (v) mobilization of financial resources, (vi) making markets work more effectively for sustainable development and (vii) multi-stakeholder approaches and public participation. Although there are 17 references to 'climate change' in CSD-9's outcome report, the issue of climate change was only referenced in conjunction with the decision on the protection of the atmosphere (Decision 9/2 but not at all in the decision on energy [Decision 9/1 – 'Energy for Sustainable Development']). It is worth pointing out this that in 2001 CSD, all aspects of the energy sector in relation to sustainable development including advanced fossil fuel and nuclear energy along with energy access and renewables were discussed, but the global discussion and debate did not result in agreed UN energy for sustainable development targets or goals. What was also noteworthy about CSD-9 was that the Chair's summary of the High-Level Segment Report, but not the actual CSD decisions themselves contained the first ever mention of a global target related to energy access, stating that 'some proposed that' the UN 'should adopt a target of cutting by half by 2015 the proportion of people without access to clean fuels and electricity' (UN 2001, p. 46). Interestingly, it would take a decade before this target was broadly referenced by a UN agreement.

Despite the acclaim accorded to the 2015 SDGs being transformative and ambitious, an argument could be made that SDG 7, represented a case of diminished expectations when contrasted to the far-reaching 2002 UN Department of Economic and Social Affairs (DESA) background paper entitled 'A Guide for Potential Partnerships on Energy for Sustainable Development'. This background report prepared for the consideration by the 2002 WSSD included the first ever articulation of specific targets for renewable energy and curbing of fossil fuel air pollution and also included a wide range of innovative partnership proposals with an implementation date of 2015 including but not limited to:

- '200 million households by 2015 to have access to modern efficient cooking fuels and systems with a particular focus on reducing gender inequities.
- Progressively increase the contribution of renewable energy in the energy mix of countries - 105 by 2015.
- Provision of renewable energy powered (vaccine refrigerators, water pumps and other allied health systems) that can service 100,000 primary health care centers.
- Low NOx burners and particulate matter pollution controls on all new coal-fired power plants.
- Implementation of a 6 mega-cities sustainable transportation initiative focused on megacities in developing countries' (UNDESA 2002, p. 7–16).

This background report served as a template for a more comprehensive UN Water, Energy, Health, Agriculture and Biodiversity (WEHAB) consultative process involving all

UN agencies and major global and regional institutions. The 2002 WEHAB framework reports were not consensus UN documents but were intended to inform global action at the 2002 WSSD via a series of indicative targets in the case of sustainable energy. What is relevant here is that the indicative targets identified in the 2002 Energy Report were based on the idea that the world was ready to take bold global action back in 2002! More than a decade of intergovernmental negotiations later, it is sobering to note that the so-called transformative nature of SDG 7 represents nothing other than watered-down energy for sustainable development targets proposed back in 2002.

The array of sustainable energy targets referenced in the UN's 2002 Energy Report, 'Framework for Action on Energy', are worth examining 20 years later in comparison to those contained within SDG 7. The indicative targets listed back in 2002 serve as evidence of the lack of specificity accorded to current global targets related to SDG 7. Additionally, the 2002 Report remains the first and only UN commissioned report on energy and sustainable development that examined 'Energy' in connection to the four other WEHAB areas (UN/WEHAB 2002). A glaring shortcoming of the WEHAB reports was that climate change was not identified as one of the four cross-cutting areas, but again this can be traced back to the fact the implementation mandate for climate change was not seen as residing with UN agencies or even the UNGA, but instead within the UNFCCC process. Put simply, global climate change was separately negotiated from energy for SDGs within the UN. But, the 2002 Energy Report nevertheless emphasized climate change as priority concern, and it provided a framework for action on energy by focusing on five challenge areas: Accessibility, Energy Efficiency, Renewable Energy, Advanced Fossil Fuel Technologies and Energy and Transport. It also clearly linked energy with poverty reduction, and, most importantly, it focused on the grave health and pollution impacts associated with heavy reliance on solid fuels. The paper was blunt about the linkages between lack of access to clean energy and air pollution: 'Current patterns of energy supply and consumption are clearly unsustainable. Nearly one-third of the world has no access to electricity, and another third has only poor access. . .The social issues linked to energy use include poverty alleviation, opportunities for women, the demographic transition and urbanization. Overall, limited access to energy services marginalizes poor people and seriously limits their ability to improve their living conditions. The poor typically spend a greater portion of their income than the rich do on indispensable energy services, such as cooking. At the same time, they frequently forgo or compromise on services like lighting and space heating. Modern energy services can be a vital entry point for improving the position of women in households and societies. It is mainly women who do the cooking, so they and their children are most vulnerable to indoor air pollution from cooking fires' (2002, p. 7). A comprehensive list of indicative targets in each of the five challenge areas along with action areas and indicative targets have been identified (Table 2.1).

What is critical to highlight is that the 2002 Energy Report and its indicative targets were not agreed upon within the context of the global negotiations that contributed to the JPOI. In this regard, the JPOI reflected the UN's sustainable development policy reality whereby sustainable energy was not assigned immediate priority, nor were there explicit references prioritizing policy intersections between sustainable energy, poverty eradication and climate change.

Table 2.1 Challenge area and indicative targets listed in WEHAB energy report.

Challenge area	Indicative targets to be achieved by 2015
Access to energy and modern energy services	To achieve the MDG of reducing the proportion of people living in extreme poverty by half, commensurate decreases in the number of people without access to electricity and clean cooking fuels are required. This implies targeting 800 million to 1 billion people to be provided with modern energy services by 2015. This corresponds to half of the estimated number of people currently living in extreme poverty. The 400 million households that currently depend on traditional fuels need access to modern efficient cooking fuels and systems. This will contribute to addressing gender inequity at the household and community level. Substantially increase access to modern energy services from an estimated base-line situation of 10% of the population in rural sub-Saharan Africa. Develop appropriate institutional and regulatory frameworks.
Energy efficiency improvements	In order to realize the potential of end-use energy efficiency improvements, which are estimated to be in the range of 25–40% in residential and commercial buildings, industry, agriculture and transport sectors in all countries, appropriate targets for every five years are needed. Substantially increase the application of appropriate energy efficiency standards and labeling programmes from the current coverage in about 30 developing countries to a much larger number. To improve the efficiency of converting fuels to power from the current low levels, it is necessary to substantially increase the share of modern electricity generation technologies, such as natural gas-based combined cycle, in national supply mixes.
Renewable energy	Progressively increase the contribution of renewable energy in the global primary energy mix from the current baseline of 2% for modern renewables. For example, at the current rate of expansion, wind energy is expected to increase from the existing generating capacity of 25,000 MW to 100,000–150,000 MW in the next decade. Targets are required to generate similar trends in other forms of renewable energy such as biomass, solar, hydro and biofuels. At 1–2 kw per health care centre, 100–200 Mw capacity is required for 100,000 health care centres (vaccine refrigerators, water pumps and other allied health systems). At 500 w per school, 100,000 schools require 50 Mw capacity. (Particular focus on rural and remote areas) To support the achievement of the MDG on reducing under-five mortality by two-thirds, provide all vaccine and immunization programmes and centres with appropriate renewable energy systems (to suit local conditions). The average vaccine refrigerator requires 250–500 w. To support the achievement of the MDG to reduce by half the proportion of people who do not have access to safe drinking water, it would be necessary to reach 500 million people with 40 litres per capita, which would require 1 million water pumps. (One pump is expected to serve on average 500 people in a community.)

(Continued)

Table 2.1 (Continued)

Challenge area	Indicative targets to be achieved by 2015
Advanced fossil fuel technologies	Assuming that capital stock renovation of 5–10% a year can be achieved, the entire energy system can be upgraded with advanced technology options in the next 20–30 years if performance criteria are explicit.
	Given current technology availability and trends, starting from 2005, 12 Gw per year of clean coal technologies in the next 10 years is feasible.
	Phased retrofitting of existing coal-fired power generating plants and introduction of new plants with low nitrogen oxide burners and particulate pollution controls will be required in order to achieve local, regional and global environmental benefits.
	Progressively increase the share of modern energy technologies as a means to support economic productivity and development.
Energy and transport	Phasing out of lead in gasoline, reduction of sulphur and benzene in fuels and reduction of particulates in vehicle exhaust in all countries.
	Implementation of sustainable transport in mega-cities focused on cleaner fuels, technology advancement and modal shifts, particularly in developing countries.
	Progressively increase the share of new technologies in transport, including three-wheelers or buses, through expanded use of new fuels, compressed natural gas vehicles, electric and electric hybrids and fuel cell vehicles.

Source: UN/WEHAB 2002, pp. 16–24.

The JPOI recognized the UNFCCC as the 'key' instrument for addressing climate change and called for the timely ratification of its KP (UN 2002, para. 38 (f), p. 36), but the locus for action on climate change was seen within the context of intergovernmental negotiations conducted primarily under the aegis of the UNFCCC and therefore held distinct from broader UN negotiations on sustainable development. Following the adoption of the JPOI, there continued to be a lack of attention on sustainable energy which is hard to reconcile from a global policy perspective, because in addition to the 2002 Energy Report, the well-regarded 2005 UN Millennium Project Report explicitly stated that: 'Failure to include energy considerations in national MDG strategies and development planning frameworks will severely limit the ability to achieve the MDGs'. The report had 10 clear recommendations on priority energy issues that were seen as critical for achieving the MDGs but none were immediately acted on (UN 2005, pp. 2–5).

By 2006–2007, CSD-14 and CSD-15, respectively, included a focus on energy for sustainable development and climate change, but once again, tense negotiations and the reports of these meetings provide conclusive evidence of the political challenges within the UN of linking energy for sustainable development and climate change. For the first time ever in the history of the CSD's deliberations, there was no consensus-based decision agreement that emanated at the conclusion of CSD-15 in 2007, and, consequently, only a Chairman's summary was agreed upon which stated: 'Delegates achieved near unanimity on the industrial development and air pollution/atmosphere themes, but remained divided on key points related to climate change and energy' (UN 2007, pp. 2–3).

Sustainable energy became a focal issue only in 2010, after then UN Secretary General Kofi Annan proposed a series of initiatives envisaged to create targeted voluntary partnerships involving governments, multilateral and regional institutions, civil society organizations and for-profit enterprises including the following:

- Every Woman Every Child
- Sustainable Energy for All (SEforAll) (discussed below)
- Global Education First Initiative
- Zero Hunger Challenge
- Scaling Up Nutrition Movement
- Call to Action on Sanitation

After the failure of CSD-15 to arrive at consensus-based sustainable energy targets, the topic of sustainable energy re-emerged outside of the intergovernmental negotiations but within the UN via the Secretary General's 2010 launch of the Advisory Group on Energy and Climate Change (AGECC). AGECC was the first high-level UN group that was explicitly 'convened to address the dual challenges of meeting the world's energy needs for development while contributing to a reduction in GHGs' (AGECC 2010, p. 3). AGECC was not created as the result of an intergovernmental agreement, but its report called on the UN system and member states to commit themselves to two complementary goals:

- 'Ensure universal access to modern energy services by 2030;
- Reduce global energy intensity by 40% by 2030'.

The report defined universal energy access as 'access to clean, reliable and affordable energy services for cooking, heating, lighting, communication and productive uses' (AGECC 2010, p. 13). More notably, it pointed out that investments in energy access and energy efficiency could be mobilized through funds made available by the Fast Track Funding of $30 billion committed at the COP-15 in Copenhagen. Here, the idea that financing for sustainable energy access for all and energy efficiency were to be accessed via the shared framework of climate financing should be highlighted because currently there are two separate SDGs with unclear financing streams, no integrated partnership framework on financing. By 2011, the AGECC was replaced by the Sustainable Energy for All (SEforAll), which is discussed in the section that follows. SEforAll was launched as a joint initiative of the UN and the World Bank – and with three interlinked objectives to be achieved by 2030:

- Ensure universal access to modern energy services.
- Double the global rate of improvement in energy efficiency.
- Double the share of renewable energy in the global energy mix (UN 2011).

Sustainable energy also came to fore with the UNGA declaration of 2012 as the International Year of Sustainable Energy. But again, despite the importance of energy as a critical lever in addressing poverty eradication and climate change, in keeping with past UN sustainable development negotiations, energy and climate change were referenced within separate sections of the 2012 FWW resulting in separate and distinct SDGs – SDG 7 (Box 2.2) and SDG 13 (Box 2.3).

SDG 7's targets related to ensuring access to affordable, reliable, sustainable modern energy for all are, however, not directly associated with specific sustainable energy policy measures, guidance and/or focused programmes by any single UN entity, instead they fall

Box 2.2 SDG 7 – targets and indicators.

SDG 7: Ensure access to affordable, reliable, sustainable and modern energy for all

Targets	Indicators
7.1 **By 2030, ensure universal access to affordable, reliable and modern energy services.**	7.1.1 Proportion of population with access to electricity.
	7.1.2 Proportion of population with primary reliance on clean fuels and technology.
7.2 **By 2030, increase substantially the share of renewable energy in the global energy mix.**	7.2.1 Renewable energy share in the total final energy consumption.
7.3 **By 2030, double the global rate of improvement in energy efficiency.**	7.3.1 Energy intensity measured in terms of primary energy and GDP.
7.a **By 2030, enhance international cooperation to facilitate access to clean energy research and technology, including renewable energy, energy efficiency and advanced and cleaner fossil fuel technology, and promote investment in energy infrastructure and clean energy technology.**	7.a. 1 Mobilized amount of United States dollars per year starting in 2020 accountable towards the $100 billion commitment.
7.b **By 2030, expand infrastructure and upgrade technology for supplying modern and sustainable energy services for all in developing countries, in particular least developed countries, small island developing States, and land-locked developing countries, in accordance with their respective programmes of support.**	7.b.1 Investments in energy efficiency as a percentage of GDP and the amount of foreign direct investment in financial transfer for infrastructure and technology to sustainable development services.

Source: UN SDG Knowledge website (2021).

under the purview of SEforAll and overlap with mandates of different agencies. Additionally, the use of broader, proportional indicators in place of targeted policy indicators included in the 2002 Energy Report references a lack of programmatic focus. The lack of specificity related to energy financing in targets 7a and 7b constitutes a challenge particularly in terms of implementing effective and timely financing for sustainable energy via partnerships or global partnership mechanisms which are themselves conceptually and programmatically hard to track. More importantly, while all the other SDGs were formally

Box 2.3 SDG 13 – Targets and indicators.

SDG 13: Take urgent action to combat climate change and its impacts	
Targets	**Indicators**
13.1 **Strengthen resilience and adaptive capacity to climate related hazards and natural disasters in all countries.**	**13.1.1** Number of countries with national and local disaster risk reduction strategies. **13.1.2** Number of deaths, missing persons and persons affected by disaster per 100,000 people.
13.2 **Integrate climate change measures into national policies, strategies and planning.**	**13.2.1** Number of countries that have communicated the establishment or operationalization of an integrated policy/strategy/plan which increases their ability to adapt to the adverse impacts of climate change, and foster climate resilience and low greenhouse gas emissions development in a manner that does not threaten food production (including a national adaptation plan, nationally determined contribution, national communication, biennial update report or other).
13.3 **Improve education, awareness-raising and human and institutional capacity on climate change mitigation, adaptation, impact reduction and early warning.**	**13.3.1** Number of countries that have integrated mitigation, adaptation, impact reduction and early warning into primary, secondary and tertiary curricula. **13.3.2** Number of countries that have communicated the strengthening of institutional, systemic and individual capacity-building to implement adaptation, mitigation and technology transfer, and development actions.

(Continued)

Box 2.3 (Continued)

13.a **Implement the commitment undertaken by developed-country parties to the United Nations Framework Convention on Climate Change to a goal of mobilizing jointly $100 billion annually by 2020 from all sources to address the needs of developing countries in the context of meaningful mitigation actions and transparency on implementation and fully operationalize the Green Climate Fund through its capitalization as soon as possible.**	**13.a. 1** Mobilized amount of United States dollars per year starting in 2020 accountable towards the $100 billion commitment.
13. b **Promote mechanisms for raising capacity for effective climate change-related planning and management in least developed countries and small island developing States, including focusing on women, youth and local and marginalized communities.**	**13.b.1** Number of least developed countries and small island developing States that are receiving specialized support, and amount of support, including finance, technology and capacity-building, for mechanisms for raising capacities for effective climate change-related planning and management, including focusing on women, youth and local and marginalized communities.

Source: UN SDG Knowledge website (2021).

adopted in September 2015 by the UNGA, there was a place holder put with regard to the language of SDG 13, which was only finalized within the separate global negotiating context of the December 2015 PA. According to the UN's Sustainable Development Knowledge Platform (which contains information on all the SDGs): 'Climate change presents the single biggest threat to development, and its widespread, unprecedented impacts disproportionately burden the poorest and most vulnerable. Urgent action to combat climate change and minimize its disruptions is integral to the successful implementation of the Sustainable Development Goals' (UN SDG Knowledge website 2021).

Logically, this clarion call of climate action being integral to the successful implementation of the SDGs necessitates that partnerships on SDG 13 be integrated into the remaining SDGs, or at the very least be linked with partnerships related to SDG 7 given that energy is the shared driver in both SDGs. But SDG 13's clear acknowledgment that the UNFCCC 'is the primary international, intergovernmental forum for negotiating the global response to climate change' reveals a distinction as to the UNGA not being the locus for intergovernmental action for climate change unlike all the other SDGs.

Accordingly, the question that needs to be raised is whether the goal silos of SDGs 7 and 13 allow for integrated partnership action on sustainable energy and climate action, especially if the intergovernmental purview of the implementation of SDG 13 lies primarily within the UNFCCC, and outside of the UNGA?

The following sections are informed by the research finding that: 'The examination of key UN-led global outcomes on sustainable development and climate change reveals that, for decades now, the two agendas have consistently been separated out with few consistent attempts at programmatic and policy linkages between energy for sustainable development and climate change objectives . . . Simply separating out the programmatic and linkages between energy access for the poor and climate change mitigation perpetuates the idea that energy for sustainable development and climate change should somehow continue to be considered as separate silos within the UN context. It is time to take stock that these separate negotiating silos have been known to exist despite global rhetoric and evidence that they represent dual and intersecting global concerns, and that a shared and universal post-2015 sustainable development agenda will be much harder to accomplish if these silos persist' (Cherian 2015, p. 233). Segregated, separate UN-led negotiations silos and SDGs on sustainable energy and climate change are hard to justify in view of the fact that the UN's 2030 Agenda hinges expressly upon being an integrated agenda. Energy is unmistakably the shared driver in global responses to air pollution and climate change. Increasing access to clean energy and climate action goals are inherently linked to the UN's overarching globally agreed priority to eradicate poverty worldwide by 2030. It appears illogical to expect that the implementation of clean energy and clean air targets can be segregated from climate change goals when access to clean energy is the common denominator in reducing poverty and air pollution. The global policy problem that needs to be confronted is that in addition to the intergovernmental silo outcomes – represented by separate SDGs on sustainable energy and climate change and an entirely separate negotiation track on the PA – there is a considerable lack of definitional clarity on the concept and application of partnership-driven actions which are key to the nexus between climate, clean air and clean energy action.

2.3 Delving into the UN Acronym Soup on Partnerships for Sustainable Development

There is a variegated world of UN-affiliated partnerships ranging from PSDs, Multi-Stakeholder Partnerships (MSPs), International Cooperative Initiatives (ICIs), Voluntary Initiatives (VIs) and Global Partnerships on Sustainable Developments (GPSDs). Voluntary partnership initiatives are therefore not easy to keep track of within the UN context, because in addition to PSDs, MSPs and VIs, there are partnerships that fall under the purview to the UN Office of Partnerships (UNOP). Within the world of UN acronyms, the UNOP should not be confused for the UN Office of Project Services (UNOPS) designated to help UN agencies implement their mandates. The sheer range of partnership modalities and the ambiguity over how the diverse array of partnerships arrangements are tracked within the UN system have made any comprehensive analysis of partnerships and their results incredibly hard. To make matters more confusing, UN-affiliated PSDs, MSPs, ICIs, VIs and GPSDs involving non-state actors do not necessarily conform to the broader category of cooperative partnerships termed 'public private partnerships' (PPPs) used by governments. PPPs are widely referenced in national development and infrastructure efforts within all countries; but typically PPPs involve a more formalized contractual arrangement between state (public) and non-state (private) stakeholders, whereas PSDs with the UN context and MSPs do not necessarily even need to involve state actors and are characterized by entirely voluntary arrangements amongst the partnerships stakeholders. But even here there is ample definitional confusion, as Colverson and Perera noted, since PPPs are a 'generic term being used' for a variety of different types of contractual agreements 'between the State and the private sector for the purpose of public infrastructure development and services provision'. They go on to state that there is 'no one single, concise definition of PPPs' and point out that defining a PPP accurately 'is problematic because by nature it is a contextual concept, responding to the institutional, legal, investment, and public procurement settings of different jurisdictions, while also considering the contextual nature of an individual agreement' (2012, p. 2). The definitional ambiguity associated with PPPs also extends into the programmatic analysis of PSDs and the GPSD within the UN context.

To begin with, the UN intergovernmental context cannot be faulted for not generating a dizzying array of partnership modalities, but the real challenge is to decipher and track the gamut of modalities. As the previous section highlighted, PSDs have been heralded as key for the implementation of the UN's agenda on sustainable development for decades. The role of GPSDs are also clearly referenced within two of the most historic and ambitious UN global agreed outcomes universally agreed to by all UN member states – the SDGs and the PA with an implementation deadline of 2030. There is a long negotiation-related history to the concept of PSDs that can be traced back to the UN's 1992 Agenda 21 agreement. The opening paragraph of the preamble of the most voluminous UN agreements ever negotiated, the 40-chapter Agenda 21, universally adopted in 1992 stated: 'Humanity stands at a defining moment in history. We are confronted with a perpetuation of disparities between and within nations, a worsening of poverty, hunger, ill health and illiteracy, and the continuing deterioration of the ecosystems on which we depend for our well-being. However, integration of environment and development concerns and greater attention to them will lead to the fulfillment of basic needs, improved living standards for all, better protected and managed ecosystems and a safer, more prosperous future. No nation can achieve this on its own; but together we can - in a global partnership for sustainable development' (UN 1992, Agenda 21, preamble: para. 1.1). A decade later, the UN's 2000 Millennium Development Declaration included Goal 8, which called for a different kind of global partnership – the GPD. Two decades later, in 2015, the UNGA adopted a new 'plan of action for people, planet and prosperity' – *2030 Agenda for Sustainable Development* – which called for mobilizing 'the means required to implement this Agenda through a revitalized Global Partnership for Sustainable Development . . . focused in particular on the needs of the poorest and most vulnerable and with the participation of all countries, all stakeholders and all people' (UNGA 2015a, preamble). Notwithstanding the capitalization, and the addition of 'revitalized', is it euphemistic policy evolution or déjà vu policy confusion if two landmark UN global sustainable development agreements invoked the need for global partnership more than 20 years apart? There is also ample room for definitional confusion as the GPD referenced as 2000 MDG 8 is different from the 2015 UN SDA's call for 'a revitalized and enhanced global sustainable development partnership' (GSPD) – currently enshrined as SDG 17.

The programmatic challenge is that 2015 variant of the GPSD is not, however, the only global partnership focused on within the SDGs. There is also the Global Partnership for Effective Development Cooperation (GPEDC) which states that: 'The Global Partnership for Effective Development Co-operation is a multi-stakeholder platform to advance the effectiveness of development efforts by all actors, to deliver results that are long-lasting and contribute to the achievement of the Sustainable Development Goals (SDGs)' (Effective Cooperation. Org 2021). The GPEDC which was originally launched in 2011 is expected to contribute directly to the strengthening of the 2015 GPSD, but there is no clearly agreed global guidance that specifies exactly how accountability for the implementation of shared principles and differentiated commitments will be tracked across these two interlinked partnership mechanisms; nor is it immediately clear how the framework for actions related to implementing the GPEDC will be accounted for within the GPSD. Additionally, it is also unclear how the delivery records of the GPSD and the GPEDC can be evaluated given their differing conceptual reach, governance structures and mandates despite their ostensible overlapping objectives to 'contribute to the achievement of the SDGs'.

Interestingly, the GPEDC is listed as being based on the four shared principles of effective development co-operation (agreed to by more than 160 countries in 2011) via the Busan Partnership for Effective Development Cooperation which included:

- '*Ownership of development priorities by developing countries*-Partnerships for development can only succeed if they are led by developing countries, implementing approaches that are tailored to country-specific situations and needs.
- *Focus on results*-Development efforts must have a lasting impact on eradicating poverty and reducing inequality, and on enhancing developing countries' capacities aligned with their own priorities.
- *Inclusive development partnerships*-Openness, trust, mutual respect and learning lie at the core of effective partnerships, recognising the different and complementary roles of all actors.
- *Transparency and accountability to each other*-Mutual accountability and accountability to the intended beneficiaries of development co-operation, as well as to respective citizen, organisations, constituents and shareholders is critical to delivering results. Transparent practices form the basis for enhanced accountability' (emphasis included, OECD 2011, p. 3).

By the logic of the GPEDC, the onus for successful partnerships resides with developing countries, and effective partnerships are result based, inclusive, transparent and mutually accountable. But the problem with these core agreed-upon principles is that currently the UN-affiliated global partnership framework is not driven by the ownership priorities of developing countries nor is there a comprehensive, transparent, fully accessible and accountable partnership tracking methodology that cuts across the SDGs. And then there is a partnership implementation challenge that is associated with the fact that the 2015 GSPD is a stand-alone SDG and not seen as cross-cutting all the SDGs in terms of delivering results.

The concept of NNSA-driven PSDs – separate from the GPSD referenced above – can also be traced back to the Agenda 21's call for NNSAs – 'major groups' (civil society stakeholders) 'moving towards real social partnership in support of common efforts for sustainable development' (Agenda 21, Chapter 23, para. 23.4). But, a decade later, the major 2002 UN agreement, the JPOI, which officially heralded PSDs as a significant outcome, interestingly made no mention of any GPSD. The lack of reference to a GPSD is at odds with the JPOI's overall focus on PSDs. A crucial organizational feature of the JPOI was its identification of the UN's CSD as the focal point for partnerships that promote sustainable development – PSDs (JPOI, para. 130 b). UN's CSD was the intergovernmental forum where energy for sustainable development but not climate change which was and remains separately negotiated under the aegis of the UNFCCC. The NNSA-driven PSDs of the JPOI were classified as 'Type 2 partnerships', to distinguish them from the 'intergovernmental commitments' or 'Type 1 partnerships'. According to the UN Guidelines, Type 2 partnerships were entirely new VIs anticipated to be additional or complementary to Type 1 outcomes, yet also contribute directly to the implementation of Agenda 21 and the JPOI (UN 2003a). The fact that Type 2 partnership outcomes were envisaged as being supplemental to, and also non-substitutable for Type 1 – governmental commitments by definition – necessitated the need for a reliable partnership monitoring and tracking system; but there was a lack of policy guidance on a comprehensive, globally accessible partnership verification and monitoring framework in the JPOI.

Paradoxically, despite the 2002 JPOI's affirmation of PSDs, and its guideline list of do's and don'ts on partnerships, the UN's benchmark definition of the term 'partnerships' was established only in 2005 within the context of the UNGA Resolution entitled 'Towards Global Partnerships' which stated that: 'Partnerships are voluntary and collaborative relationships between various parties, both public and non-public, in which all participants agree to work together to achieve a common purpose or undertake a specific task and, as mutually agreed, to share risks and responsibilities, resources and benefits' (UN 2006, p. 3). This 2005 UNGA definition of partnerships also clearly recognized the 'importance of the contribution of voluntary partnerships to the achievement of the internationally agreed development goals', but it emphasized that these voluntary partnerships 'are a complement to, but not intended to substitute for, the commitments made by Governments with a view to addressing these goals' (UN 2006, p. 3). It is exactly this distinction between the voluntary, complementary and non-substitutable nature of Type 2 partnerships that sets them apart and distinct from Type 1 intergovernmental effort that still remains a policy challenge in terms of monitoring, measuring and verifying tangible partnership results.

Measuring implementation results of different types of partnership initiatives within the UN context over time has proven to be a thorny problem to solve. Despite a 20-year trajectory of partnerships ranging from the GPSD, to PSDs and MSPs, it was only in 2015 via a UNGA Resolution entitled 'Towards Global Partnerships: A Principle-based Approach to Enhanced Cooperation Between the United Nations and All Relevant Partners', that the series of principles constituting an operational framework for UN-affiliated partnerships with all major groups was outlined. This resolution specifically called for a 'discussion on the best practices and ways to improve, inter alia, transparency, accountability and the sharing of experiences of multi-stakeholder partnerships and on the review and monitoring of those partnerships, including the role of Member States in review and monitoring' (UN 2016, p. 5). But, the 2016 UNGA resolution's definition of partnerships is an identical match to the 2005 benchmark definition, and the obvious question is why did it take more than a decade for a global discussion on 'best practices and way to improve transparency and accountability' on partnerships deemed critical to achieving the UN's sustainable development goals?

To be clear there has been no shortage in partnership announcements. The Rio+20 Conference saw the emergence of over 700 voluntary commitments. However, the lack of a comprehensive and fully accessible framework for tracking a diverse array of PSDs, VIs, ICIs needs to be flagged as a major global concern. More specifically: 'From the early inception of the concept of partnerships, including both PSDs and the GPSD, there was no stated attempt to provide or devise a framework, database or registry that could keep account of these varied partnership actions in terms of the agreement contained in either Agenda 21 or the JPOI. Agenda 21 contained 53 different references to databases and one to a land registry, but *none* of these related to the issue of partnerships or the implementation of the GPSD. Meanwhile the JPOI contained *no mention* of any kind of registry, and included only three references to the need for databases, including water resources related databases, integrated databases on development hazards in relation to health, and accurate databases for developing countries in relation to environmental protection. Despite a 20 year trajectory of voluntary efforts, including both PSDs and PPPs, focused on a range of sustainable development issues and an ostensible GPSD, the first mention of a registry is contained in the 2012 FWW'. And while the 2012 FWW outcome 'includes three references to a registry on VCs in its concluding paragraph (para 238)' there is 'no reference to a partnership registry in the FWW' (emphasis added, Cherian 2015, p. 203).

The UN's Rio+20 Conference underscored the need for VIs from NNSA stakeholders complementing the initiatives to be undertaken by governmental action. The outcome agreement – the FWW – called for the establishment and maintenance of 'a comprehensive registry' of partnerships. More specifically, para. 283 of the FWW outcome document stated:

'We welcome the commitments voluntarily entered into at the United Nations Conference on Sustainable Development and throughout 2012 by all stakeholders and their networks to implement concrete policies, plans, programmes, projects and actions to promote sustainable development and poverty eradication. We invite the Secretary-General to compile these commitments and facilitate access to other registries that have compiled commitments, in an Internet-based registry. The registry should make information about the commitments fully transparent and accessible to the public, and it should be periodically updated' (UNGA 2012, para. 283). Notwithstanding this globally agreed upon declaration, what exactly does an evaluation of the UN's online partnership platform reveal?

Clough et al. (2019) studied the online UN partnerships, as well as the results via an online survey of partnership initiatives. Two of their noteworthy conclusions are:

- 'Engagement with the UN Platform is *clearly limited*, as evidenced by reporting rates. This *should prompt reflection on the purpose of the platform and its future role*: should it function as a more passive repository of commitments and searchable directory of initiatives (and if so, how could it fulfil this function better)? or, is there more scope for making it an active platform, encouraging coordination and learning? This reflection, in turn, should be situated in terms of the role *the UN is willing and able to play* in cultivating an "ecosystem" of partnerships for the SDGs.

- This analysis has reported on the "state of play" for these SDG partnerships – but what should the basis for evaluation be, or the standard against which to assess these findings? – *it is not easy to highlight gaps, or areas of strength, without understanding the wider context of the "global partnership for sustainable development"*. There is a clear need to better understand how these initiatives and partnerships function, and the different kinds of roles diverse partnerships play in implementation of the SDGs' (emphasis added, 2019, p. 7).

What this lack of programmatic and implementation clarity means is that challenge of differentiating between the partnerships with the GPSD and the GPDEC also extends into the function of the UN online partnership platform. And, there is also the additional challenge of distinguishing between the UN's institutional locus on partnerships as reflected by the fact that there are a number of different UN entities focused on implementing 'partnerships'.

There is the UN Office of Partnerships (UNOP) which 'serves as a global gateway for catalysing and building partnership initiatives between public and private sector stakeholders including civil society organizations, businesses, philanthropy, trade unions, academia and the United Nations in furtherance of the Sustainable Development Goals (SDGs)' (UNOP website 2021). While UNOP facilitates partnerships within the UN, it also 'serves as the administrative hub for the United Nations Democracy Fund (UNDEF)' and 'as the Secretariat for the Secretary-General's Sustainable Development Goals (SDG) Advocates' who are 17 eminent global personalities from business, sport, entertainment, academia and advocacy, appointed by the Secretary-General to raise awareness of the SDGs and accelerate efforts to achieve the Goals by 2030. But, UNOP is not to be conflated with UN Office of Project Services (UNOPS) which 'provides infrastructure, procurement and project management services to help build the future' and also 'support the achievement of the Sustainable Development Goals' by responding to 'partners' needs' (UNOPS website 2021). In addition to the UNOP and the UNOPS, to make matters more confusing, there is also the United Nations Fund for International Partnerships (UNFIP) established in 1998 to serve as the interface between the United Nations Foundation and the UN system. According to the UNFIP website, the UN Foundation collaborates with the UN to foster innovative partnerships, campaigns and initiatives that contribute to the achievement of the 2030 Agenda for Sustainable Development and the PA under the UNFCCC to secure better and healthier lives for people across the globe by focusing on Global Health; Women, Girls and Population; Energy and Climate; Capacity and Development; Fiduciary Agreements; Advocacy and Communications; and Multi-stakeholder Alliances and other development-related issues.

Additionally, a number of UN-affiliated private sector investment vehicles have been created or retooled to catalyze private sector-based 'partnership' investment towards the 2030 Agenda. These include the 'Strategic Framework and Action Plan for Private Investment in the SDGs', which was proposed by the Geneva-based United Nations Conference on Trade and Development (UNCTAD). There is also 'The UN-supported Principles for Responsible Investment (PRI)', which has over 1300 signatories globally, '. . . with combined assets under management of approximately US$ 45 trillion. These signatories – asset owners, investment managers and service providers – have agreed to six principles that are all consistent with the SDGs. The PRI Initiative supports signatories in implementing the principles by providing guidance documents, case studies, and forums for dialogue and exchange of best practices. Through its Policy and Research Work Stream, it helps to mobilize investors to engage in public policy and brings long-term investors' perspectives into policy decision-making processes at the UN and at the international and regional levels' (PSISD 2015, p. 9).

The UN-related acronym soup on partnerships also includes opportunities for partnership-driven platforms and knowledge-sharing networks amongst developing countries themselves – via the vehicle of South-South Cooperation (SSC) that has been doing work but arguably not harnessing economies of scale because the tangled morass of competing UN and other global institutional agendas currently impedes scaling up sustainable energy and climate partnerships. For instance, the Southern Climate Partnership Incubator (SCPI) initiative is implemented by the United Nations Office for South-South Cooperation (UNOSSC) in cooperation with other UN agencies but is completely different from UNOP. The precise genesis of SSC can be traced back to the 1978 Buenos Aires Plan of Action on SSC and to the reaffirmation of the principles of South-South Cooperation by the Group of 77 and China (the largest bloc/grouping of countries within the UN) in September 2009. The entity responsible for implementing climate partnerships under the aegis of the UNOSSC is the UN Climate Partnerships for the Global South which is also referred to as the Southern Climate Partnership Incubator. According to the UN:'SCPI is an UN Secretary-General-led initiative that is designed to incubate, facilitate, and support cross-country partnerships. The initiative also supports countries' implementation of the historic Paris Agreement and achievement of the Sustainable Development Goals of the 2030 Agenda (SDGs). The initiative is both an online and offline multi-stakeholder platform' (UN.org /SCPI 2021). But definitional confusion appears once again because according to a 2017 UN report entitled 'Climate Partnerships for a Sustainable Future', developing countries have continually '. . . stressed that South-South cooperation is a complement to, and not a substitute for, North-South development cooperation. It differs from ODA in that it is characterized as a "partnership among equals, based on solidarity" and is guided by the principles of respect for national sovereignty and ownership, free of any conditionality. It aims to utilize capacities and experience available in developing countries. There have been initiatives in the United Nations system in support of the strengthening of South-South cooperation, leading to operational definitions about South-South cooperation that are used within the United Nations system. Nevertheless, it is worth pointing out that *there is no standardized definition* of South-South cooperation' (emphasis added, UNOSSC/South Centre 2017, p. 29). A search of the SCPI site which states that the SCPI is both an online and offline incubator for climate partnerships reveals that the last time a high-level partnership event was held was on 29 November 2017 at the Global South-South Expo, while all the case-studies listed relate back to July 2016 (UN.org website/SPCI 2021).

The bottom line is that UN database on climate and clean energy partnerships are confusing to access and lack an integrated approach as many different UN entities report on such partnerships. A detailed examination of the UN PSD and GSPD databases reveals that currently there is still no comprehensive reporting or accountability framework that can allow for consistent tracking and comparing between and amongst a diverse range of VIs on sustainable development. The lack of both a clear tracking framework and programmatic clarity on PSDs and GPSDs poses verification challenges for evaluating the NNSA-driven partnerships on clean energy and climate change.

2.4 Conclusion: Confusion Rather Than Clarity Prevails with Segregated Silos and Partnerships on Sustainable Energy and Climate Change

The existence of separate SDG silos on sustainable energy and climate, combined with a lack of specificity in terms of concrete targets, and the urgent need to scale up progress in relation to both SDGs remain a challenge. The UN itself summarized 'progress of Goal 7' in 2018 as follows:

- 'From 2000 to 2016, the proportion of the global population with access to electricity *increased* from 78 per cent to 87 per cent, with the absolute number of people living without electricity dipping to just below 1 billion.
- In the least developed countries, the proportion of the people with access to electricity *more than doubled* between 2000 and 2016.
- In 2016, *3 billion people (41 per cent of the world's population)* were still cooking with polluting fuel and stove combinations'

(UN Sustainable Development Goals Report 2018).

The lack of integration between SDG 7 and SDG 13 targets and indicators is layered upon programmatic and implementation related confusion when it comes to tracking partnerships. The extent of policy confusion associated with the implementation of PSDs, GPSDs, VCs etc. therefore extends into the tracking of partnerships between these two SDGs and SDG 17 – the GPSD. Currently, there is no comprehensive, fully accessible tracking framework for verifying and evaluating partnerships on integrated global issues such as energy and climate change. This lack of a globally integrated framework for all partnerships on climate and sustainable energy stems largely from the fact that currently the overall mandate for the implementation of the UN's SDG 7 (sustainable energy for all) and SDG 13 (climate change) appears to lie within the UNGA but climate related action falls within the separate institutional purview of the PA process. From an implementation viewpoint, SDG 13 – climate change – stands apart from the other SDGs, as it does not have the same institutional locus as SDG 7 or other SDGs, because SDG 13 was formally and fully agreed to only after the December 2015 PA was adopted within the context of the UNFCCC negotiating process. In other words, unlike SDG 7, the PA and the UNFCCC provide the intergovernmental venue for negotiating all climate goals including the delivery of progress on SDG 13, and, consequently, the overall institutional mandate for implementation of SDG 13 is located under the aegis of the UNFCCC and the PA. This in turn raises two challenges that are directly relevant to developing countries where poverty eradication and access to affordable, clean and modern energy remain a compelling priority: (i) How can the tracking of UN-affiliated global partnerships related to these two SDGs and any future climate change goals occur within the context of two distinctly separate intergovernmental negotiating silos? and (ii) How can an integrated 2030 UN agenda with its overall priority on poverty eradication that is supposed to facilitate linked partnerships involving a range of NNSAs be expected to deliver results via segregated SDGs and partnership modalities?

A comprehensive 2020 report tracking SDG 7 entitled 'Energy Progress Report' jointly produced by 'the 5 custodian agencies' responsible for SDG 7, namely, IEA, International Renewable

Energy Agency (IRENA), UN Statistics, World Bank and WHO referenced that global efforts fall well short of the scale required to reach SDG 7 targeted goals by 2030. The 2020 SDG 7 Tracking Report highlighted the following with regard to specific targets (see also Figure 2.3):

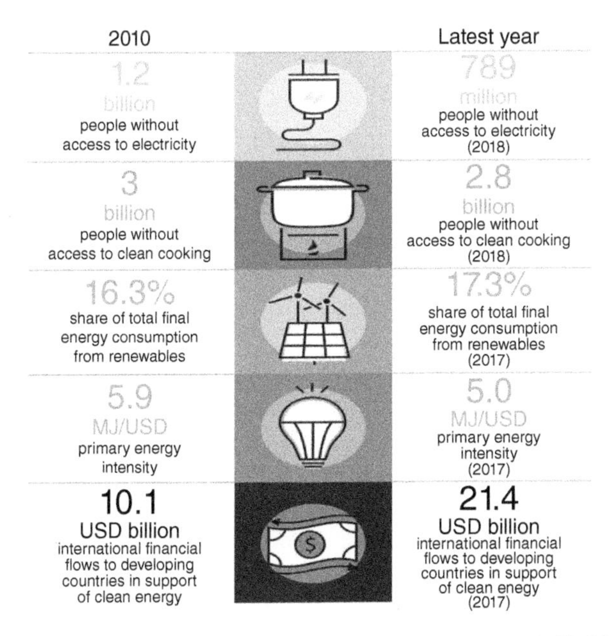

Figure 2.3 Primary indicators of global targets towards SDG 7. *Source:* IEA et al. (2020, p. 2).

- **Target 7.1.1** – Proportion of population with access to electricity: Recent years have seen rapid growth in access to electricity after an accelerated deployment of affordable electrification options, including on- and off-grid solutions. As a result, *the global population lacking access to electricity dropped* to 789 million in 2018, from 1.2 billion in 2010.
- **Target 7.1.2** – Proportion of population with primary reliance on clean fuels and technology: In contrast, the global population without access to clean cooking solutions *remained largely unchanged* during the same period, standing at close to 3 billion. The *rate of increase in access to clean cooking* has *even decelerated since 2012*, falling behind population growth in some countries. *Increased efforts are needed to ensure universal access to both electricity and clean cooking,* consistent with SDG target 7.1.
- **Target 7.2** – Increase the share of renewable energy in the global energy mix: 'The share continued to increase in 2017 (+ 0.1 percentage points), although at a slower pace than the year before (+ 0.2 percentage points), reaching 17.3 percent of total final energy consumption (TFEC) in 2017, up from 17.2 percent in 2016 and 16.3 percent in 2010. Solar PV and wind are key drivers behind the fast-growing share of renewables in the generation of electricity. But *renewables' share in the heating and transportation sectors lags far behind its* potential. An acceleration of renewables in all sectors will be needed to achieve target 7.2'.
- **Target 7.3** – Double the global rate of improvement in energy efficiency by 2030: 'Global primary energy intensity, defined as total primary energy supply per unit of GDP, reached 5.0 megajoules per USD dollar in 2017, equivalent to a 1.7 percent rate of improvement

from 2016—the lowest since 2010. Preliminary estimates for 2018 (1.3 percent) and 2019 (2 percent) suggest that the improvement rate would reach approximately 2.1 percent between 2010-2019 which is lower than the required 2.6 percent annual target rate for the years between 2010 and 2030. Consequently, achieving the goal will require an energy intensity improvement rate of at least 3 percent per year from now through to 2030—a challenging proposition'.

- **Target 7.A** – Promote access to technology and investments in clean energy: 'With target 7.A.1 focusing on international public financial flows to developing countries in support of clean and renewable energy. Total flows reached USD 21.4 billion in 2017, double the level of 2010. Although this is a promising increase, *only 12 percent of financial flows in 2017 reached the least-developed countries,* which are the *furthest from achieving* the various SDG 7 targets. *Increased efforts are needed to make sure finance reaches the countries most in need*' (emphasis added, World Bank, IEA 2020, pp. 1–3).

Currently, SEforAll is the only institutional locus albeit with quasi-UN affiliation that is aimed at delivering results on SDG 7 via 'partnerships'. In spite of the central importance of sustainable energy as a shared driver in addressing several SDGs and in particular SDG 13, the decision was made to have the SEforAll institutional structure not become a UN entity. Instead, it was created as an international not-for-profit organization, headquartered in Vienna under Austrian law with a satellite office in Washington, DC. According to the SEforAll website, it 'works in partnership with the United Nations and leaders in government, the private sector, financial institutions, civil society and philanthropies to drive faster action towards the achievement of Sustainable Development Goal 7 (SDG 7) – access to affordable, reliable, sustainable and modern energy for all by 2030 – in line with the Paris Agreement on climate'. But even here, there is conceptual differentiation associated with the use of the term 'partnership' as both a modality and a mechanism, which is demonstrated in the case of the Renewable Energy and Efficiency Partnership (REEP) and SEforAll. Unlike SEforAll, which began as an ad hoc intergovernmental committee within the UN, REEP was launched initially as a government-driven partnership by the UK in the 2002 WSSD and continued following the Type 2 partnership process of stakeholder consultation till 2004, when it obtained formal status as an international NGO. In 2016, REEP was granted status as a *quasi-international organization* located in Austria. While the transformation of partnerships into quasi-international organizations focused on sustainable or renewable energy offers the possibility of harnessing global attention, the challenge for the UN-affiliated global partnerships remains the overall difficulties in distinguishing between globally relevant partnership actions and partnering networks. But, SEforAll is not the only operational UN-affiliated partnership initiative focused on increasing clean energy access for the poor as the 2016 UN Secretary General's Report also referenced the role of Global Alliance for Clean Cookstoves (GACC).

GACC can be traced back to the 2002 launch by the US EPA of a Partnership for Clean Indoor Air (PCIA) at the UN's WSSD. The PCIA was transformed in 2010, when the US Department of State along with the US EPA announced the GACC. Back in 2015, the GACC website listed its mission which was 'to mobilize high-level national and donor commitments foster the adoption of clean cookstoves and fuels in 100 million households by 2020 and clearly identified a 10 year business plan comprising of 3 phases: Phase One (2012-2014)-Launch global and in-country efforts to rapidly grow the sector; Phase Two (2015-2017)-Drive investments, innovations and operations to scale; and Phase Three (2018-2020)-Establish thriving and sustainable global market for clean cookstoves' (Cherian 2015, p. 244). Since then the GACC has been renamed as the Clean Cooking Alliance (CCA), and its website contains no reference to the target of 10 million households by 2020 and instead states that the CCA 'works with a global network of partners to build an inclusive industry that makes clean cooking accessible to the three billion people who live each day without it'. CCA's work is referenced as being built around three core pillars:

- 'Driving consumer demand for cleaner, more modern stoves and fuels by supporting behavior change and awareness-raising interventions;
- Mobilizing investment to build a pipeline of scalable businesses capable of delivering affordable, appropriate, high-quality clean cooking products; and
- Fostering an enabling environment for industry growth by advocating for effective and predictable policies, providing trusted, relevant data, and serving as the convener and champion of the clean cooking sector' (CCA website 2021).

But with only a few years left to achieve its previously stated goal, can the CCA scale up action in those countries and communities in Sub-Saharan Africa and South Asia where energy access issues are most relevant based on the list of partners it currently has including: CCAC; Clinton Global Initiative; Cooking for Life; Energy for All; Every Woman Every Child, Global Bioenergy Partnership; Global LPG Partnership; Roundtable on Sustainable Biomaterials; Safe Access to Fuel and Energy; Sustainable Energy for All; World Access to Modern Energy – Milan Expo 2015 and the World LPG Association (WLPGA). A current search of the UNFIP website reflects the fact: 'As at 31 December 2017, the cumulative allocations approved by the UN Foundation for UNFIP projects for implementation by UN partners reached approximately $1.47 billion. Since its inception, 641 projects have been implemented or are under implementation by 48 United Nations entities in 128 countries. During 2016, the Foundation disbursed funding in the amount of $26.1 million, focusing on polio eradication, malaria, HIV/AIDS and measles prevention, humanitarian relief, disease surveillance, reproductive healthcare, girls empowerment, sustainable energy development and protection of the environment, and other issues' (UN.org/UNFIP 2021). The UNFIP website does not provide any more specific breakdown of information as to funding allocated towards sustainable/clean energy access or climate change.

There is an urgency in providing access to non-polluting energy services for all. The UNDESA published 'Global Sustainable Development Report: The Future is Now' clearly highlighted both the urgency in linked action on climate change and clean energy access: 'Supplying a world population in 2050 of an estimated 9 to 10 billion people with energy mainly provided by fossil fuel sources is simply incompatible with meeting global climate targets. Providing clean and efficient energy for all in a climate-friendly way is economically and socially desirable, and is technically feasible. *The benefits are predicted to outweigh the costs of transforming our energy systems by a factor of three.* It is estimated that for *every dollar spent on shifting to a sustainable energy system, the transition would generate between $3 and $7,* including through *savings from reduced air pollution, improved health and lower environmental damage* arising from the transition to clean energy sources' (emphasis added, UNDESA 2019, p. 78).

The question is what do 'partnerships' look like within the context of the SEforAll which is billed as the primary UN-affiliated entity for promoting and delivering on SDG7? The SEforAll website currently lists 66 different types of partners, which are further categorized into the following groups:

- **Funding Partners (17 in total):** Australian Development Cooperation; Bloomberg Philanthropies; Climate Emergency Collaboration Group; Climate Works Foundation; Denmark Ministry of Foreign Affairs; Iceland Ministry of Foreign Affairs; IKEA Foundation; Kigali Cooling Efficiency Program; Mott Foundation; Rockefeller Foundation; Shell Foundation; Swedish Postcode Foundation; Swiss Agency for Development Cooperation; Transforming Energy Access; UK Aid; UN Foundation and Wallace Global Fund
- **Civil Society Partners (10 in total):** Alliance for Rural Electrification; CCA; Clinton Foundation; Global LPG Partnership; Initiative for Sustainable Energy Policy; Power for All; REN21; SNV; Terrawatt Initiative and the UN Foundation
- **Private Sector Partners (18 in total):** ABB; ACOB Lighting Technology; Associação Lusófona de Energias Renováveis (ALER); Arçelik A.Ş.; Danfoss, EDP, Enel, ENGIE PowerCorner; EURELECTRIC; the Global Off-Grid Lighting Association (GOGLA); Iberdrola; International Gas Union (IGU); International Hydropower Association (IHA); Johnson Controls; Schneider Electric; SIGNIFY; TOTAL and WLPGA
- **Multilateral Partner (1 in total):** International Solar Alliance (ISA)
- **Hubs and Accelerators (14 in total):** 'Hubs' billed as those which connect international partners with knowledge and experience on the ground include five SEforAll Hubs – Africa Hub, Asia-Pacific Regional Hub, Latin-America and Caribbean Hub, the Knowledge Hub and the Middle East Hub, as well as the Energy Efficiency Hub (Denmark) and Energy Efficiency Facilitating Hub (Japan) and 'Accelerators' billed as bringing together networks of public and private sector leaders to advance progress on the energy transition by sector including the Gender and People Centered Accelerator (SEforAll), Appliances and Equipment Accelerator (UNEP and International Copper Association), Building Efficiency Accelerator (Johnson Controls and World Resource Institute), District Energy in Cities Initiative (Danfoss and UNEP), Global Fuel Economy Initiative (FIA Foundation and UNEP), Industrial Energy Accelerator (Carbon Trust and UN Industrial Development Organization) and Lighting Accelerator (SIGNIFY and UNEP)

(SEforAll website 2021).

In terms of NNSAs – private sector involvement in SEforAll, there is an issue of distinguishing between those that are private sector/corporate entities versus those that are non-profit industry or association groups. Major energy private sector entities are absent from the list. Of the current list: ALER is a non-profit association with 'the mission to promote renewable energies in Portuguese speaking countries'; EURELECTRIC is 'the sector association representing the common interests of the electricity industry at pan-European level'; GOGLA is a not-for-profit global association founded in 2012 for 'off-grid solar energy'; IGU founded in 1931 is a 'worldwide non-profit organisation' whose mission is to 'advocate gas as an integral part of a sustainable global energy system, and to promote the political, technical and economic progress of the gas industry'; IHA is 'a non-profit membership organization formed in 1995 whose mission is to advance sustainable hydropower' and WLPGA 'is the authoritative voice of the global LPG industry representing the full LPG value chain' (SEforAll website 2021). Arguably, improving results on SDG 7 requires a more full-scale involvement of renewable and clean energy private sector partners responsible for innovation, design, production, distribution and dissemination of sustainable energy services, systems and technologies directly related to the relevant targets of SDG 7 that increase access to clean energy for all especially the 2.8 billion who still lack access to non-polluting energy to this day.

In addition to the entities such as SEforAll and GACC, it is also important to briefly review various UN-affiliated partnership mechanisms/entities such as the UN Global Compact (UNGC) to see how energy and climate private sector driven partnerships and initiatives have evolved over time. The UNGC was launched at a time when businesses were largely excluded from UN-led intergovernmental discussions that were taking place then on charting a sustainable development future. The conceptual idea for the UNGC came from UN Secretary-General Kofi Annan's 1999 address to the WEF in which he proposed that '. . . the business leaders gathered in Davos, and . . . the United Nations, initiate a global compact of shared values and principles, which will give a human face to the global market' (Secretary-General/Press Release SG/SM/6881). As the global emphasis shifted towards ensuring that the private sector contributes its fair share to sustainable development, the UNGC has now billed itself the 'world's largest corporate sustainability initiative', whose mission is 'a call to all companies' to:

- 'Do business responsibly by aligning their strategies and operations with Ten Principles on human rights, labour, environment and anti-corruption; and
- Take strategic actions to advance broader societal goals, such as the UN Sustainable Development Goals, with an emphasis on collaboration and innovation' (UNGC website 2021).

In 2004, the UNGC launched its first Leaders' Summit – initially a triennial gathering of top executives of Global Compact companies and other stakeholder groups – but from 2016 on, these Summits are annual gatherings held within the context of the annual UNGA opening. Additionally, the UNGC also launched an annual high-level expressly private sector-focused event – the Private Sector Forum (PSF) – in 2008 during the annual opening of the UNGA to, among other things, demonstrate:

> . . . that the United Nations and the private sector have embarked upon a new phase of collaboration – with leaders from business, civil society, Governments, and the UN committing to deeper levels of partnership . . . [new frameworks] for UN-business collaboration . . . designed to more effectively mobilize private sector engagement in the work of the United Nations . . . [and] support of specific Millennium Development Goals
>
> (Private Sector Forum Meeting Report 2008, p. 3).

Based on the metric of membership, the UNGC could be determined to be a modality for private sector involvement in the SDGs. It has definitely ushered in an age of media announcements made by business at the typically one-day high-level events held in the margin of the annual opening sessions of the UNGA, but deciphering tangible partnership outcome-related results is hard as there is still no comprehensive, time-verifiable database that allows for tracking UNGC SDG partnership results.

The list of thematic/focal areas discussed by the UNGC and PSF is impressive and ranges from 2007 launch of 'Caring for Climate' to the 2019 PSF 'Transformative Business Leadership for a 1.5° C Future'. However from a tracking perspective, the reality is that it is difficult to definitively ascertain whether and to what extent the commitments and pledges for action made in these fora were actually met given the absence of a globally accessible, time-bound, partnership monitoring framework aimed at verifying the varied partnership and initiatives undertaken by the private sector. Despite the existence of the UNGC's early push to embrace the private sector through a series of high-profile fora, it was only in 2010 that the UNGA formally recognized a private sector role in delivering on the MDGs: 'The United Nations Global Compact, in which the world's major companies are committed to global social responsibility, will this year take on the Millennium Development Goals as a central focus of its participating companies. These companies will share technologies, business models, outreach strategies and skilled managers towards the scaling-up of Millennium Development Goal initiatives in many parts of the world' (UNGA Report A/64/665 2010, para. 114, p. 31). The UNGC has also been criticized for not embracing the larger poverty eradication goal-oriented MDG agenda which was seen to be ultimately short-changed and, more importantly, for enabling companies to '. . . in effect use their association with the UN to enhance their image without changing their corporate behaviour (the so-called "bluewash" issue where several major companies which are members have been accused of engaging in practices that are contrary to the principles of the Compact)' (Fortin and Jolly 2015).

In terms of tracking private sector driven partnerships on clean energy and climate change, a search of the UN Business Action Hub was referenced as being:'An interactive online platform designed to match business and UN in order to facilitate action to advance UN goals and the SDGs.' (UNGC/Business Action Hub 2021). This UN-Business Action Hub was developed as a joint effort of the UNGC and others. As it turns out, one of the main functions of the 'platform' was previously listed as a means to allow businesses to search for and interact with potential partners to scale the impact of their projects. However, a search of the prior online avatar of UN business'business.un.org' site now directs traffic to the UNGC site which lists the launch of the 'Climate Ambition Accelerator' that is referenced as comprising a massive list of 13,043 companies, 160 countries and 79,597 public reports. But, the site itself does not provide a searchable database or any detailed information as to any specific partnerships.

The closest approximation of a means by which to track NNSA-related partnerships aimed at the SDGs is the UN SDGs Partnerships Platform (SDGPP) for SDGs' which is billed as: 'The global registry of voluntary commitments and multi-stakeholder partnerships, facilitating global engagement of all stakeholders in support of the implementation of the Sustainable Development Goals' (UN SDGPP 2021). When the search filter of SDG 7 (sustainable energy) was applied on the online platform, as of 22 February 2021, 688 partnerships are listed, which included many partnerships related to fashion. The platform allowed for searches that are related to the specific targets associated with SDG 7 and when the target 'By 2030, ensure universal access to affordable, reliable and modern energy services' was selected, as of 22 February 2021, 181 partnership initiatives were listed, including a number that had nothing to do with sustainable energy access such as those related to ocean health, eradication of water hyacinths, bioacoustics songs amongst them (UN SDG Partnerships Platform 2021).

When the search filter of SDG 13 (climate change) was applied, 1072 partnership initiatives were listed and included amongst them are again a number of the same fashion-related partnerships. What is difficult to explain is why the application of targets reveals an overlap of partnerships that have little to do with addressing climate needs of the poorest and most vulnerable communities. For instance, the filter of SDG 13's Target 13.b: 'Promote mechanisms for raising capacity for effective climate change-related planning and management in least developed countries and small island developing States, including focusing on women, youth and local and marginalized communities' as well as the filter of SDG 7's Target 7.b: 'By 2030, expand infrastructure and upgrade technology for supplying modern and sustainable energy services for all in developing countries, in particular least developed countries, small island developing States, and land-locked developing countries, in accordance with their respective programmes of support' both indicate partnerships that have no connection to LDCs, SIDS and developing countries. Searches using these two separate targets of the two SDGs for example list 'NYC Junior Ambassadors programme' listed as 'created in 2015 by the NYC Mayor's Office for International Affairs' (UN SDG Partnerships Platform). What is surprising to note is that as of 12 December 2021, a search of the UN SDGPP which is the only online database for UN affiliated SDG partnership reveals the following opening clause: '*15 September 2021: Our website is currently experiencing technical issues which prevent (sic) users from registering new initiatives/commitments, updating existing ones, submit (sic) progress reports and accurately using the search filters.* We are working tirelessly to fix these issues. Please keep checking this website for updates' (emphasis added, UN SDGPP 2021). In essence, the UN has acknowledged that its primary database for registering, updating and tracking SDG partnerships has not been functioning for atleast three months as evidenced above. There is an arguably urgent global policy imperative to achieve the 17 SDG and in particular SDGs 7 and 13. But, the silo existence of partnership actions aimed at achieving SDGs that are inherently linked as in the case of SDGs 7 and 13, and the absence of a clear UN definitional, implementation and tracking framework for partnerships poses a massive global challenge in measuring progress. The review of key agreed UN sustainable development outcomes related to partnerships indicates confusion and overlap rather than policy and programmatic clarity. A related concern that needs to be flagged is that there is no comprehensive tracking framework for evaluating partnerships on integrated global issues like energy and climate change.

Currently, there are a plethora of well-intentioned UN-affiliated partnership mechanisms but a clear lack of definitional coherence and policy guidance as to points of engagement with NNSA like the business sector especially in terms of linked action on SDGs 7, 13 and 17 pertaining to the poorest and most vulnerable countries and communities. Furthermore, deciphering and measuring results of UN-affiliated global partnerships is complicated by the fact that ad hoc references to GPSDs and other variants of PSDs are scattered throughout key UN global outcomes on sustainable development. A conclusive finding is that the existence of two separate SDGs combined with the absence of a consistent UN-affiliated globally accessible partnership framework for distinguishing and tracking the role of the private sector in global partnership mechanisms over a 20-year period thwarts a comprehensive evaluation of diverse partnerships results. Various invocations of global partnerships with oftentimes overlapping mandates and/or indistinguishable frameworks are not ideally positioned to track results envisaged to meet the specific needs of the poorest countries and communities.

Programmatic and implementation rather than clarity prevails as to tracking partnerships results as the UNGC does not currently have a publicly accessible database of business-related partnerships, while the UN SDGs Knowledge Platform lists partnerships that are completely unrelated to issues of clean energy and climate change. Resolving concerns related to the overall lack of definitional coherence and tracking of partnership results are particularly relevant and urgent in a resource-constrained global context where the poorest, developing countries/communities are unable to access and leverage scaled-up resources.

References

AGECC (2010). *Energy for a Sustainable Future: Summary Report and Recommendations*. New York: UN.

Alston, M. (2014). Gender mainstreaming and climate change. *Women's Studies International Forum* 47: 287–294.

Banerjee, D. (2014). Toward an integrative framework for environmental justice research. *Society and Natural Resources* 27: 805–819.

Basu, A. (2016). Zika, poverty and reproductive health. *UN Foundations Blog* (29 January 2016). https://unfoundation.org/blog/post/zika-poverty-and-reproductive-health (accessed 22 February 2021).

Bouchard, M. (2015). Transportation emerges as crucial to escaping poverty. *New York Times* (7 May 2015). https://www.nytimes.com/2015/05/07/upshot/transportation-emerges-as-crucial-to-escaping-poverty.html (accessed 25 April 2021).

Bullard, R.D. (ed.) (1993). *Confronting Environmental Racism: Voices from the Grassroots*. Boston: South End Press.

Bullard, R. (ed.) (2005). *The Quest for Environmental Justice*. San Francisco: Sierra Club Books.

Bullard, R. et al. (2007). *Toxic Wastes and Race at Twenty 1987-2007*. Cleveland: UCC https://www.nrdc.org/sites/default/files/toxic-wastes-and-race-at-twenty-1987-2007.pdf (accessed 22 April 2021).

Bullard, R. and Wright, B. (2012). *The Wrong Complexion for Protection: How the Government Response to Disaster Endangers African American Communities*. New York: New York University Press.

Chasek, P. and Downie, D. (2021). *Global Environmental Politics*, 8e. New York: Routledge.

Chasek, P. and Wagner, L. (ed.) (2012). *The Roads from Rio: Lessons Learned from Twenty Years of Multilateral Environmental Negotiations*. New York: Routledge.

Cherian, A. (2012). Confronting as multitude of multilateral environmental agreements. In: *Global Environmental Issues* (ed. F. Harris), 39–61. Chichester: Wiley-Blackwell.

Cherian, A. (2015). *Energy and Global Climate Change: Bridging the Sustainable Development Divide*. Oxford: Wiley Blackwell.

Chetty, R. and Hendren, N. (2017). The Impacts of Neighborhoods on Intergenerational Mobility II: County-Level Estimates. https://opportunityinsights.org/wp-content/uploads/2018/03/movers_paper2.pdf (accessed 23 April 2021).

Clayton, S., Manning, C., Krygsman, K., and Speiser, M. (2017). *Mental Health and Our Changing Climate: Impacts, Implications, and Guidance*. Washington, DC: American Psychological Association.

Clayton, S. et al. (2021). *Mental Health and our Changing Climate: Impacts, Inequities, Responses.* Washington DC: American Psychological Association and ecoAmerica. https://www.apa.org/news/press/releases/mental-health-climate-change.pdf (accessed on 4 December 2021.

Clean Cooking Alliance website (2021). Strategic Partnerships and Alliances. https://www.cleancookingalliance.org/about/our-partners/index.html (accessed 22 February 2021).

Clough, E. et al. (2019). A Study of Partnerships and Initiatives registered on the UN SDG Partnership Platform. https://eprints.ncl.ac.uk/file_store/production/261442/A0ECA892-0298-4BD8-882B-03187A900D4E.pdf (accessed 20 February 2021).

Colverson, S. and Perera, O. (2012). *Harnessing the Power of Public-Private Partnerships: the Role of Hybrid Financing Strategies in Sustainable Development.* Winnipeg: IISD.

Detraz, N. (2014). *Environmental Security and Gender.* New York: Routledge.

Detraz, N. (2017). *Gender and the Environment.* Cambridge, UK: Polity Press.

Effective Cooperation.Org (2021). Welcome to the Global Partnership. https://www.effectivecooperation.org/ (accessed 20 February 2021).

Fortin, C. and Jolly, R. (2015) The United Nations and Business: Towards New Modes of Global Governance? May 2015 *IDS Bulletin* 46(3).

Frieden, T. (2016). Entire World Literature on Zika: Image. Twitter. https://twitter.com/CDCDirector/status/698162506137784321 (accessed 23 April 2021).

Gubler, D. (2012). Economic burden of dengue. *The American Journal of Tropical Medicine and Hygiene* 86 (5): 743–744.

Haas, P. (2004). When does power listen to truth? A constructivist approach to the policy process. *Journal of European Policy* 11 (4): 569–592.

Hartmann, B. (2010). Rethinking climate refugees and climate conflict. *Journal of International Development* 22 (2): 233–246.

Hayes, K. and Poland, B. (2018). Addressing mental health in a changing climate: incorporating mental health indicators into climate change and health vulnerability and adaptation assessments. *International Journal of Environmental Research and Public Health* 15 (9): 1806.

IASC (2008). *Mental Health and Psychosocial Support: Checklist for Field Use.* Geneva: IASC.

IEA et al. (2020). Tracking SDG 7: The Energy Progress Report 2020. https://trackingsdg7.esmap.org/data/files/download-documents/tracking_sdg_7_2020-full_report_-_web_0.pdf (accessed 22 February 2021).

IPCC (2014). AR 5 Climate Change 2014: Impacts, Adaptation and Vulnerability. https://www.ipcc.ch/site/assets/uploads/2018/02/WGIIAR5-PartA_FINAL.pdf (accessed 23 April 2021).

Kanie, N. (2007). Governance with multilateral environmental agreements: a healthy or ill-equipped fragmentation? In: *Global Environmental Governance: Perspectives on the Current Debate* (ed. W. Hoffman and L. Swart), 67–86. New York: Center for UN Reform Education.

Kanie, N. and Haas, P. (2004). *Emerging Forces in Environmental Governance.* Tokyo: UN University Press.

Lancet Editorial (2016). Zika virus: a new global threat for 2016. *The Lancet.* https://www.thelancet.com/journals/lancet/article/PIIS0140-6736%2816%2900014-3/fulltext?elsca1=etoc&elsca2=email&elsca3=0140-6736_20160109_387_10014_&elsca4=Public%20Health%7CInfectious%20Diseases%7CHealth%20Policy%7CInternal%2FFamily%20Medicine%7CGeneral%20Surgery%7CLancet (accessed 23 April 2021).

Leach, M. (ed.) (2015). *Gender Equality and Sustainable Development.* New York: Routledge.

Martinez-Alier, J. (2002). *The Environmentalism of the Poor: A Study of Ecological Conflicts and Valuation*. Cheltenham UK: Edward Elgar.

Martinez-Alier, J. (2014). The environmentalism of the poor. *Geofourm* 54: 239–241.

Mohney, G. (2016). Health officials say mosquitoes that spread Zika now in 30 states. *ABC News*. https://www.yahoo.com/gma/health-officials-mosquitos-spread-zika-now-30-states-190823478--abc-news-health.html (accessed 23 April 2021).

Moon, S. et al. (2015). Will Ebola change the game? Ten essential reforms before the next pandemic-the report of the Harvard-LSHTM independent panel on the global response to Ebola. *The Lancet* 386: 2204–2221. https://www.thelancet.com/pdfs/journals/lancet/PIIS0140-6736%2815%2900946-0.pdf (accessed 23 April 2021).

Najam, A. (2005). Developing countries and global environmental governance. *International Environmental Agreements* 5 (3): 303–321.

OECD (2011). Busan Partnership for Effective Development Cooperation. https://www.oecd.org/dac/effectiveness/49650173.pdf (accessed 23 April 2021).

Oxfam Press Release (2021) *Oxfam's verdict on the COP26 outcome*. 13 November 2021. https://www.oxfam.org/en/press-releases/oxfams-verdict-cop26-outcome (accessed 23 April 2021)

Pellow, D. and Brulle, R. (ed.) (2005). *Power, Justice and the Environment: A Critical Appraisal of the Environmental Justice Movement*. Cambridge, MA: MIT Press.

Rehman, I.H. et al. (2017). Accelerating access to energy services: the way forward. *Advances in Climate Change Research* 8 (1): 57–61.

Sachs, J. (2015). *The Age of Sustainable Development*. New York: Columbia University Press.

SEforALL website (2021). Partnerships. https://www.seforall.org/partnerships/partners (accessed 22 February 2021).

Shiva, V. (2008). *Soil not Oil: Environmental Justice in an Age of Climate Crisis*. Berkeley: North Atlantic Books.

Simeone, R.M. et al. (2016). Possible Zika virus infection among pregnant women – United States and Territories. *CDC MMWR* **2016**: 65. https://www.cdc.gov/mmwr/volumes/65/wr/mm6520e1.htm (accessed 23 April 2021).

Susskind, L. (1994). *Environmental Diplomacy: Negotiating More Effective Global Agreements*. Oxford, UK: Oxford University Press.

UCC (1987). *Toxic Wastes and Race in the United States*. UCC Commission for Racial Justice. New York: UCC. https://www.nrc.gov/docs/ML1310/ML13109A339.pdf (accessed on 22 April 2021)

UN (1987). Report of the World Commission on Environment and Development. A/42/427 Our Common Future: Report of the Word Commission on Environment and Development. http://www.un-documents.net/wced-ocf.htm (accessed 22 April 2021).

UN (1992). *United Nations Conference on Environment and Development (UNCED)*, Agenda 21. New York: UN. A/Conf.151/26. https://sustainabledevelopment.un.org/content/documents/Agenda21.pdf (accessed 1 February 2021).

UN (2001). *Commission on Sustainable Development: Report of the Ninth Session. E/CN.17/2001/19*. New York: UN.

UN (2002). *Johannesburg Programme of Implementation*. New York: UNDESA.

UN (2003a). *Partnerships for SD Brochure: 2003. Elaborated by Interested Parties in the Context of the World Summit on Sustainable Development*. New York: UN http://www.un.org/esa/sustdev/partnerships/guiding_principles7june2002.pdf (accessed 1 February 2021).

UN (2003b). Follow-up to Johannesburg and the Future Role of the CSD – The Implementation Track E/CN.17/2003 /2. http://www.un.org/esa/sustdev/csd/csd11/sgreport.pdf (accessed 1 February 2021).

UN (2005). *Millennium Project Report.* New York: UN.

UN (2006). *Resolution adopted by the General Assembly on 22 December 2005: Towards Global Partnerships.* A/RES/60/215. New York: UN.

UN (2007). *Commission on Sustainable Development: Report on the Fifteenth Session. E/CN.17/2007/15.* New York: UN.

UN (2008). *MDG Gap Task Force Report: Delivering on the Global Partnership for Achieving the Millennium Development Goals.* New York: UN.

UN (2011). *Sustainable Energy for All – A Vision Statement by Ban Ki Moon Secretary General of the United Nations.* New York: UN.

UN (2015a). MDG Gap Task Force Report 2015: Fact sheet: Where are the Gaps? http://www.un.org/millenniumgoals/pdf/MDG_Gap_2015_PR_Fact_Sheet_English.pdf (accessed 2 March 2021).

UN (2015b). *Millennium Development Goals Report.* New York: UN https://www.un.org/millenniumgoals/2015_MDG_Report/pdf/MDG%202015%20rev%20(July%201).pdf (accessed 25 April 2021).

UN (2016). Report of the Secretary General: Enhanced cooperation between the United Nations and all relevant partners, in particular the private sector, A/70/296. http://www.un.org/ga/search/view_doc.asp?symbol=A/70/296 (accessed 22 April 2021).

UN (2018). *Sustainable Development Goals Report: 2018.* New York: UN https://unstats.un.org/sdgs/files/report/2018/TheSustainableDevelopmentGoalsReport2018-EN.pdf (accessed 22 April 2021).

UNGC website (2021). https://www.unglobalcompact.org/ and https://www.unglobalcompact.org/library/search?search%5Bissue_areas%5D%5B%5D=211 (accessed 19 April 2021).

UN Press release (2017). Amid Humanitarian Funding Gap, 20 Million People across Africa, Yemen at Risk of Starvation, Emergency Relief Chief Warns Security Council. https://www.un.org/press/en/2017/sc12748.doc.htm (accessed 22 March 2021).

UN SDG Knowledge website (2021). Sustainable Development Knowledge Platform. https://sustainabledevelopment.un.org/index.html (accessed 22 April 2021).

UN SDG Partnerships Platform (2021). 'Goal 7' and 'Goal 13'. https://sustainabledevelopment.un.org/partnerships/goal7 https://sustainabledevelopment.un.org/partnerships/goal13 (accessed 22 February 2021).

UN SDGPP (2021) Sustainable Development Goals Partnership Platform. https://sustainabledevelopment.un.org/partnerships/ (accessed 12 December 2021).

UN.org website. (2021a). Our Work – What We Do. https://www.un.org/en/our-work (accessed 24 April 2021).

UN.org website. (2021b). UN Millennium Development Goals. https://www.un.org/millenniumgoals/reports.shtml (accessed 23 April 2021).

UN.org/SCPI (2021). About the UN Climate Partnerships for the Global South. https://www.un.org/sustainabledevelopment/scpibackground/ (accessed 24 November 2021).

UN.org/UNFIP (2021). What is UNFIP. https://www.un.org/partnerships/content/what-unfip (accessed 24 November 2021).

UN/South Centre (2017) Climate Partnerships for a Sustainable Future. https://www.un.org/sustainabledevelopment/wp-content/uploads/2017/11/Report-on-Climate-Partnerships-for-a-Sustainable-Future.pdf (accessed 23 April 2021).

UN/WEHAB (2002). *A Framework for Action on Energy*. New York: UN http://www.gdrc.org/sustdev/un-desd/wehab_energy.pdf (accessed 24 April 2021).

UNDESA (2002). *A Guide for Potential Partnerships on Energy for Sustainable Development*. New York: UNDESA Background Paper No 3.

UNDESA (2019). *Global Sustainable Development Report: The Future is Now*. New York: UNDESA https://sustainabledevelopment.un.org/content/documents/24797GSDR_report_2019.pdf (accessed 24 November 2021).

UNEP (2007). *Glossary of Terms for Negotiators of Multilateral Environmental Agreements*. Nairobi, Kenya: UNEP.

UNEP (2010). Auditing the Implementation of Multilateral Environmental Agreements (MEAs): A Primer for Auditors. https://wedocs.unep.org/bitstream/handle/20.500.11822/17290/Auditing_Implementation_of_MEAs.pdf?sequence=1&%3BisAllowed= (accessed 21 February 2021).

UNEP (2019). *Environmental Rule of Law*. Nairobi: UNEP https://wedocs.unep.org/bitstream/handle/20.500.11822/27279/Environmental_rule_of_law.pdf?sequence=1&isAllowed=y (accessed 20 April 2021.

UNESCAP (2019). *The Disaster Riskcape: Asia Pacific Disaster Report 2019*. Bangkok: UNESCAP https://www.unescap.org/sites/default/d8files/knowledge-products/Asia-Pacific%20Disaster%20Report%202019_full%20version.pdf (accessed 23 April 2021).

UNGA (1997). *Program for the Further Implementation of Agenda 21*. A/RES/S-19/2. New York: UN.

UNGA (2010). Keeping the promise: a forward-looking review to promote an agreed action agenda to achieve the Millennium Development Goals by 2015. A/64/665. https://undocs.org/pdf?symbol=en/A/64/665 (accessed 22 January 2021).

UNGA (2012). *The Future We Want*. New York: UN. A/RES/66/288. http://www.un.org/ga/search/view_doc.asp?symbol=A/RES/66/288&Lang=E (accessed 22 January 2021).

UNGA (2015a). *Transforming our World: 2030 Sustainable Development Agenda*. New York: UN. A/Res/70 (15 October 2015). https://www.un.org/ga/search/view_doc.asp?symbol=A/RES/70/1&Lang=E (accessed 23 April 2021).

UNGA (2015b). Towards Global Partnerships: A Principle-Based Approach to Enhanced Cooperation Between the United Nations and All Relevant Partners A/Res/70/224. http://www.un.org/en/ga/search/view_doc.asp?symbol=A/RES/70/224 (accessed 20 February 2021).

UNGA (2016). UN SG Report on UN Office of Partnerships (2016) 15 July 2016. A/Res/71/159. http://www.un.org/ga/search/view_doc.asp?symbol=A/71/159 (accessed 12 March 2021).

UNGC, UNCTAD et al. (2015). Private Sector Investment and Sustainable Development (PSISD). https://3blmedia.com/media/other/Private_Sector_Investment_and_Sustainable_Development.pdf (accessed 24 March 2021).

UNOPS (2021). About us: UNOPS. https://www.unops.org/ (accessed 24 November 2021).

US EPA (2013). Enviromental Justice Related Terms. https://www.epa.gov/sites/default/files/2015-02/documents/team-ej-lexicon.pdf (accessed 22 April 2021).

Vittor, A. (2016). To tackle the Zika virus, eliminate urban poverty. *New York Times* (29 January 2016). https://www.nytimes.com/roomfordebate/2016/01/29/

how-to-stop-the-spread-of-zika/to-tackle-the-zika-virus-alleviate-urban-poverty (accessed 23 April 2021).

Watts, N. et al. (2018). Countdown on health and climate change: from 25 years of inaction to a global transformation for public health. *The Lancet.* 10 February 391 (10120): 581–630.

Weiss, E.B. (1989). *In Fairness to Future Generations: International Law, Common Patrimomy, and Intergovernmental Equity.* Dobbs Ferry, NY: Transnational Publishers.

Weiss, E.B. (1992). In fairness to future generations and sustainable development. *American University International Law Review* 8 (1): 19–26.

White, G. (2015). *Stranded: Hhow America's Ffailing Ppublic Ttransportation increases inequalities. The Atlantic* https://www.theatlantic.com/business/archive/2015/05/stranded-how-americas-failing-public-transportation-increases-inequality/393419/ (accessed 23 April 2021).

WHO (2018). *COP-24: Special Report – Health and Climate Change.* Geneva: WHO https://apps.who.int/iris/handle/10665/276405 (accessed 22 April 2021).

WMO (2020). *State of the Climate in Africa.* Geneva: WMO https://library.wmo.int/doc_num.php?explnum_id=10421 (accessed 23 April 2021).

World Bank, IEA et al. (2020). *Tracking SDG 7: Energy Progress Report 2020.* Washington DC: World Bank https://trackingsdg7.esmap.org/data/files/download-documents/tracking_sdg_7_2020-full_report_-_web_0.pdf (accessed 22 February 2021).

World Vision (2019). Typhoon Haiyan: Facts, FAQs and How to Help. https://www.worldvision.org/disaster-relief-news-stories/typhoon-haiyan-facts (accessed 22 April 2021).

3

Looking Beyond the Global Climate Change Negotiations Silo

Examining UN Climate Change Outcomes for Linked Action on Clean Air and Clean Energy for All

3.1 Time to Look Beyond Tense Annual Climate Negotiations: Developing Countries' Urgent Needs

The UNFCCC entered into force on 21 March 1994, as the universally agreed-upon framework for international cooperation to combat climate change. It remains one of the most historic and globally consequential UN agreements with the broadest ratification – 197 Parties (196 UN member states and 1 regional economic integration organization) (UNFCCC website 2021a). The Preamble to the 1992 UNFCCC which to date is the only climate agreement globally agreed to by all UN member affirms that '. . .responses to climate change should be coordinated with social and economic development in an integrated manner with a view to avoiding adverse impacts on the latter, taking into full account the legitimate priority needs of developing countries for the achievement of sustained economic growth and the eradication of poverty' (UNEP/WMO/IUC 1994, p. 4). From a development perspective, the UNFCCC further emphasized the differences between developed and developing economies and the need to integrate energy sector actions with climate change goals by: 'Recognizing that all countries, especially developing countries, need access to resources required to achieve sustainable social and economic development and that in order for developing countries to progress towards that goal, *their energy consumption will need to grow* taking into account the possibilities for achieving greater energy efficiency and for controlling greenhouse gas emissions in general, including through the application of new technologies on terms which make such an application economically and socially beneficial' (emphasis added, UNEP/WMO/IUC 1994, p. 4). Since poverty eradication is an overarching principle, and access to sustainable energy for all is also crucial to the UN's SDA, it stands to reason that progress towards integrating poverty eradication, sustainable energy and climate change should not be stymied.

The UNFCCC also recognized that developing countries have development imperatives, which they must pursue as a priority, and that their ability to undertake climate action depends on the extent of support they receive from developed countries. The UNFCCC's framing of the concept of 'common but differentiated responsibilities' of developed and developing countries with respect to climate change and their respective commitments under the UNFCCC and its related instruments (such as the KP and the PA) are particularly important because it provides the legal and policy architecture for the multilateral

Air Pollution, Clean Energy and Climate Change, First Edition. Anilla Cherian.

climate change policy regime.As dicussed in Chapter 1, the UN-led global community has long known the developing countries, especially SIDs and LDCs are on the frontline in terms of battling climate change despite being the least responsible for causing historic and current escalations in GHGs. Back in 2003, Barnett and Adger pointed out that the physical impacts on small island nations could push their socio-political systems into collapse. In 2011, Nicholls et al. concluded that the small population size and associated costs of adaptation to sea-level rise to island nations are such that forced abandonment is a likely outcome even for small changes in sea-level rise. Against this backdrop, the protracted climate change policy regime predicated on intergovernmentally-negotiated outcomes arrived at via 26 annual climate conferences convened in cities across the globe serves as inequitable delaying mechanism. It is time to acknowledge that addressing the climate needs of smallest and most vulnerable communities and countries has not been comprehensively addressed as a result of annual word-parsing intergovernmental exercises.

It is also time to recognize that the role of NNSAs which are crucial to the effective resolution of climate change has been circumscribed within these negotiated outcomes in the context of the UN's SDGs and the PA. There has been a documented history of NNSAs – namely fossil fuel companies actively engaged with funding climate denial research and/or pushing political interests in delaying climate mitigation within the US and within the context of UN negotiations. In his 1997 book 'The Heat Is On', Pulitzer Prize–winning journalist Ross Gelbspan laid bare the carefully orchestrated 'campaign' by oil and coal groups combined with conservative political interests to confuse the public by claiming that disruptive weather patterns do not equate to global climate change. But, addressing global climate change is a massive challenge not just in the climate denial executive suites of oil, coal and insurance companies but also because it requires confronting huge, existing inequities faced by the global poor who ironically contribute the least in terms of per capita GHG emissions, and are also the most vulnerable to adverse climatic impacts. As Gelbspan noted, this 'further complication' in the context of climate change was that 'the consequences of global economic inequality are as critical as the carbon in the air. Any solution to climate change must address this economic gulf' (1997, p. 11). Optimistically, in spite of the massive global disinformation campaign, Gelbspan saw promise back in 1997 in 'an unlikely coalition' of '*small island nations whose fear of being flooded out of existence by rising sea levels and increasingly severe hurricanes'* have found 'common cause with Germany, Britain, Ghana and other nations that fear the disastrous consequences of unchecked climate change. And they are beginning to force the world's negotiators to confront the problem' (1997, pp. 12–13).

Now more than thirty years after the adoption of the globally agreed precautionary principle on climate change enshrined as Article 3.3 of the UNFCCC, the collective foot dragging on climate change by private sector has been signalled as a problem by the CEOs of Amazon and Black Rock amongst a whole slew of other major business/industry leaders (Fink 2020; Bezos 2021). What is striking about Jeff Bezos' 2021 letter to his shareholders ahead of his departure as CEO is his blunt reference that climate change is as real as gravity: 'You don't have to say that photosynthesis is real, or make the case that gravity is real, or that water boils at 100 degrees Celsius at sea level. These things are simply true, as is the reality of climate change' (Bezos 2021). But perhaps an equally big eye opener is Bezos' reference to the Climate Pledge (launched in 2019), which in less than two years now includes 53 companies representing almost every economic sector – all committed to achieve net-zero carbon in their worldwide businesses by 2040. According to the Climate Pledge website, Amazon cofounded the Climate Pledge in 2019, and became the first signatory, the Pledge is a billed as 'a cross-sector community of companies, organizations, individuals and partners working to crack the climate crises and solve the challenges of decarbonizing our economy'. Signatories to the Pledge which include companies such as IBM, Infosys and Microsoft amongst others are required to:

- 'Measure and report on GHG emissions on a regular basis;
- Implement decarbonization strategies in line with the Paris Agreement through real business change and innovations, including efficiency improvements, renewable energy, materials reductions, and other carbon emission elimination strategies;
- Neutralize any remaining emissions with additional, quantifiable, real, permanent, and socially-beneficial offsets to achieve net zero annual carbon emissions by 2040' (Climate Pledge website 2021).

Today, there is an unmistakable global urgency in looking for innovative global partnerships that engage both UN member states and NNSAs to work in concert to mitigate and adapt to the climate crisis. However, the question is whether the UN-led global community is effectively focused on the pressing concern about PM pollution problem that negatively impacts millions of poorer and more vulnerable lives.Timely and effective global partnerships that simultaneously mitigate climate change and energy related pollution can ameliorate grave risks of irreversible impacts for some of the poorest regions, countries and communities in the world. But, what does an examination of the agreed outcomes of the decades-long intergovernmental climate change negotiations on two key partnership-driven mechanisms – the UNFCCC's KP's CDM and the UNFCCC partnership process – reveal regarding actions aimed at jointly addressing climate, increase access to sustainable energy, poverty eradication and SLCPs? What if anything does the PA specify in terms of increasing access to clean energy and reducing air pollution related SLCPs?

The chasm is growing between those countries, communities and households who bear the brunt of impacts and those who can afford to escape. The 2019 Klynveld Peat Marwick Goerdeier (KPMG) Change Readiness Index (now its fourth year of bi-annual publication) studied 140 countries on how effectively they prepare for, and respond to, major change events. The report recognized that 'from geopolitics to technology to climate, the world is changing at a rapid pace. Which countries seem prepared for the opportunities? And which appear ill equipped to manage the risks?'. These questions were answered by measuring each country across three key pillars of capability: enterprise, government and people and civil society. Not surprisingly, Switzerland and Singapore top the list, but when it comes to climate change and developing countries, the 2019 Index reveals that: 'Countries most susceptible to climate risks are mostly low income and lower-middle income countries. Less mature economies like Chad, South Sudan and Afghanistan are the worst performing in climate resilience, as are countries in Sub-Saharan Africa and South Asia. The majority of higher income economies are considered low risk, high readiness countries'. The report offers a bleak prognosis pointing out that 'poorer countries face double jeopardy when it comes to climate change: a higher risk from the negative impacts of climate change and a lower capacity to implement climate-ready policies and institutions' (KPMG 2019, Executive Summary). The 'double jeopardy' faced by poorer countries provides a sense of the climate and poverty threat multiplier, but when exposure to fossil fuel energy related air pollution is factored in, the outcomes are grim. How exactly have the UNFCCC and its PA addressed the issue of SLCPs associated with the world's greatest environmental health risks – air pollution?

KPMG's 2019 reference about the 'double jeopardy' faced by the poorest and most vulnerable lives within the smallest and most vulnerable countries can be traced back to the numerous warnings issued by developing country negotiators from SIDS and LDCs. The late Ambassador John Ashe who served as Chair for both Subsidiary Bodies of the UNFCCC and Co-Chair of the Rio+20 Summit pointed out that climate change negotiators from SIDs and LDCs recognized early on that their countries were sadly the real 'canaries' in the minefield of adverse climatic impacts (Ashe Personal notes 2016). In a 2014 article entitled 'Seizing the Moment' in UNEP's Our Planet journal, Ashe highlighted the following: 'The expected impacts of the adverse effects of climate change only serve to add a particularly nefarious layer of vulnerability, which seems all the more cruel when one considers that the SIDS share of global emissions is very low, that they have limited adaptive capacity and options, and they face very significant costs for adaptation' (p. 12). More than 30 years after the adoption of the first UNGA resolution on climate change, the prognosis for poorer, smaller and more vulnerable countries has only gotten worse, but not all countries, cities and communities are anticipated to bear the brunt of climate change equally. As the authors of a 2018 scientific comparison of climate models put it: '*The countries that have contributed least* to climate change, and *are most vulnerable to extreme events,* are *projected to experience the strongest increase in variability*. These changes would therefore *amplify the inequality* associated with the impacts of a changing climate' (emphasis added, Bathiany et al. 2018, p. 1).

The policy reality about climate change negotiations is that from their inception they have consistently been the forum for contentious debate coalesced around the need for consensus-based textually derived negotiations between developed and developing countries. Reflecting on the early history of the climate change negotiations, Gupta argued that climate change negotiations ran the risk of deadlock where neither industrialized countries nor developing countries were motivated to take further action on the grounds that the other party would then be perceived as getting a free ride (1997). 'A Review of the Effectiveness of Developing Country Participation in the Climate Change Convention Negotiations' by the Overseas Development Institute highlighted that: 'Analysis of the outcomes of the UNFCCC indicates that developing countries have generally been losers, and see themselves as being cheated. A classic example was the way the Brazilian "Clean Development Fund", introduced as a polluter-pays tax on non-compliance, was turned around at the end of the negotiation process at Kyoto into the "Clean Development Mechanism" (CDM) – a mechanism through which developed countries could gain carbon credits for carbon offset activities. This and other outcomes illustrate how developed countries armed with superior negotiating power and information are able to convert their interpretations of the UNFCCC into outcomes, often through forced or "non-decisions" (from the developing country perspective). The outcomes of the UNFCCC have often been fudged so that they are politically acceptable in developed countries. These factors make developing countries suspicious and defensive to developed country initiatives' (Richards 2001). Gupta pointed to a series of factors which were representative of the tension between developed and developing countries on climate change negotiations including: (i) increasing complexity of the definition of the climate change issue from an environmental to a development issue; (ii) the inability of the developed countries to reduce their own emissions and raise funds commensurate with the nature of the problem and their initial commitments; (iii) the increasing engagement of different social actors in the discussion and, in particular, the gradual use of market mechanisms in the regime; (iv) the increasing search for alternative solutions within the formal negotiations and (v) the search for solutions outside the regime (Gupta 2010).

Even before the UNFCCC formally entered into force on 21 March 1994, developing country negotiators, in particular those on the frontline of the battle against SLR and coastal zone inundation were cognizant of the UNFCCC's precautionary principle and actively aware of the UNFCCC limitations. Here, it is useful to go back to a 1998 Pew Center Report entitled 'Equity & Global Climate Change: The Complex Elements of Global Fairness' which referenced the concept of 'equity' as compared to 'justice' in terms of climate negotiations and noted that there were different philosophical approaches to equity. The report pointed out that the concept of equity as it applies to developing countries has been amply referenced in international treaties and law principles, with the International Court of Justice (ICJ) defining 'equity' as being a 'general principle directly applicable as law that is, one of many considerations that can be used to reach a solution. However, the ICJ also recognized that there are no precise guidelines for defining equity' (Claussen and McNeilly 1998, p. 9). The 1998 Report's conclusion posed this question: 'What are the criteria that climate change negotiators need to employ when considering equity?' The answer is hauntingly timely: 'If the end result of negotiations is not fair- by most governments' definitions-then it will not be implemented. Little or no mitigation of climate change is then the unfair outcome to those who will bear the brunt of the impacts' (1998, p. 20). Now that over 20 years of climate negotiations have passed, it is hard to ignore the fact that the little or no mitigation of climate change is the unfair burden borne by those who face the brunt of adverse climatic impacts and are also heavily exposed to toxic levels of solid fuel/fossil fuel air pollution.

The problem is that policy perspectives within the context of the UN-led intergovernmental climate negotiations as well as the broader sustainable development negotiations have tended to focus largely on the dichotomy between developing versus developed countries or the need for environmental norm following, or the role of effective environmental governance mechanisms. There is, however, an overall lack of research on the interlinked global environmental health risks posed by air pollution and climate change taking into account viewpoints from developing country policymakers. From an early stage onwards, policy and data on the impacts of global warming was available primarily for regions and countries in the developed world (CRU/IERL 1992; Parry and Duncan 1995). It is important to recognize, however, that global climate change negotiators from developing countries, in particular SIDS and LDCs, were active in contesting the inequities associated with adverse climatic impacts and the differential burdens of human lives lost as a result. Those relatively new to the 'heated' and convoluted three decades of climate negotiations may be surprised to hear that back when the IPCC prepared its Second Assessment Report (AR2), its Working Group II got inflamed in the controversy that a Bangladeshi or Indian life was worth much less than that of an American or Northern European life. A principal economist behind many of the IPCC's calculations for the AR2, Sam Fankhauser, worked at the World Bank's Global Environmental Facility, which then and to date, provides funding for climate change projects designed to help developing countries with responding to climate change. Fankhauser was the co-author of the IPCC's draft chapter on the social cost of climate change which calculated that there were five times as many deaths in the poor nations as in the OECD countries as a result of global warming, but that the mortality cost of the OECD deaths was more than twice as great. Writing in the New Scientist on 1 April 1995, Fred Pearce noted: 'Is the death of an overweight American from heatstroke a greater loss to the world than a Bangladeshi farmer struck down by a tropical cyclone? Economists advising the world's governments on how to cope with global warming say yes. And their answer poses a new threat to climate negotiations beginning in Berlin this week. A draft of a forthcoming report from the Intergovernmental Panel on Climate Change (IPCC), currently being circulated among economists, values a dead American or European at $1.5 million, but a dead citizen from a "low-income country" at a tenth as much, or $150 000' (Pearce 1995).

In the end, the draft of Chapter 6 of the IPCC's Working Group III of AR2 may only be remembered by grizzled veterans of the climate change negotiations process. The pitched fight related to its so-called pricing the life of the world's poor resulted in political controversy breaking out at the first COP to the UNFCCC. Late Ambassador Ashe, who was put in charge of seeking global consensus on this highly politically charged topic, recalled it was only after extremely fraught, late night, tense still negotiation meetings on the topic that compromise was reached between developed and developing countries (Ashe, personal notes 2016). As the record shows, the comprise text on AR2's Working Group III stated that: 'The literature on the subject in this section is controversial and mainly based on research done on developed countries, often extrapolated to developing countries. There is no consensus about how to value statistical lives or how to aggregate statistical lives across countries. Monetary valuation should not obscure the human consequences of anthropogenic climate change damages, because the value of life has meaning beyond monetary valuation. . ..It may be noted that, in virtually all of the literature discussed in this section, the developing country statistical lives have not been equally valued at the developed country value, nor are other damages in developing countries equally valued at the developed country value. Because national circumstances, including opportunity costs, differ, economists sometimes evaluate certain kinds of impacts differently amongst countries. . .. An approach that includes equal valuation of impacts on human life wherever they occur may yield different global aggregate estimates than those reported below. For example, equalizing the value of a statistical life at a global average could leave total global damage unchanged but would increase markedly the share of these damages borne by the developing world. Equalizing the value at the level typical in developed countries would increase monetized damages several times, and would further increase the share of the developing countries in the total damage estimate.' (IPCC 1995, p. 50). In a subsequent 1997 paper, entitled 'Social Costs of Climate Change', Fankhasuer and Tol steered clear of human valuation and concluded that while earlier studies have estimated an aggregated monetized damage equivalent to 1.5–2.0% of world GDP, the 'OECD would face losses equivalent to 1.0 to 1.5% of GDP, and developing countries 2.0 to 9.0%'and added that 'these figures are not comprehensive and highly uncertain' with 'newer studies increasingly emphasizing the power of adaptation' (1997, p. 399). However, the lack of capacities for poorer countries, communities and households to adapt to the adverse impacts of climate change layered upon the burden of energy related air pollution has not been sufficiently addressed. This lack of capacity to cope with climate change buttresses a broader argument that reducing exposure to toxic levels of PM pollution should be considered both as a means to mitigate SLCPs and as a means to ameliorate climate related health vulnerabilities.

The issue of assessing the value of human life as a result of adverse climatic impacts was deemed too politically charged for the IPCC and, by extension, the 1995 COP. The tense discussion on the valuation of human lives led to then Indian Environment Minister Kamal Nath's choosing to contrast the 'survival emissions' of developing countries versus 'luxury emissions' by developed ones in his speech to the 1995 COP. This distinction has import and carried through for more than two decades within the UNFCCC's discussions by developing countries. Nath highlighted the importance of prioritization given to poverty eradication within the UNFCCC and noted: 'The Climate Change Convention is not merely about the control of greenhouse gases. Eradication of poverty, avoiding risks to food production and sustainable development are three principles quite "explicit" in the Convention . . . How can we talk of "burden sharing"? Equitable burden sharing in emission reduction has no meaning unless it is preceded by equitable benefit sharing of environmental space. Even at a conservative estimate, the privileges enjoyed by the North for excess emissions are worth 100 billion dollars annually. This is the "environmental rent" that ought to be transferred to developing countries every year in lieu of "eating" into their environmental space' (Nath 1995). The conceptual idea of a shared global environmental space that needs to be fairly allocated in terms of burden sharing as it relates to climate change also lies at the heart of Oxfam's finding about carbon inequality and undergirds the UN Rapporteur's reference to climate apartheid. This concept also poses the question as to whether equitable burden sharing in terms of access clean air has or even can be adequately factored into climate change negotiations, and whether air pollution has been conceived as intrinsically linked to GHG emissions reductions.

In addition to the tensions between developed and developing countries in the context of intergovernmental climate negotiations, it is worth acknowledging that the shaping of policy responses to climate change within developing countries that have played and continue to play a key role remains less reflected in academic research accessible within developed countries. However, the policy reality is that Indian researchers have since the early days of UNFCCC process been articulating policy positions relevant to India, including 'Gender Issues in Energy Policy' (Parikh 1995), and 'CO_2 Emissions Reduction from the Power System in India' (Chattopadhyay and Parikh 1994). Back in 1991, in a report entitled 'Global Warming in an Unequal World', Indian environmentalists Anil Agarwal and Sunita Narain took aim at a 1990 World Resources Institute (WRI) Report 'A Guide to the Global Environment' and referenced it as 'an excellent example of environmental colonialism' which was 'based less on science and more on politically motivated and mathematical jugglery' (1991, p. 3). Agarwal and Narain took issue with WRI's methodology and emphasis on carbon dioxide and methane production figures that were correlated to deforestation, agriculture and livestock emanating from developing countries. Their proposal was that: '. . . a system of global tradeable permits should be introduced to control global greenhouse gas emissions. All countries should be given tradeable quotas in proportion to their population share and the total quotas should equal the world's natural sinks. The quantity of unused permissible emissions can be sold by low-level greenhouse gas emitting countries to high-level greenhouse gas producers at a certain fixed rate' (p. 20). Needless to say, the idea of a per capita–based permits never took off within the UNFCCC, but the core principle of energy and climate equity for those who are most vulnerable and have contributed the least to GHG emissions in particular households in LDCs and SIDS persists. Dubash (2015) addressed a range of substantive issues that shape current Indian views on global climate negotiations, the national politics around climate change and, equally importantly, the manner in which cross-cutting issues such as energy, urbanization, water and forests have been addressed.

From a developing country perspective, the analytical costing of climate change with implications for the public health costs of energy related air pollution not being adequately factored is another issue that needs to be flagged. The 2018 Nobel Prize for Economics was awarded to Professor Nordhaus whose landmark paper 'To slow of not to slow: The economics of the greenhouse effect' presented the first cost-benefit analyses of policies aimed at mitigating GHGs. Nordhaus went on to create the first integrated assessment model – the 'dynamic integrated climate-economy' (DICE) – and its initial 1992 results found that no action would result in global warming of 3 °C, but an 'optimal policy' in which the discounted value of the benefits of action was slightly more (US$199 billion) than the costs of action would slow the rate of warming marginally. Bob Ward of the Grantham Institute (UK) pointed out that the 2018 Nobel Prize had focused a 'spotlight' on the policy controversies associated with estimating the potential costs of climate change and noted that model has been updated many times, but its results 'which have changed relatively little over the past 26 years, are at odds with the scientific consensus'. Ackerman et al. (2013) for instance demonstrated that the damage function, which describes how economic costs would increase as global warming advances, in the 'default' setting of the earlier DICE-2007 version of the model does not include potentially catastrophic consequences of global warming. Meanwhile Okullo (2020) underlined the importance of factoring the social costs of climatic damages and pointed that failing to account for the interaction between damage and climate sensitivity uncertainty underestimates the cost of climate change by more than US dollars 1 trillion.

About 25 years ago, Grubb et al. (1995) argued that an important issue for climate–economy assessments could be the dynamic characteristics of energy systems. They suggested that energy systems had potential to adapt to emission constraints, but in ways constrained by their very long-lived and path-dependent nature. More recently, Grubb et al. analyzed the widely used DICE as well as the Integrated Assessment Models and demonstrated that these models assume temporal independence – that abatement costs in one period are not affected by prior abatement – and argued in favour of 'dynamic realism' which represents 'the diversity of dynamic linkages across sectors and technologies and implies the need for more nuanced policy than a single global carbon price' (Grubb et al. 2021). The point that Grubb et al. recognized more than 25 years ago about how energy systems are both adaptable and yet constrained is key to informing the overall discussion on how developing countries and cities and NNSAs can respond to the policy nexus between increasing clean energy access and reducing SLCPs.

As discussed previously, intergovernmental negotiations on climate change between developed and developing countries have been tense over the years. For example, at the

2008 COP-14 held in Poznan, developed and developing countries were at loggerheads. As noted in an article analyzing the Poznan COP by the Centre for Science and Environment (an Indian NGO): 'Developing countries had submitted proposals on mitigation, adaptation, technology transfer and the architecture of the global financial mechanism for climate change. Poznan saw little movement on these key areas. At the high-level ministerial session, then Chair of the Group of 77 and China, Ambassador John Ashe, said they expected an initial response by the developed countries to their proposals. There was none till the end'. The fact that there was no formal response to the proposal put forward by the largest intergovernmental organization of developing countries in the UN comprising 134 UN members speaks volumes as to the impasse in negotiations. At this COP, developed countries placed a precondition that developing countries had to take on binding targets to reduce their emissions. The call to differentiate amongst developing countries based on the stage of their economic development and shares of GHG emissions – a proposal by Japan – was endorsed by many developed nations. The article also reported that the Chair of the largest negotiating group of countries, Ambassador Ashe 'firmly rejected any proposal directed towards differentiating between developing countries saying such an idea was contrary to the principles of the UNFCCC. Ashe said this approach bordered on the immoral and was counterproductive' (CSE 2008).

The next year, at the 2009 COP-15 in Copenhagen, a secret draft text prepared by a small group of developed countries – the so-called Danish draft text – was leaked to the Guardian and spurred a huge wave of tension amongst developing and developed countries. The draft text was viewed as a serious departure from the UNFCCC KP's principle that developed nations, which have historically emitted the bulk of CO_2, should take on binding commitments to reduce emissions. The draft handed effective control of climate change finance to the World Bank and thereby weakening the UN's role in handling climate finance; did not allow poor countries to emit more than 1.44 tonnes of carbon per person by 2050, while allowing rich countries to emit 2.67 tonnes; sought to hold temperature rises to 2 °C; divided poor countries further by creating a new category of developing countries called 'the most vulnerable' and mentioned the sum of $10bn a year to help poor countries adapt to climate change from 2012 to 2015 and without clear indication of how this sum was possible (Vidal 2009).

In the lead up to the adoption of the entirely voluntary PA, there were contentious debates between developed and developing countries negotiators as to who should be paying the price for historical and aggregate GHG emissions, as well as the costs for non-action being borne by those countries and communities who contributed the least to the problem. In 2011, the WRI published a working paper, entitled 'Building the Climate Change Regime', that outlined three different approaches for a future climate agreement: (i) Proceeding without new legally binding commitments, (ii) immediate adoption of new, legally binding commitments, and (iii) achieving new legally binding commitments as soon as possible - strengthening the components of a legal character (Moncel et al. 2011, pp. 53–65). This in turn paved the way for the move toward a voluntary non-legally binding agreement that was signalled in the COP 2014's Lima Accord which was, as Davenport put, was not based on the threat of sanctions or legal consequences but on 'global peer pressure': 'The strength of the accord – the fact that it includes pledges by every country to put forward a plan to reduce emissions at home – is also its greatest weakness. In order to get every country to agree to the deal, including the United States, the world's largest historic carbon polluter, the Lima Accord does not include legally binding requirements that countries cut their emissions by any particular amount' (Davenport 2014). Interestingly, Greenstone in an article focused on the voluntary/non-binding PA contrasted the reluctance to move ahead with the internationally binding KP stated that: '. . . countries' commitments from Kyoto - a treaty that was binding in writing and voluntary in nature - produced surprising results. Most countries met or even exceeded their pledged reduction targets for greenhouse gas emissions from 2008-2012' (Greenstone 2015). Now with the enormous challenge of curbing energy related air pollution and SLCPs, the questionthat has not yet been answered is: Where is the global peer pressure to address the needs of climate vulnerable and energy poor?

There is broad global agreement that addressing climate change requires all countries, (economically advanced or economically disadvantaged, large, small, island or landlocked or any variation in between), to confront the direct links between fossil fuel energy which still drives the engine of socio-economic development and human well-being. The policy challenge confronting the UN-led global community is that climate change has been evidenced to have vastly differential negative impacts on lives in developing countries, and that persistent delays in global action pose a clear and present danger for those who are most vulnerable to energy related air pollution and climate adversities (WHO 2018). It is important to ask if the agreed outcomes of intergovernmental climate negotiations specifically focus on providing policy guidance on clean energy technologies, services and tools that could bridge the divide between increasing energy access and air pollution exposure. This in turn requires attention to the fact that one of the key features associated with climate change negotiations is that since their inception, there has been no formal agreement on the decision-making rules. The rules of procedure governing the UNFCCC negotiating process and its PA have not been formally adopted due to long-standing objections between developed and developing countries over the issue of Rule 42, which focuses on how to reach agreement on substantive issues (i.e. whether or not to allow a two-third majority vote as a last resort on all matters of substance). Essentially, what this has translated into over the years is that decision-making within the UNFCCC negotiations context offers only two broad end options: (i) consensus-based agreement or the adoption of a decision on the basis of no objection or (ii) unanimous vote of all Parties voting in favour.

Global expectations from climate activists were high after the PA was finally adopted as the result of extremely lengthy all-night long negotiations. 174 States and the EU formally signed the PA on Earth Day (22 April 2016), but on that date 15 states – mostly all SIDS – deposited their instruments of ratification. On 3 September 2016, China and the United States, the world's two largest GHG emitters, representing nearly 38% of GHG emissions, deposited their official instruments of ratification. On 21 September 2016, a UN High-Level Event focused on the PA's entry into force. As per Article 21, the PA also has a dual threshold to be met before its entry into force, namely, 30 days after 'at least 55 Parties to the Convention, accounting in total for at least an estimated 55 per cent of the total global GHG emissions have deposited their instruments of ratification, acceptance or acceptance' (UNFCCC 2015, p. 30). A total of 29 countries ratified the Agreement ahead of the 2016 UN Event, and 32 additional countries deposited their instruments of ratification at the event, bringing the total number to 61 ratifying countries that together represent approximately 47.79% of global GHG emissions. While the first threshold of 55 countries was reached at the 21 September UN Event, the second significant threshold of 55% of global GHG emissions was met when India committed to ratifying the Agreement on 2 October 2016 – the birth anniversary of Mahatma Gandhi. This was a pertinent example of the increasing leadership role of members of the BASIC (Brazil, South Africa, India and China) developing country group that was formed in 2009 during the UN Climate Conference in Copenhagen. With India's ratification, the Agreement formally entered into force ahead of the annual global climate change negotiations meeting – 2016 COP-22 held in Morocco.

The overwhelming preference to work towards a consensus agreement, however long or protracted the process meant that the final, plenary session of the 2018 24th UN Climate Conference in Poland for instance had to be postponed no less than six times. In the waning days of 2019, climate activists across the world were left disappointed that the 25th UN annual climate negotiations meeting ended in utter disarray. Despite extending the two-week 2019 Madrid COP for an additional two days, countries were unable to come to agreement on key aspects of the PA's implementation. After 25 consecutive annual climate conferences convened in diverse cities, countries failed to deliver essential outcomes such as setting a rulebook for the PA and designing a global carbon market. At COP-25, Parties to the PA could only agree to a 'watered-down text' which 'reflects a failure to agree on the key outcomes that were needed at the summit: setting a rulebook for the Paris Agreement and designing a global carbon market' (Keating 2019). The recent return to the PA negotiations cycle by the second-largest aggregate emitter sent a clear message about the US re-joining the PA (UN Treaties 2021). While the 2021 massive UN Glasgow COP-26 was an assemblage of global heads of state, prominent as well as self-proclaimed climate activists, COP-26 did little to address the needs of those struggling from climate vulnerabilities and exposed to toxic energy related air pollution. If climate, clean air and clean energy justice are truly goals to be acted upon, UN-led global community and business tycoons might wish to ask: What might be achieved by looking beyond the convening of costly, colossal climate conferences, and choosing instead to tackle climate change, poverty eradication and pollution reduction as integrated rather than segregated global SDGs via linked NNSA partnership-based actions.

It is hard to deny that the UN's SDA has grown in size and complexity over the past decades since the 1992 Agenda 21 – the first global blueprint for sustainable development was adopted. The UN sustainable development complex can now be characterized as a massive global juggernaut that involves a morass of meetings and technical fora convened in diverse global cities which negotiators from all UN member states travel to and from, attend in single person or large delegations depending on the size and economic status of the country the negotiators hail from. The pace and scope of UN sustainable development-related meetings is such that the one predictable intergovernmental policy constant may well be the always-expanding glossary of negotiating meeting group acronyms and related jargon. What is particularly challenging for developing countries is that the UN's SDA 'complex' has coalesced into evolving interplay amongst myriad MEAs and SDGs with associated negotiations, reporting and delivery requirements that pose serious resource and institutional constraints on UN member states where ironically the needs to eradicate poverty and meet the SDGs are most urgently felt.

In light of the UN's overarching priority on poverty eradication anchored in all of its far-reaching 17 SDGs, and given the direct impacts on public health resulting from climatic change, it is surprising to find that the 2015 PA on climate change contains only a couple of references to health and none to curbing PM air pollution. The 'right to health' is proclaimed in the preamble, but it is not clear what if any modalities are available to enable the implementation of this 'right'? The only other significant reference to health is found in a single paragraph 109 in a section dealing with enhanced action prior to 2020 which: 'Recognizes the social, economic and environmental value of voluntary mitigation actions and their co-benefits for adaptation, health and sustainable development' (UNFCCC 2015, p. 15). But again, it is unclear how the mere invocation of voluntary mitigation actions can somehow be expected to translate into concrete, real-time globally responsive measures that can for example, address VBDs associated with climatic change. Can the intersection between climate change and exposure to toxic levels of pollution and socio-economic inequalities arising lack of access to clean energy and adequate public health services be left to the purview of protracted, politicized annual COP negotiations? The current juncture in global environmental history necessitates acknowledgement not just of the embedded inertia of 26 annual climate summits that have delayed in scaling up requisite action, but also the urgency of leveraging the role of NNSAs in linked action on climate change and clean energy that improves the lives of the energy poor and climate vulnerable. The structure and format of annual two-week climate change COPs aimed at securing consensus-based intergovernmental agreement, has not served to progress the UN's commitment that the needs of the energy poor and climate vulnerable are being addressed in these fora. Since the past is often a portent of the future, and with a view towards saving lives threatened by PM air pollution, developing countries need to urgently look beyond the confines of the current PA negotiations process that does not factor in curbing SCLPs and energy-related air pollution.

3.2 Shedding Some Light on Clean Energy and Climate Action by NNSAs Within the UNFCCC: A Brief Overview of the Clean Development Mechanism

Decades of intergovernmental climate negotiation outcomes predicated on the concept of common and differentiated responsibilities have not yet yielded time-bound, legally binding climate mitigation targets that all of the world's largest aggregate emitters have agreed to abide by, and these negotiations have been proven to be arduous and politically contentious from the perspective of the energy poor and climate vulnerable. During the early days of intergovernmental negotiations, for example at the Berlin COP-1 held in 1995, developing country negotiators from SIDS and LDCs were active proponents of the need for time-bound climate mitigation commitment. After two and a half years of intensive discussions under the aegis of the eight sessions of the Ad Hoc Group on the Berlin Mandate which included arduous, all-night negotiations during the Third Conference of the Parties (COP-3) to the UNFCCC, the KP was adopted in the early hours of 11 December 1997. The KP was opened for signature on 16 March 1998. But several years passed before it entered into force on 18 November 2004 – 90 days after it was ratified by at least 55 Parties to the Convention, including Annex I Parties accounting for at least 55% of the total Annex I Parties 1990 carbon dioxide emissions (UNFCCC 1997). The KP's entry into force referred to a 'double threshold or double trigger' gave de facto withholding power to large aggregate emitting countries like the United States which never ratified and distinguished the KP's entry into force from that of the UNFCCC. The KP was finally ratified on 16 February 2005, but it never included the full spectrum of Parties that signed on the UNFCCC, including notable exceptions like the US and Australia, while Canada withdrew from the KP effective from December 2012. The Protocol's first commitment period started in 2008 and ended in 2012. The second commitment period began on 1 January 2013 and ended in 2020. The KP called for the establishment of a CDM, and, more recently, the 2015 PA referenced a different kind of partnership mechanism – its Article 6 Mechanism – which is intended to 'promote the mitigation of greenhouse gas emissions while fostering partnerships with the private sector' (UNFCCC 2015, Annex p. 24).

The KP's goal was to add targets to the principal objective of the UNFCCC, namely, to reduce GHG concentrations. The KP's adoption of legally binding quantified emission limitation and reduction commitments (QELRCs) for industrialized countries and countries with economies in transition (Annex I Parties to the UNFCCC) represented a step in fulfilling the provisions of the Berlin Mandate adopted at the first COP in 1995. The heart of the KP is Articles 3.1 and 3.2 which stated:

'3.1 The Parties included in Annex I shall, individually or jointly, ensure that their aggregate anthropogenic carbon dioxide equivalent emissions of the greenhouse gases listed in Annex A do not exceed their assigned amounts, calculated pursuant to their quantified emission limitation and reduction commitments inscribed in Annex B and in accordance with the provisions of this Article, with a view to reducing their overall emissions of such gases by at least 5 per cent below 1990 levels in the commitment period 2008 to 2012.

3.2 Each Party included in Annex I shall, by 2005, have made demonstrable progress in achieving its commitments under this Protocol' (UNFCCC 1997, p. 4).

The list of the six GHGs and a number of 'sector/source categories' listed in Annex A are provided below, while the list of countries included in the KP are provided in Box 3.1:

- 'Carbon dioxide (CO_2)
- Methane (CH_4)
- Nitrous oxide (N_2O)
- Hydrofluorocarbons (HFCs)
- Perfluorocarbons (PFCs)
- Sulphur hexafluoride (SF6)' (UNFCCC 1997, p. 22).

The KP listed a number of 'sector/source categories' and as shown below the broad sectors were then broken down into specific source categories:

'Energy

Fuel combustion

Energy industries

Manufacturing industries and construction

Transport

Other sectors
Other
Fugitive emissions from fuels
Solid fuels
Oil and natural gas
Other
Industrial processes
Mineral products
Chemical industry
Metal production
Other production
Production of halocarbons and sulphur hexafluoride
Consumption of halocarbons and sulphur hexafluoride
Other
Solvent and other product use
Agriculture
Enteric fermentation
Manure management
Rice cultivation
Agricultural soils
Prescribed burning of savannas
Field burning of agricultural residues
Other
Waste

Solid waste disposal on land
Wastewater handling
Waste incineration
Other' (UNFCCC 1997, pp. 22–23)

In his analysis of comparing the voluntary and non-binding PA versus the 'internationally binding' KP, Greenstone made a compelling case that even so-called binding agreements are voluntary and that: 'Ultimately domestic measures- laws, policies and regulations- are necessary to make international commitment'. He pointed to the unexpected positive results achieved by the KP despite the fact that the US never ratified the KP, and Canada recognizing that it was 31% away from meeting its KP commitment withdrew from the KP in 2011: 'Overall the world reduced emissions 25 percent more than had been pledged in the treaty' (Greenstone 2015). What is particularly noteworthy from the perspective of understanding national motivations to address climate change and air pollution in a linked manner, Greenstone referenced that '. . . in China and perhaps in India as well, growing awareness that people are being exposed to deadly air pollution is playing an important role in the formulation of policy on greenhouse gases. More broadly, all over the world, national decisions are being swayed by evidence of damage caused by climate events, and, of course, by the varied responses of the local political systems and its citizens' (Greenstone 2015). This point about the localization of responses to climate change based on increased recognition of air pollution is essential to the discussion in the following chapters which focus on the role of cities and especially the linkages between PM pollution and access to clean energy in a country like India.

Box 3.1 Countries included in Annex B to the Kyoto Protocol and their emissions targets.

	Annex B
Party	**Quantified emission limitation or reduction commitment**
	(% of base year or period)
Australia	108
Austria	92
Belgium	92
Bulgaria[a]	92
Canada	94
Croatiaa	95
Czech Republic[a]	92
Denmark	92
Estonia[a]	92
European Community	92
Finland	92
France	92
Germany	92
Greece	92
Hungary[a]	94
Iceland	110
Ireland	92
Italy	92
Japan	94
Latvia[a]	92
Liechtenstein	92
Lithuania[a]	92
Luxembourg	92
Monaco	92
Netherlands	92
New Zealand	100
Norway	101
Poland[a]	94
Portugal	92
Romania[a]	92
Russian Federation[a]	100

Slovakia[a]	92
Slovenia[a]	92
Spain	92
Sweden	92
Switzerland	92
Ukraine[a]	100
United Kingdom of Great Britain and Northern Ireland	92
United States of America	93

[a] Countries that are undergoing the process of transition to a market economy.
Source: UNFCCC (1997, p. 24).

In addition to establishing legally binding emissions targets for industrialized countries, what made the KP different from the current PA, which has still not yet fully defined its Article 6 mechanism, was that the KP also included innovative market-driven mechanisms that involved NNSAs – the private sector. To help Annex I Parties manage and/or reduce the cost of meeting their emission reduction commitments, the KP permitted these Parties to use three market-based mechanisms, also known as 'flexible mechanisms': Emissions Trading (ET), Joint Implementation (JI), and the CDM, defined in Article 12 of the Protocol. Of these three flexible mechanisms, the CDM was the only one that was predicated on the idea of a broad partnership and included technology transfer between developed (Annex 1 of the UNFCCC) and developing countries that was expressly focused on linking sustainable development objectives with emissions reductions. The extent to which sustainable development objectives were fulfilled by the CDM serves as the basis for an improved understanding of any ongoing or future role of the NNSAs such as the private sector within the PA. Article 12.2 of the KP specifically defines the CDM as follows: 'The purpose of the clean development mechanism shall be to assist Parties not included in Annex I in achieving sustainable development and in contributing to the ultimate objective of the Convention, and to assist Parties included in Annex I in achieving compliance with their quantified emission limitation and reduction commitments under Article 3' (UNFCCC 1997 p. 16). Put simply, the CDM allowed developed countries with emission reduction commitments under the KP to implement emission reduction projects in developing countries and thereby gain emission reduction credits. CDM projects have three overall criteria: (i) Projects must be voluntary; (ii) Projects must be able to show long-term climate change mitigation benefits and (iii) Projects must contribute to emissions reductions above and beyond business as usual (so-called additionality criteria) (CSDA/FIELD/WRI 1998). Any viable CDM projects could therefore earn saleable certified emission reduction (CER) credits, each equivalent to 1 tonne of CO_2, which can be counted towards meeting a developed country (i.e. Annex I Parties) KP targets. But, as noted by the UNFCCC, there was another important feature of the CDM, which was that the CDM was 'the main source of income for the UNFCCC's Adaptation Fund, which was established to finance adaptation projects and programmes in developing country Parties to the Kyoto Protocol that are particularly vulnerable to the adverse effects of climate change'. As per the site, the Adaptation Fund was financed by a 2% levy on CERs issued by the CDM (UNFCCC website 2021b).

The role of the CDM as a partnership mechanism between developed and developing countries, in particular its role as a financing mechanism for adaptation projects, along with an active role for NNSAs gave the CDM its novel appeal. What made the CDM the first of its kind from the UN perspective was that although both public and private entities were eligible to develop CDM projects, the CDM was heavily tilted to and actively encouraged participation by NNSAs-the private sector. The basic underlying operating premise of the CDM was that private companies, usually from – but not limited to – developed countries, would identify and ultimately fund projects in developing countries that reduce GHG emissions, on the basis that such projects met the developing countries' sustainable development criteria, and had what was called an 'additionality' requirement, meaning that the projects' proposed emissions reductions would be 'additional' to what would have been possible without the proposed CDM funding. However, this problem of 'additionality' would prove to be a challenge for developing countries as noted by Michaelowa and Pallav (2007). From a business/private sector perspective, the challenges of engaging with the CDM were twofold in that CDM emission reduction projects had the potential for lowest cost reduction options globally available, but these lower costs could only be realized if the transaction costs associated with administering the CDM offsets were kept at a minimum, and yet the CDM also needed to simultaneously ensure high-quality, verifiable emissions reductions. Key challenges associated with the CDM such as ambiguous certification procedures and the ability of private sector entities in developing countries to ensure that the criteria and modalities related to project activities benefit national sustainable development objectives were highlighted since the CDM's early inception (Cherian 1998).

The KP's embrace of market-based flexibility mechanisms as cost-effective tools in mitigating emissions, in particular the CDM's role for the private sector, allowed for the first ever partnership mechanism between developed and developing countries. Werksman very early on identified the private sector as being essential to the effective operation of the CDM, arguing that speed in implementation, low transaction costs and policy certainty were prized by the private sector in engaging with the CDM (Werksman 1998). According to a 2011 Pew Center Report, entitled 'The Clean Development Mechanism: Review of the First International Offset Program', the CDM was seen as representing 'The primary international offset program in existence today and, while not perfect, it has helped to establish a global price on GHG emission reductions. Further, it has managed to establish—in just eight years—a fairly credible, internationally-recognized, carbon offset market that is worth $2.7 billion with participation from a large number of developing countries and private investors' (Gillenwater and Seres 2011, p. 35). Since the US never ratified the KP, it meant that the EU countries were the largest users of CERS. While China and India led the way in CDM projects, and 111 countries participated in the record of the CDM's implementation, the sustainable development benefits for the smallest, most vulnerable and poorest countries which could not participate in the larger offset programs were less clear. What made the CDM different was its duality of application related to its twofold objectives: (i) to help developing (non-Annex 1) countries achieve their sustainable development priorities by focusing private sector investment into emission reduction projects; and (ii) to assist developed or industrialized (Annex I) countries in meeting their emission reduction commitments under the KP through CERs. What it did not envisage was an opportunity for any reverse transactions or bi-directional project operational activities whereby CERs would be gained by large aggregate emitting developing countries for private sector investment in Annex I countries. And the simple reason for this was that the KP was a binding agreement only on Annex 1 developed countries. Now with the universal and voluntary nature of the PA, the premise of how potential partnerships to offset GHG emissions might work needs a different operational configuration. According to the UN's 2018 report entitled 'Achievements of the Clean Development Mechanism: 2001-2018': 'The CDM has exceeded everyone's expectations. Since 2001, over 8,000 projects and Programmes of Activities in 111 countries which have reduced or avoided 2 billion tonnes of CO2 equivalent and sparked investment of close to USD 304 billion in climate and sustainable development projects' (2018, p. 5).

CDM projects required independent verification by the CDM's governing body – the Executive Board – which then awarded these projects CERs, each equivalent to one ton of carbon dioxide. These CERs, also called 'offset credits' because they 'offset' the developed countries' emissions with reductions in developing countries, could then be sold by the project proponent(s) to developed countries, which then used them to meet a part of their own national reduction commitments under the KP. The CDM also clearly required that project proposals include an explanation of how the individual project objectives contribute to sustainable development. The 2011 Pew Center Report highlighted that although the CDM required that project proposals include such an explanation, the actual implementation process was less clear: 'Developing countries have few incentives to apply stringent criteria for sustainable development since they are effectively competing for CDM projects with other developing countries. More stringent standards could raise the cost of projects and deter potential investors. The CDM review process does include procedures to ensure projects do not cause adverse economic, social, or environmental harm. Yet, few procedures exist to ensure that projects produce social and environmental sustainable development benefits. Until more specific sustainable development criteria— at the global or national level—are introduced, it is difficult to objectively assess to what degree the CDM contributes to this goal' (2011, p. 30). Reiterating this concern, the first Chair of the CDM, who also served as Chair of the UNFCCC Subsidiary Body on Implementation (11th–15th sessions) and Co-chair of the Rio+20 Summit Ambassador John Ashe observed: 'The CDM was specifically meant to stimulate sustainable development and emission reductions, while giving industrialized countries some flexibility in how they meet their emission reduction or limitation targets but its record of providing sustainable development and/or emissions offsets in the smallest and poorest countries remains a concern. More significantly, in the absence of clear sustainable development guidelines for CDM projects which serves as a competitive mechanism, developing countries are less likely to implement stringent sustainable development criteria since they are competing for CDM projects with other developing countries' (Ashe personal notes 2016).

The complex and somewhat unresolved nature of the CDM with sustainable development objectives was best summarized by an article published in the Guardian entitled 'What Is the CDM?' According to the Guardian's analysis: 'CDM projects are not without their controversies. Questions surround the sustainable development credentials of certain projects, particularly in the case of industrial gas projects. HFC-23 projects, for example, seem to create perverse incentives to continue to produce the ozone depleting gas HCFC-22 in order to destroy the waste gas by-product HFC-23. Indeed in response to this, starting in May 2013, the EU has banned companies covered by the EU Emissions Trading Scheme from using CERs from HFC-23 and N2O adipic acid industrial gas projects . . . Meanwhile, the geographical distribution of CDM projects, over 80% of which originate in China and India, calls into question the ability of the CDM to drive broad engagement with sustainable development across developing countries. What's more, critics would suggest a more fundamental flaw in the CDM is that it is impossible to prove the "additionality" of a project in comparison to a hypothetical baseline' (Guardian Environment Network 2011). Also in 2011, an article in Reuters reported about a Honduran CDM project that raised concern about human rights abuses: 'The Aguan biogas project in Honduras is applying for registration under the United Nations' Clean Development Mechanism (CDM), which rewards firms for investing in emissions-cutting schemes in developing countries. Environmental watchdog group CDM Watch claims the project, which is being developed by Grupo Dinant subsidiary Exportadora del Atlantico, is linked to human rights violations including the killing of more than 20 peasants in the past year who were occupying the land' (Chestney 2011). In contrast to the above, Chichilinsky and Sheeran saw 'magic' in the KP's carbon market and posited: 'As we will soon see, Kyoto's carbon market can avert global warming at no cost to the global economy. It can foster sustainable development and can close the global divide between rich and poor nations. And it can do all this at no cost to the taxpayer, while providing incentives to create and implement clean technologies of the future . . . The suspense in this drama is whether Kyoto will survive its critics - whether it can be saved. Stay tuned' (2009, p. 104).

The role of clean energy businesses in the KP, which was never ratified by the US does serve as an example of the interest by NNSAs – private sector in the nexus between climate and energy, but notably can be contrasted to the role of the private sector in the Montreal Protocol on Substances That Deplete the Ozone Layer. From the outset, it is important to point out that integrated energy sector involvement in the implementation of the UNFCCC and the PA process has never been as comprehensive as the role of the private sector in the Montreal Protocol (MP). In his 1991 book, Ozone Diplomacy, US Ambassador Richard Benedick who played a key role in the adoption of the MP, provided a comprehensive historical analysis of how the international chemical industry 'vigorously denied any connection' between the condition of the ozone layer and increasing sales of anthropogenic chemicals – the chlorofluorocarbons (CFCs) in the mid-1970s. But, as Benedick noted the unique collaboration between scientists, policymakers and industry meant that by 1989, the MP entered into force with 'ratification from 29 countries and the European Commission together accounting for an estimated 83 percent of global consumption of CFCs and halons' (1991, p. 117). Benedick shed valuable light in noting that the chemical industry estimated back in 1989 that 30% of CFC demand could be replaced inexpensively via conservation and recycling, 30% demand could be met by replacing CFCs in aerosol sprays, cleaning agents and plastic foam blowing, but the remaining 30 and 10% could only be met by hydrochlorofluorocarbons (HCFCs) and HFCs, respectively, with the latter being seen as outside the scope of the MP (pp. 135–136).

Benedick also pointed out that US chemical industry alliance groups such as the Alliance for Responsible CFC Policy advocated more forcefully than their European counterparts in calling for 'time certain phaseout dates within the period of 2030-2050 with longer lived HCFC phased out earlier', and this advocacy position was 'fully consistent' with established science (p. 137). Although HFCs were initially not part of the MP, in 2016 during the 28th Meeting of the Parties to the MP in Kigali/Rwanda, more than 170 countries agreed to amend the MP. The Kigali Amendment aims for the phase-down of HFCs by cutting their production and consumption. Given their zero impact on the depletion of the ozone layer, HFCs are currently used as replacements of HCFCs and CFCs; however, HFCs are powerful GHGs and SLCPs (UNIDO website 2021). The role of the Kigali Amendment to the MP.

Unlike the MP, the KP did not survive its critics. Notwithstanding its demise, one of the biggest overall accomplishments of the CDM from the perspective of those countries most vulnerable to the adverse impacts of climate change was that it was the main source of income for the UNFCCC's Adaptation Fund. This Adaptation Fund was first established in 2001 to finance adaptation projects and programmes in developing country Parties to the KP that are particularly vulnerable to the adverse effects of climate change. Developing country negotiators from SIDS, in particular who were well aware of the double burden of being most vulnerable to adverse climatic impacts including SLR and least responsible for causing the problem of climate change, were instrumental in ensuring that Adaptation Fund is financed by a 2% levy on CERs issued by the CDM. As of 3 January, 2021, which is the date listed for the most current update on the CDM site, the share of proceeds from the CDM project activities for the Adaptation Fund totals US $ 40,827,393 (UNFCCC website 2021b).

Despite its challenges, the partnership aspect of the CDM represented a win–win for the private sector and governments and was also its greatest asset as noted by Streck, who also outlined the flaws of the CDM process: 'The CDM does not work efficiently. . .. The mechanism has failed to develop a regulatory due process to guarantee fundamental fairness, justice, and respect for property rights . . . Private economic actor firms will invest time and resources in generating, monitoring, and certifying emissions reductions only if they are assured a reasonable degree of regulatory certainty. The CDM governance will have to be put on the right track for the second commitment period, enhancing the predictability of its decisions and private-sector confidence in the system' (Streck 2010, pp. 71–73). It is useful to keep in mind that the UNFCCC as well as the KP's CDM emerged within an intergovernmental negotiating context where there was global consensus that the principal responsibility for climate mitigation resided with those historically responsible aggregate GHG emitters – the industrialized/developed countries. According the 2011 Pew Center Report, the CDM 'while not perfect' was nevertheless the main international offset programme which '. . . *helped to establish a global price on GHG emission reductions' and* 'managed to establish—in just eight years—a fairly credible, internationally-recognized, carbon offset market that is worth $2.7 billion with participation from a large number of developing countries and private investors' (2011, p. 35). More importantly, looking past the end of the commitment period of the KP, the World Bank went on to use the CDM for results-based finance through the Carbon Initiative for Development (CI-Dev). According to

CI-Dev, CDM has been used as 'the methodological framework' to secure approximately $76 million in emission reductions from 13 energy access projects, 12 of which are in Sub-Saharan Africa and one in South Asia. But CI-Dev also noted that after 2020 the KP will be replaced by the PA which will sponsor the need for the global community to look for new mechanisms (CI-Dev website 2021). The UNFCCC Secretariat in its Annual Report 2019 provided the most to update schematic on the CDM (see Figure 3.1).

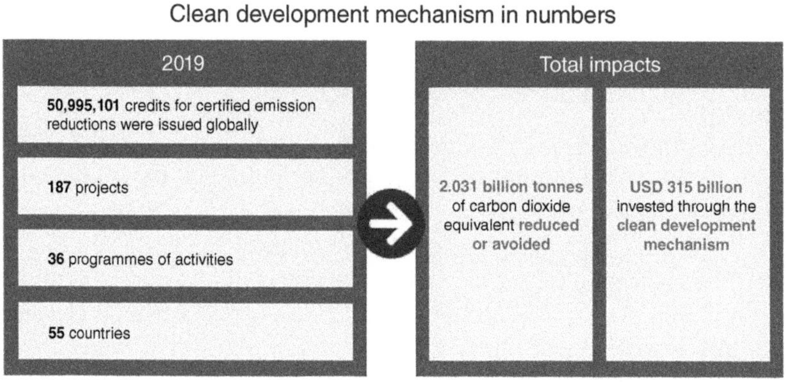

Clean development mechanism in numbers

2019

50,995,101 credits for certified emission reductions were issued globally

187 projects

36 programmes of activities

55 countries

Total impacts

2.031 billion tonnes of carbon dioxide equivalent reduced or avoided

USD 315 billion invested through the clean development mechanism

Figure 3.1 CDM in 2019 (Schematic overview by UNFCCC secretariat). *Source:* UNFCCC (2020, p. 22).

The negotiations for the PA rule book and the full articulation of its Article 6 mechanism have been significantly delayed, and COP-26 has failed to fully deliver. The fossil fuel energy sector is a principal driver of GHG emissions, but notably, the PA contains no specific guidance on increasing access to clean energy in developing countries and has just one cursory reference to 'energy' in Article 16 and that is in connection to the International Atomic Energy Agency. The PA contains no reference whatsoever to the term 'particulate matter' pollution nor does it address the issue of reducing SLCPs. The final Paris Climate Conference outcome document has one solitary reference to increasing access to clean energy in developing countries where these needs are the greatest: 'Acknowledging the need to promote universal access to sustainable energy in developing countries, in particular in Africa, through the enhanced deployment of renewable energy' (UNFCCC 2015, p. 3). What is interesting is that the very next reference is that UN member states agreed universally to 'uphold and promote regional and international cooperation in order to mobilize stronger and more ambitious climate action by all Parties and non-Party stakeholders, including civil society, the private sector, financial institutions, cities and other subnational authorities, local communities and indigenous people' (UNFCCC 2015 p. 3). It is precisely this reference aimed at mobilizing cooperation by all non-Party stakeholders – NNSAs – that is central to air pollution reduction efforts in two of the largest aggregate GHG emitters where curbing SLCPs poses a public health crisis. The following sections focus on the role of NNSAs within the context of the PA and provides a backdrop for the argument that NNSAs are vital to the future of climate mitigation, including curbing SLCPs and increasing access to clean energy for all.

3.3 Role of NNSAs in the PA: Explosive Growth but Still Operating from the Margins of the UNFCCC Negotiations Framework

It is sobering to keep in mind that in spite of all the global media and political attention paid to climate change, the much-ballyhooed PA depends completely on individual countries scaling up their voluntary commitments. Three decades of nation-state-driven intergovernmental negotiations have still not resulted in a legally binding, comprehensive global climate change agreement. Given the urgency of the linked climate change, poverty eradication, lack of access to sustainable energy and pollution reduction crises, it is useful to therefore look beyond the vagaries of national political election cycles towards a more-inclusive global partnership agenda that involves the active engagement of NNSAs – all civil society actors.

Climate and clean air responsive policies need to be framed differently, and acted upon urgently by integrating socio-economic development with reducing vulnerability and financing the transition to clean energy for all. NNSAs ranging from CEOs of global companies to the world's foremost religious leaders have spoken out on the challenge of climate change, but none have drawn as a clear a linkage to the triple threats of poverty and climate change and energy related air pollution as Pope Francis. On 24 September 2015, the Pope's address to the largest gathering of world leaders at the UN's assembly of nations – the UNGA – urging the adoption of a transformative and universal SDA harnessed massive global attention. Ahead of addressing the UNGA, the Pope's 24 May 2015 encyclical on environment and human ecology, 'Laudato Si: On Care of Our Common Home', included a section entitled 'Pollution and Climate Change' which stated: 'Exposure to atmospheric pollutants produces a broad spectrum of health hazards, especially for the poor, and causes millions of premature deaths. People take sick, for example, from breathing high levels of smoke from fuels used in cooking or heating'. On the subject of climate change, Paragraph 25 notes: 'Climate change is a global problem with grave implications. . . It represents one of the principal challenges facing humanity in our day. Its worst impact will probably be felt by developing countries in coming decades' (Pope Francis 2015). But it was really paragraphs 49 and 54 of the encyclical that laid bare the inequities of poverty and exclusion in the face of decades of global summits and intergovernmental negotiations: 'Paragraph 49: It needs to be said that, generally speaking, there is little in the way of clear awareness of problems which especially affect the excluded. Yet they are the majority of the planet's population, billions of people. . .. Today, however, we have to realize that a true ecological approach always becomes a social approach; it must integrate questions of justice in debates on the environment, so as to hear both the cry of the earth and the cry of the poor'. Paragraph 54 went on even further and focused explicitly on the 'failure' of global summits: 'The *failure of global summits on the environment* make it plain that our politics are subject to technology and finance. There are too many special interests, and economic interests easily end up trumping the common good and manipulating information so that their own plans will not be affected. . .. Consequently the most one can expect is *superficial rhetoric, sporadic acts of philanthropy and perfunctory expressions of concern for the environment,* whereas any genuine attempt by groups within society to introduce change is viewed as a nuisance based on romantic illusions or an obstacle to be circumvented' (Pope Francis 2015).

Broadly speaking, NNSAs activities related to the UNFCCC negotiating process can be categorized as attempting to shape the parameters of the global climate change agenda that has been set on the basis of intergovernmental, consensus-driven textual agreement. The political process of textual word-parsing that has been accepted as the norm within the UN context has made it challenging for defining clear rules of partnership engagement with NSAs related to the energy sector and even harder to track partnership results which have seen an exponential growth as evidenced in Chapter 2. Civil society engagement including the increased commitments shown by cities and local municipalities are increasingly the real frontline for responsive action. While negotiations between countries may have stalemated over the years, there has been a dynamic shift towards NNSAs actively involved in addressing climate change and clean energy.

Two decades ago, the 1997 annual climate summit – COP-3 – which resulted in the adoption of the KP Treaty to the UNFCCC, was viewed as the first large, intergovernmental climate meeting bringing together close to 10,000 participants. 'Rising Voices against Global Warming', a compilation of civil society stakeholders inputs noted that COP-3 participants included 2273 government delegates, 3712 media representatives, 3663 NGO representatives and 79 representatives from intergovernmental organizations, and added a nod to the role of virtual world: 'As a sign of our modern and virtual reality, COP 3 set the stage for the biggest live broadcast ever on the Internet' (Taalab 1998, p. xiii). Back in 2001, Raustiala argued that '. . . as the climate change issue has risen in salience and complexity, the NGO community surrounding it has evolved considerably'. He noted that whereas environmental NGOs were active with time, 'increasing numbers of business NGOs have become involved in the policy process', and 'business groups with a wide variety of stakes in the policy outcome have mobilized to monitor and influence the proceedings. The major interested industries include, among others, fossil fuels (coal, oil, natural gas), automobiles, insurers, power generation, and alternative energy suppliers (hydroelectric, solar, wind)' (2001, p. 7). NNSAs' interest has grown exponentially through the years. The 2019 annual climate summit – COP-25 – held in Madrid was attended by over 26,000 participants. The schematic diagram in Figure 3.2 excerpted from the UNFCCC Climate Change Annual Report 2019 provides evidence how of the annual COPs have grown, with the additional reference that the UNFCCC Secretariat supported and held over 20 high-level events during COP-25 (2020, p. 25).

As Chapter 2 evidenced, NNSAs have been referenced as essential to meeting global SDGs. What is not often reflected on, however, is the fact that NNSAs, in particular environment-focused NGOs, have long demanded a more formal role in the negotiating process for climate change and sustainable development starting with the CSD. As noted by the Sustainable Development Issue Network publication focused on the CSD, late Ambassador Ashe, chair of the thirteenth session of the CSD said that he and his Bureau 'would love to go down in history for innovative thinking at this CSD, but that it is not possible to climb a mountain and get to the other side in one shot'. However, he

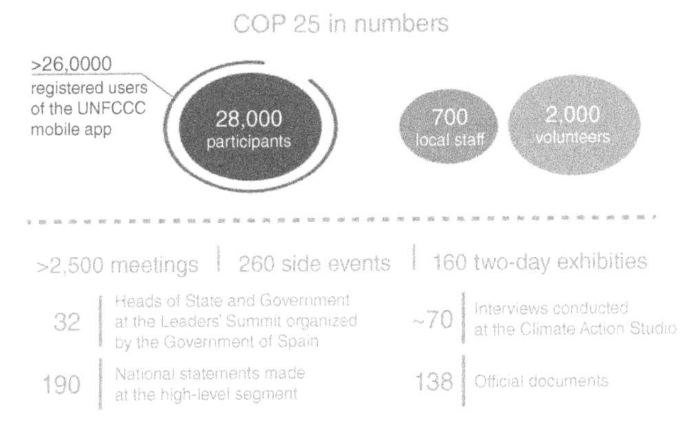

Figure 3.2 Schematic diagram of COP-25 (Excerpted from UNFCCC 2019 Climate Change Annual Report). *Source:* UNFCCC (2020, p. 25).

encouraged civil society to try and emphasized the important role that NGOs play in the process: 'You have a role as lobbyists and watchdogs, and I hope you keep up that role . . . It's up to you to hold governments' feet to the fire. Make sure they do what they say they will do' (SDIN 2005, p. 1). Notwithstanding, NNSAs interests in engaging with the intergovernmental negotiating process in both climate and energy for sustainable development call for more active engagement by NNSAs, particularly environmental NNSAs, have been stoutly resisted by some UN member states who have argued that nation-states are the only legally recognized entities responsible for ironically in this case, the entirely voluntary and non-legally binding submissions of NDCs under the aegis of the PA. The general resistance by some state actors within the UNFCCC process should give pause and has led to an examination of the impacts NNSAs have had on the UNFCCC process itself. In their study of the effectiveness of (environmental) NGOs at the UNFCCC negotiations, Busby et al noted that:

The negotiating process magnifies the number of actors, establishes a high bar for agreement, aims to encompass all the relevant issues in a single agreement, and encourages very close attention to legal language by the focus on treaty-based commitments. By design, that elevates the political stakes of any agreement but also encumbers the *negotiations with structural features that favor deadlock and broad but shallow agreements when such inertia is overcome* . . . observation of recent meetings suggests that there is a growing skepticism – originating from both within the NGO community and from states themselves – regarding the value of formal participation of civil society groups in climate governance around the UNFCCC process' (emphasis added, Busby et al. 2015, pp. 89–90.

In spite of the lack of a formalized process for NNSAs inclusion within the UNFCCC process, the overall global interest in responding to climate change resulted in a focus on the role of NNSAs in the lead-up to the 2015 PA. The call for responsive actions from NNSAs unleashed ideas on how such a fourth pillar which is how NNSAs role was envisaged and, more importantly, how it could be made part of the PA itself. In its 'white paper' on the subject, the Yale Climate Change Dialogue (YCCD) envisaged an 'alliance', which 'requires participation from people and organizations at all levels of society, from government to the private sector. Real success in Paris will require an agreement architecture that both builds on and goes beyond the traditional nation-state-dominated framework of targets and timetables for emissions reductions driven by top-down mandates. . .. Implementing this great alliance demands not only a policy framework that provides for broader engagement, but also a new structure of metrics and reporting. An innovative metrics framework encourages action by subnational jurisdictions and other non-state actors while offering structure, accountability, and recognition for city, state, corporate, and civil society mitigation efforts' (YCCD 2015, p. 3). Recognizing that climate change has captured the attention of a wide array of NSA stakeholders ranging from city mayors to CEOs, the YCCD went on to point out that: 'Climate change poses a systemic problem of global scope that requires solutions across many scales and sectors. Unlike many past issues that have been the focus of international agreements — for example, arms control, navigation, and land mines — national or central governments have no monopoly of control over the actions necessary to address global climate change. Indeed, the *actions required to reduce greenhouse gas emissions lie much more within the ambit of mayors, regional leaders including governors/premiers, and CEOs. Thus, the global community needs a 21st century climate change framework that can deliver new mechanisms to connect, aggregate, organize, support, and order crosscutting actions taken by a broad assortment of leaders'* (emphasis added, YCCD 2015, p. 5). And it is exactly this – a twenty-first century framework for action – that allows diverse actors – NNSAs- organized around the implementation rubric of curbing fossil fuel air pollution, access to clean energy and climate action within cities and regions that is essential.

What is hard to refute is that NNSAs play a key role in current and future climate responsive actions both within and outside of the confines of the PA. Gordon pointed out that in terms of responding to climate change goals and targets, 'cities have come to see themselves as generating collectively meaningful global governance' and 'seek to be globally accountable, and are doing so through practices of measurement, reporting, and disclosure' (2016, pp. 83–84). Widerberg and Pattberg noted that, in the build-up to the December 2015 meeting in Paris: 'The dynamic and experimental approach taken by many NSAs could inspire a ratcheting up of national action or even fill some of the ambition gap left between the cumulative Intended Nationally Determined Contributions (INDCs) and the emission scenarios putting us on safe path towards limiting warming to 2 degrees Celsius . . . in short, a vast number of NSAs hold ample mitigation potential for contributing to reducing global GHG emissions. How do we harness all this potential and what do we actually know about the currently active NSAs?' (2015, p. 2). They then identified several existing obstacles to harnessing the contribution of NSAs towards the overall goal of climate change mitigation within the PA that are relevant to keep in mind for the future efficacy of climate action including:

1) 'Targets of many climate initiatives are announced in vague terms such as "promoting information sharing," "engage in capacity building," and "encourage politicians." Moreover, most initiatives put forward voluntary commitments and hence make it difficult if not impossible for accountability and compliance to be enforced.

2) Patchwork of available Monitoring, Reporting and Verification (MRV) standards inhibits streamlined assessment and review of the commitments made by NSAs. The *Non-State Actor Zone (NAZCA) for Climate Action portal* launched at the UN climate change conference in Lima and registers commitments to action by companies, cities, subnational regions, and investors to address climate change) database, the largest database to date on non-state climate actions, explicitly refers to this problem under its FAQ section on why there are no aggregate numbers on tonnes of GHG mitigated.
3) *In sum, fuzzy targets in combination with incoherent MRV standards increase the risk for an actor to engage in "green washing," i.e. reaping the good-will of announcing mitigation action without ever changing behavior.* The importance of NSAs to accomplishing additional mitigation successes, on top of promises made in the INDCs, heightens the urgency of harnessing the potentials they have and avoid green washing'

(*Source:* emphasis added, and adapted from Widerberg and Pattberg 2015, p. 4).

With the enticing prospect that NNSAs could play a significant role in the global effort to reduce GHG emissions, and many already coming forward with their 'commitments', it was agreed at COP-20 in Lima, Peru, to launch NAZCA portal to allow for NSA to register commitments on a global scale under the Lima–Paris Action Agenda. When the 2015 PA was agreed to, NNSA leadership was viewed as a linchpin for future implementation of scaled-up climate action. While the recognition given by the Parties to NAZCA in the lead-up to PA should be seen as a plus, the larger goal of formal inclusion of the efforts and contributions of non-state actors in the Agreement itself remained unmet. Instead, at COP-21, NSAs undertook the 'Paris Pledge for Action (L'Appel de Paris)', which was an initiative of the COP-21 French Presidency: According to the L'Appel de Paris website, in the week the UN member states adopted the PA, NNSAs ranging from cities, regions, companies and investors representing 150 million people and US$11 trillion supported the L'Appel de Paris – Paris Pledge for Action (PPA). The French Minister Laurent Fabius who played an influential role in securing global agreement on the PA reflected on the role of NNSAs in the PPA: 'Non-state actor leadership is key to the success of COP21 and to the effective transition to a low-emissions and climate-resilient future. The world needs you to step up and rise to the challenges of climate change and sustainable development. This is why I strongly encourage you to take bold actions and make ambitious commitments, both individually and collectively, register them on NAZCA, and sign on to the Paris Pledge for Action, to make sure the commitments made in Paris by governments are achieved or even exceeded'. But, as of 1 Jan 2022, the Paris Pledge for Action/ L'Appel de Paris website states that it is 'now closed for future signatories' (Paris Pledge for Action /L'Appel de Paris website 2021).

The PA was seen as signalling a massive shift towards low-and zero-carbon emissions technologies and services that governments and NNSAs (including small, medium and large businesses) and city and municipal leaders would voluntarily undertake. However, there was, and remains a disconnect between the pledge to spur NNSAs and actual integrative action within the formal PA process. While the process leading up to the PA's adoption generated an unprecedented level of interest by NNSAs, the intergovernmental negotiation process was unable to arrive at a consensus that NNSAs should be provided any juridical status within the PA, which would require a rethinking – if not redesigning – of decades of established UN legal practice and parlance that accords primary legal status to sovereign nation-states – 'Parties' to the UNFCCC and its PA. Additionally, one of the thorniest challenges has been how to track the GHG reduction commitments of any NNSA such as an energy sector company that cannot sign on to implement what is entirely a voluntary-driven PA. There is also the problem of a plethora of pledge with in the Climate Pledge requiring industry/business signatories to commit to reporting on their GHG emissions reductions but a disconcerting lack of a globally accessible and verifiable framework that allows for the real-time tracking of GHG reduction commitments.

The Pledge was not a formal part of the PA and not legally linked to the COP decision. NAZCA, on the other hand, as previously mentioned, was welcomed in the PA related COP decision. In a nod to the role of private sector partnerships, Article 6 of the PA contains a number of provisions to facilitate ET through market and non-market mechanisms, but it was and remains the subject of contentious debate during the negotiations ever since. As Marcu observes: 'It was one of the last issues to be agreed, in the last night of COP21 . . . This "midnight-hour" handling of Article 6 can be attributed to a number of factors [including] the ideological opposition of some Parties to include any provision that referred to markets or could be seen as facilitating markets in the Paris Agreement. The markets provisions, contained in Article 6 . . . [should] be seen as a major success and a minor miracle. Throughout [the negotiations], and during COP 21 itself, the prediction was for a very small reference to anything related to markets, or possibly even the total omission of any such reference in the text' (Marcu 2016, p. 1).The inherent illogic of an intergovernmental negotiation process that dodges around key concepts like 'markets' and 'access to clean energy for all' needs to be referenced. The antipathy within the PA neogtiations to the word 'markets' almost proved to be a deal breaker for what eventually became Article 6. In the final consensus agreement on the PA, what emerged was an 'Article 6 that creates the following three frameworks: i) one for cooperative approaches to allow the linking of emissions trading systems, ii) one for a new "mechanism to contribute to the mitigation of greenhouse gas emissions and support sustainable development" to replace the Kyoto's flexible mechanisms, and iii) one non-market mechanism to promote "integrated, holistic and balanced non-market approaches"' (Carbon Market Watch 2017).

Despite the antipathy to 'markets' expressed by some in the PA negotiations, it is clear that the international carbon market will need to and is anticipated to play a crucial role in the post-2020 climate regime (Koakutsu et al. 2016). The private sector was in fact well represented at the COP-21 negotiations for the PA. A report by Baker&McKenzie provided a summary of key civil society, business and investment initiatives announced at COP-21:

Major global companies made individual commitments to climate action such as: 'Philips' commitment to become carbon neutral by 2020; and Goldman Sachs' commitment to invest $150 billion in clean energy projects and technology'. Major coalitions of companies made announcements such as:

> 'We Mean Business Coalition, made up of 363 companies which made a number of commitments, including internal carbon pricing and conversion to renewable energy;

The Science Based Targets initiative which enlisted 114 global companies, with the aim of setting strict science based emissions targets; and

The Green Freight Action Plan, which includes Deutsche Post, HP, Ikea and Volvo as members as well as 24 nation-states with the aim of enhancing green freight options'.

- Institutional investors and banks also made a number of announcements at the COP regarding the way climate change would shape their investments. This included:

  'The Paris Green Bonds Statement of representing USD 11.2 trillion worth of assets; and expansion of the Portfolio Decarbonization Coalition, a group of investors who have committed to systemically integrate carbon information into their portfolios, bringing the value of the Coalition's assets under management to USD 600 billion'

  (*Baker & Mckenzie 2016, p. 19*).

In addition to the above, there were three significant PPPs or coalitions of NSAs launched at the margins of Paris Climate Conference. The first announcement was made by 28 of the world's richest men and women – a veritable alphabet A for Ambani to Z for Zuckerberg soup of philanthropic billionaires – who agreed to pool their massive assets and experience to launch the 'Breakthrough Energy Coalition' (BEC). The second announcement was the official launch of 'Mission Innovation' (MI) aimed at dramatically accelerating global clean energy innovation by 20 countries (including the world's most populous and the largest oil, gas and renewable energy producers such as Brazil, China, India, France, Germany, Saudi Arabia and the United States). The third announcement was the 30 November 2015 launch of the 'International Solar Alliance' (ISA) jointly announced by the leaders of India and France and aimed at boosting the use of solar energy in developing countries.

The ISA was designed as a 'treaty based international organization' seeking 'to mobilize more than USD 1,000 billion into solar power by 2030' and 'to accelerate the development and deployment of over 1,000GW of solar generation capacity in member countries'. The Framework Agreement for ISA was officially launched at COP-22 in Morocco in 2016. The ISA has been billed *as* 'a coalition of solar-resource-rich countries (which lie either completely or partly between the Tropic of Cancer and the Tropic of Capricorn) to address their special energy needs'(ISA, 2019, pp 6-7). According to the ISA website, the amendment of its Framework Agreement in 2020 allows all member states of the UN to join the ISA: "At present, 101 countries are signatories to the ISA Framework Agreement, of which 80 countries have submitted the necessary instruments of ratification to become full members of the ISA (ISA website, 2021). Headquartered in Haryana, India, with initial seed capital of $16 million from the Government of India, and an additional $1 million each from Coal India, Indian Renewable Energy Development Agency and other Indian entities to a corpus fund, the ISA was the first global organization of its sorts (ISA 2019, p. 12). According to its website, the ISA 'has been conceived as be an action-oriented, member-driven, collaborative platform for increased deployment of solar energy technologies to enhance energy security and sustainable development, and to improve access to energy in developing member countries'. With '122 sun-belt countries' seen as prospective member countries, and a current membership of 86 countries, the ISA specifically does not aim 'to duplicate or replicate the efforts that others (like International Renewable Energy Agency (IRENA), Renewable Energy and Energy Efficiency Partnership (REEEP), International Energy Agency (IEA), Renewable Energy Policy Network for the 21st Century (REN21), United Nations bodies, bilateral organizations etc.) are currently engaged in, but will establish networks and develop synergies with them and supplement their efforts in a sustainable and focused manner'. ISA website 2021).

Tracking the role of entities like the ISA in engendering NNSAs partnership is therefore a key factor in evaluating the future of solar energy applications, financing and use in developing countries. The future of initiatives like BEC, MI and the ISA hinges on the vibrancy of the symbiotic relationship between UN member states and NSAs. But the diverse array of pledge platforms masks the global policy reality which is that tracking and verifying GHG emissions reduction commitments within a universally accessible framework has still not been realized.

3.4 Moving Beyond the Hype: Need for a Twenty-First Century Inclusive Framework on NNSAs Partnership on Climate and Clean Energy

There is a dizzying range of climate summits, pledging platforms and networks that almost daily showcase NNSA involvement in addressing climate change but there is no analog in terms of reducing air pollution or SLCPs. The WHO has unequivocally declared air pollution to be the world's single largest environmental health risk (WHO 2018). The lack of global and regional attention being paid to the linked nexus between poverty eradication, curbing fossil fuel air pollution and increasing access to clean energy for the poor is a much more depressing situation. Given the synergy between SLCPs and air pollution, and the recognition that air pollution poses a massive disease and morbidity burdens on poorer households within cities particularly in Asia and Africa, a detailed examination of climate change related partnerships and pledges should at least show some traction and progress. But as the section below evidences, while there are no shortages of climate pledges, what is missing is concrete verifiable action towards mitigating SLCPs and increasing access to clean energy for the poor.

A year after the PA was adopted, on 12 December 2017 the French President hosted the 'One Planet Summit', expressly aimed at providing additional impetus to the PA. The 2017 US withdrawal from the PA under the Trump administration notwithstanding, the Parisian effort to secure global momentum towards the PA billed as 'adaptation, mitigation, mobilization' was promoted by a triumvirate of global co-hosts – Antonio Guterres (UN Secretary General), Emmanuel Macron (President of France) and Jim Yong (World Bank President). The Summit was convened without any high-level governmental attendance from the second-largest aggregate GHG emitter, and approximately 60 representatives of governments were represented at the Summit. The Summit demonstrated that the US withdrawal from the PA had not dampened climate responsive enthusiasm, and there were plenty of pledges and announcements of climate action and green financing. In his analysis of the 2017 One Planet Summit, Keating points out that 'governments have committed to only

$100 billion per year' and 'unless countries cough up more climate finance and more ambitious emissions-reduction plans in the next year, the conclusion may have to be that such high-level summits are not working' (Keating 2017). Ironically, this pledge still remains incomplete to date. A critically important outcome of the Summit from the perspective of climate–clean energy linkages was that the President of the World Bank announced that Bank will 'no longer finance upstream oil and gas, after 2019'. The World Bank statement added: 'In exceptional circumstances, consideration will be given to financing upstream gas in the poorest countries where there is a clear benefit in terms of energy access for the poor and the project fits within the countries' Paris Agreement commitments' (World Bank 2017). This reference as to the dual challenges of climate and energy access being grappled with by the poorest developing countries and communities is key. Although the global challenges of resource mobilization and implementation constraints faced by developing countries in South Asia and Sub-Saharan Africa where the majority of the world's energy poor and climate vulnerable are located still remain largely unaddressed, this World Bank announcement is an example of a global marker signalling a shift in the role of the fossil fuel energy sector in the fight against climate change.

Since the 2017 One Planet Summit, there have been a series of other such summits, but the analytical challenge of tracking projects and partnerships remains as the website for the One Planet Summit only provides a brief snapshot as to 'First One Planet Summit' and no longer references the specific projects announced in 2017. See Box 3.2 for an excerpt of the One Planet Summit process and outcomes (2021). Instead of the term 'partnerships', the One Planet Summit site references the term 'coalitions': 'One Planet coalitions help create synergies and develop concrete actions at the local or global level. They mobilize states and non-state actors on all continents and in all sectors' (One Planet site 2021). Table 3.1 references the 10 'Coalitions' within the searchable database of the One Planet website relation to the search term of 'Energy'; and Table 3.2 references the only 2 'Coalitions' with the searchable database of the One Planet website related to the search term of 'Climate' (2021).

Box 3.2 Three years of the One Planet Summits.

Three years and three summits later, some forty coalitions and very concrete initiatives have been created. The organizations that carry them have committed themselves to achieving specific objectives and to reporting on them.

Key figures:

- 3 'One Planet' summits since 2017;
- 121 countries involved in the coalitions;
- 5 public-private partnerships set up to mobilize more than $2 billion to finance the transition in developing countries ($472M available to date);
- 308 concrete research projects conducted by 3305 researchers worldwide.

Source: One Planet Summit website 2021.

Table 3.1 'Energy' related 'Coalitions' within the One Planet process.

Global alliance for smart cities in Africa

Objective: enable access to clean energy and connectivity through a smart streetlights program.

Projects: as part of an agreement with the Rwanda Green Fund, construction of a model, climate change resilient house and a series of smart streetlights (solar power supply, LED lighting, 3G, Wi-Fi, sockets and direct connection to the hospital in Kigali for remote consultations).

Launch: One Planet Summit, Nairobi, 14 March 2019.

Members: R20, Solektra, DiCaprio Foundation, Société Générale, JCDecaux and SIGNIFY.

Contribution of cities under the global covenant of mayor: 'Global Urbis' Initiative

Objective: Mobilize private investment in high-risk sectors.

Encourage cities to make ambitious public commitments to stimulate investment, public procurement and political decisions (One Planet Charter campaign).

Investments totalling 4.3 billion euros have been mobilized.

Projects: 25 new cities committed to zero waste and 19 to developing low carbon buildings; innovate4cities programs (September 2018) with Google and Bloomberg Philanthropies, to provide 10,000 cities worldwide with the data they need (transport, buildings, air quality) to create and implement climate action plans.

Launch: One Planet Summit, Paris, 12 December 2017.

Members: European Union (EU), European Investment Bank (EIB), European Bank for Reconstruction and Development (EBRD), cities.

The International Solar Alliance (ISA)

Objective: Rapidly scale up solar energy projects across the world, in conjunction with the Paris Agreement. To increase global installed solar capacity by 1000 GW in 2030. The estimated cost would be 1000 B$.

Projects: The ISA seeks to develop the knowledge, expertise, financing, capacities and regulatory frameworks for optimizing solar energy deployment. To this end, it has initiated several projects including:

- Star-C: A global network of technical experts and academics for developing knowledge of solar energy and training local actors at every level of solar projects implementation. Star-C will launch a program in the Pacific in 2021.
- Sustainable Renewables Risk Mitigation Initiative (SRMI): A dedicated financial instrument to address financial risks linked to the implementation of solar projects, notably in developing countries. It is currently being developed by the World Bank and the Agence Française de Développement (AFD), with support from the International Renewable Energy Agency (IRENA).
- The Lomé Initiative: A joint initiative by six West African countries, seeking to establish a common regulatory framework for solar projects.
- Infopedia: An online platform dedicated to the dissemination of information, best practices and knowledge on Solar Energy.

The ISA has also identified several dedicated areas of work, such as solar applications for agricultural use, Affordable finance at scale, mini-grids, solar rooftops, solar E-mobility and storage.

Launch: COP-21 (December 2015). It was launched by France and India at COP-21.Originally restricted to countries located between the two tropics, it will soon be open to every UN member state.

Members: 70 member countries (December 2020). 18 countries in the process of admission.

Table 3.1 (Continued)

Nairobi call for action on the preservation and conservation of forests in Africa

Objectives: Contribute to stopping deforestation; map critical situations; develop global strategies made up of workable, concrete solutions.

Launch: Presidents of Kenya and France at the One Planet Summit, Nairobi, 14 March 2019.

Members: National governments (France, Kenya, Madagascar, Democratic Republic of Congo), international organizations (World Bank, United Nations, WWF) and companies.

Biarritz Pledge for fast action on efficient cooling

Objective: With the Biarritz Pledge for fast action on efficient cooling, the signatory countries commit to improving the energy efficiency of cooling equipment in parallel with the reduction of HFC refrigerants. The pledge is associated with the efficient cooling initiative. The aim is to raise awareness on the importance of improving the energy efficiency of air-conditioning and refrigeration equipment, in parallel with the reduction of HFC refrigerants (in accordance with the Kigali amendment), in order to maximize climate benefits (up to 0.8 °C of global warming could be avoided by 2100 with a dual approach).

Launch: Led by France and introduced during the G7 Environment meeting in Metz, under the Climate and Clean Air Coalition (CCAC) Biarritz, 24–26 August 2019.

Members: Around 15 countries (Germany, Canada, Japan, United Kingdom, Australia, Chile, Rwanda, India and United Arab Emirates).

Sustainable Actions for Innovative Low Impact Shipping (SAILS)

Objectives: The SAILS charter brings together shipowners engaged in voluntary initiatives to reduce their environmental impacts. Aims are to preserve biodiversity, oceans and the climate and share innovations on a larger scale.

Projects: Limiting speed in order to reduce CO_2 emissions (up to 20% reduction per vessel); systems to electrify boats at quay; on-board radars to protect marine mammals and filtering air pollutant emissions.

Launch: G7, Biarritz, 24–26 August 2019.

Members: About 10 European shipowners.

Subnational climate funds for Africa

Objective: The SnCF Africa fund works to invest for climate in the areas of waste management and access to renewable energies. Aims to finance sustainable development projects led by local authorities (regions, large cities) in 15 African countries.

Launch: One Planet Summit, Nairobi, 14 March 2019.

Members: R20 network, BlueOrchard.

Breakthrough energy venture

Objective: Expand the Breakthrough Energy Coalition (BEC) to include a diversified global network of leading banks, funds, energy producers and technology companies. Breakthrough Energy Venture (BEV) is a 1-billion-dollar investment fund. Support and create cutting-edge companies that help stop climate change.

Launch: One Planet Summit, Paris, 12 December 2017.

Partners: Canada, Mexico, the United Kingdom, France, European Commission.

Powering past coal alliance

Objective: This coalition offers support to countries seeking to reduce their dependence on coal and steel and improve their citizens' health. Comprising 58 actors including 8 governments and 24 businesses, the alliance is committed to speeding up the elimination of conventional coal-fired power stations in a sustainable and economically inclusive manner and to imposing a moratorium on new conventional coal-fired power stations without carbon capture and storage systems. France has committed to ending all electricity production from coal by 2022.

(Continued)

Table 3.1 (Continued)

Launch: Launched by the United Kingdom and Canada during COP-23 and at One Planet Summit.
Members: Governments and companies.

Caribbean climate-smart accelerator

Objective: This initiative's main areas of intervention are: infrastructure and sustainable cities, reducing vulnerability to natural risks, developing clean vehicles and renewable energies and protecting the oceans. It aims to: make the Caribbean the first Climate-Smart Zone; identify projects; seek funding and monitor implementation.

Projects:

- Install a solar pumping system to facilitate access to drinking water in Jamaica.
- Protect the oceans and the Belize reefs.
- Create a start-up to tackle the threats surrounding the Sargasso Sea.
- Finance marine conservation and global warming adaptation projects through a debt swap mechanism.

Launch: One Planet Summit, Paris, 12 December 2017. The Inter-American Development Bank, the Business Development Bank of Canada, the World Bank and Virgin United have already committed 2.8 billion dollars.

Members: 26 national governments, 40 private sector partners.

Source: One Planet Summit site (2021).

Table 3.2 'CLIMATE' related 'COALITIONS' within the One Planet process.

Climate Action 100+

Objective: Climate Action 100+ brings together investors who are committed to helping companies transform their practices. Its objective is to encourage the world's 100 most polluting companies to implement concrete measures and revise their governance model to address climate risks.

Projects: Progress report (September 2019): Ten of the targeted companies have agreed to change their lobbying practices; 12% have accepted an approach compatible with the 2 °C objective.

Launch: One Planet Summit, Paris, 12 December 2017.

Members: 373 investors (representing 35 billion dollars in assets) working with in 161 companies.

Initiative KIWA

Objective: The initiative aims to build up resilience to climate change and help preserve biodiversity in 18 Pacific Island countries and territories. It totals 31 million euros, including 13 million from the French Development Agency (AFD) and 10 million from the European Union. It seeks to optimize access to financing for nature-based solutions for civil society, local or national authorities and regional organizations.

Launch: One Planet Summit, New York, 26 September 2018.

Members: France, European Union, Australia, New Zealand and Canada.

Source: One Planet Summit site (2021).

In contrast to the PPA pledges, the NAZCA is part of the PA process and as such there is a UN-affiliated NAZCA portal for partnerships. But the entirely voluntary nature of these NNSA 'commitments' and the lack of any universally recognized MRV standard(s) to assess them have raised valid concerns related to tracking NNSAs overall contributions to global mitigation efforts. The registration of NNSA actions within the NAZCA portal has been seen as the vehicle to track the actions and mobilization needed to implement the agreement and raise global ambition amongst NSAs to address mitigation of and adaptation to climate change. So, what exactly does a search of the NAZCA portal reveal in terms of increasing access to clean energy, reducing SLCPs, curbing air pollution and reducing climate vulnerabilities?

To begin with, the NAZCA portal is a complex array of multiple searchable themes, filters and categories. There is also the issue of double counting of search filters and terms which are quite apparent as evidenced below. The portal currently allows for the search filter/category of 'actors': '*Countries, Cities, Regions, Companies, Investors and Organizations*', but no definitions or clear explanations are provided on the portal site as to the distinguishing features for the search categories differentiating amongst actors such as companies versus investors and/or organizations. It also allows for a search of the following broad filters entitled 'themes', which include:

- ***Land Use***
- ***Oceans and Coastal Zones***
- ***Water***
- ***Human Settlements***
- ***Energy***
- ***Industry***

In addition to these broad 'themes', there are lists of additional filters, and each additional filter has its own subcategories:

- '*Cross-cutting themes:* Decent work, Disclosure, Education, Finance, Gender, Health, Innovation, Long-term Strategy, Mitigation, Other, Policy and Resilience;
- *Types of Action:* Bonds Issuance, Carbon Price Establishment, Emissions Reductions, Energy Efficiency, Investment, Policy Establishment, Renewable Energy and Resource Consumption
- *SDGs:* Affordable and Clean Energy, Clean Water and Sanitation, Climate Action, Decent Work and Economic Growth, Gender Equality, Good Health and Well Being, Industry Innovation and Infrastructure, Life below Water, Life on Land, No Poverty, Partnerships to Achieve the Goal, Peace and Justice Strong Institutions, Quality Education, Reduced Inequality, Responsible Consumption and Production, Sustainable Cities and Communities, Zero Hunger.
- *Announcements:* Only Announcements, Exclude Announcements
- *Time Frame:* Short Term, Medium Term, Long Term and No Target'

(emphasis added, Global Climate Action/NAZCA website 2021).

Box 3.3 provides a summary of the search results of the NAZCA portal using a variety of search filters which are all spelled out.

Box 3.3 Summary of key search results of global climate action/NAZCA portal (as of on 22 April (Earth Day) 2021).

A search of the NAZCA portal referenced a TOTAL of: '19,372 ACTORS representing 28,383 ACTIONS across ALL '*THEMES*'

SDG 1: '*NO POVERTY*' – SEARCH FILTER RESULTS

When the search filter of SDG – '*No Poverty*' was selected, the portal referenced a TOTAL of '*3 actors representing 3 actions*'

As evidenced by the screenshot of the portal using this search filter, the portal did not provide any specific information as to the 2 countries and 1 organization listed as undertaking actions. A further search of all 'ACTORS' – Countries, Cities, Companies, Investors and Organization referenced the term: *'no matching actions'*.

No matching actions found.

Filter by **ACTORS**

2 **Countries**

0 **Cities**

0 **Regions**

0 **Companies**

0 **Investors**

1 **Organizations**

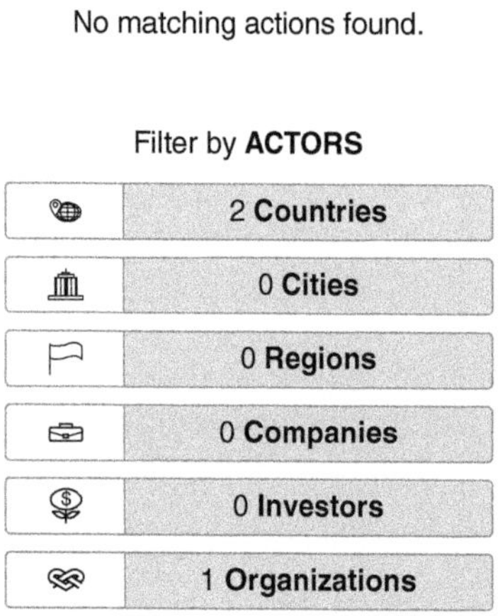

SDG 7: '*AFFORDABLE AND CLEAN ENERGY*' – SEARCH FILTER RESULTS

When the search filter of 'Affordable and Clean Energy' was selected, the portal referenced a TOTAL of: '461 actors representing **542** actions'

As evidenced by the screenshot of the portal using this search filter, the portal did not provide any information as to any specific actors and actions.
A further search of all 'ACTORS' – Countries, Cities, Companies, Investors and Organization referenced the term *'no matching actions'*

No matching actions found.

Filter by **ACTORS**

120 **Countries**

9 **Cities**

0 **Regions**

210 **Companies**

60 **Investors**

62 **Organizations**

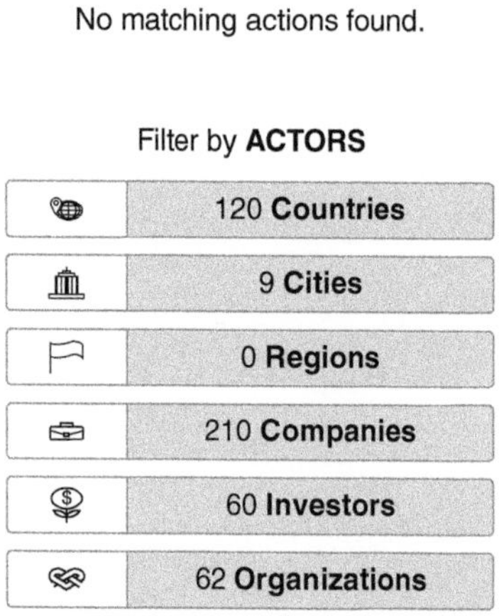

THEME OF '*ENERGY*': CONTRADICTORY SEARCH FILTER RESULTS

In contrast to SDG 7, when the search filter by the '*theme*' of '*ENERGY*' was selected, the portal referenced a TOTAL of '14,859 actors representing **16,743** actions'

As evidenced by the screenshot below, the portal referenced the following set of 'ACTORS' associated with the filtered theme of 'ENERGY'

Filter by **ACTORS**

	182 **Countries**
	10567 **Cities**
	76 **Regions**
	2902 **Companies**
	331 **Investors**
	801 **Organizations**

A further search of All ACTORS yielded the following results when:
The 'CITIES' search filter was used, the portal referenced: '**10,567** actors representing **11,075** actions' A list of municipalities was provided.

The 'COUNTRIES' search filter was used, the portal issued two contradictory references: '**182** actors representing **796** actions' and 'No matching actions found'.

The 'COMPANIES' search filter was used, the portal issued two contradictory references: '**2902** actors representing **3452** actions' and 'No matching actions found'.

The 'INVESTORS' search filter was used, the portal issued two contradictory references: **331** actors representing **412** actions and 'No matching actions found'.

The 'ORGANIZATIONS' search filter was used, the portal issued two contradictory references: '**801** actors representing **876** actions' and 'No matching actions found'.

SDG 13: '*CLIMATE ACTION*' – SEARCH FILTER RESULTS

When the search filter of 'Climate Action' was selected, the portal referenced a TOTAL of: '19,300 actors representing **28,030** actions'

As evidenced by the screenshot of the portal using this search filter, the portal referenced the following set of 'ACTORS' associated with the theme of 'CLIMATE ACTION'.

Filter by **ACTORS**

	191 **Countries**
	10748 **Cities**
	243 **Regions**
	5005 **Companies**
	1154 **Investors**
	1959 **Organizations**

A further search of All 'ACTORS' yielded the following results when:

- The 'COUNTRIES' search filter was used, the portal issued two contradictory references: **191** actors representing **1355** actions and 'No matching actions found'.
- The 'CITIES' search filter was used, the portal referenced: '**10,748** actors representing **11,970** actions'. A list of municipalities was provided.
- The 'REGIONS' search filter was used, the portal referenced: '**243** actors representing **668** actions'. A list of regions was provided.
- The 'COMPANIES' search filter was used, the portal referenced: '**5005** actors representing **9454** actions. A list of companies was provided.
- The 'INVESTORS' search filter was used, the portal referenced: '**1154** actors representing **2364** actions'. A list of companies was provided.
- The 'ORGANIZATIONS' search filter was used, the portal referenced: **1985** actors representing **2295** actions'. A list of organizations was provided.

SDG 11 – SUSTAINABLE CITIES AND COMMUNITIES– SEARCH FILTER RESULTS

When the search filter of 'Climate Action' was selected for SDG 11, the portal referenced a TOTAL of: **'494** actors representing **954** actions'

As evidenced by the screenshot of the portal using this search filter, the portal referenced the following set of 'ACTORS' associated with the theme of 'SUSTAINABLE CITIES AND COMMUNITIES'

A further search on All ACTORS yielded the following results when:

- The 'COUNTRIES' search filter was used, the portal issued two contradictory references **'56** actors representing **63** actions' and 'No matching actions found'
- The 'CITIES' search filter was used, the portal referenced: '**336** actors representing **653** actions'. A list of municipalities was provided.
- The 'REGIONS' search filter was used, the portal referenced: '**69** actors representing **205** actions'. A list of regions was provided.
- The 'COMPANIES' search filter was used, the portal issued two contradictory references: '**7** actors representing **7** actions and No matching actions found'.
- The 'INVESTORS' search filter was used, the portal issued two contradictory references: '**3** actors representing **3** actions and No matching action found'.
- The 'ORGANIZATIONS' search filter was used, the portal issued two contradictory references: '**23** actors representing **23** actions and No matching action found'.

Source: Global Climate Action/NAZCA portal (2021).

The search results of NAZCA portal revealed some major problems in terms of how the UN-managed portal is tracking and scaling up NNSA actions related to poverty eradication, increasing access to affordable and clean energy and sustainable cities. Given the central role of poverty eradication within all the SDGs and its overarching importance within the SDA and the PA, it is nothing short of alarming to see that when the search filter of SDG – 'No Poverty' was selected, the portal revealed a total of only '3 actors representing 3 actions'. What was also troubling was that the search of the NAZCA portal in relation to this SDG evidenced absolutely no information as to the 3 actors (which were only referenced as being 2 countries and 1 organization). It is also important to underscore that the results of the search filters for 'SDGs' versus 'Themes' pose a major data management and tracking challenge. For instance, the search for SDG 7 – 'Affordable and Clean Energy' versus the search filter theme of 'Energy' revealed a vastly different number of actors and actions – '461 actors representing 542 actions' versus '14,869 actors representing 16,473 actions'. This differentiation can be viewed as indicative of the lesser scope and scale of NNSAs engagement with SDG 7, which itself poses a problem in terms of the progress being made to achieve the goals of increasing access to clean and non-polluting energy for the poor. But, the even more serious concern with the NAZCA results in relation to SDG 7 was the fact that absolutely no specific information was provided on any actions by any actors in relation to this SDG within the portal. By contrast, and given that the NAZCA portal was focused on 'Climate Action', it is not at first glance surprising to see that the vast majority of actors and actions should be focused on SDG 13. But, what the search revealed is that out of '19,372 actors representing 28,383 actions' across all themes and SDGs, SDG 13 had '19,300 actors representing 28,030 actions' – in other words a mere 72 actors and 353 actions are focused on all other themes and SDGs. There is also a discrepancy as to the role of NAZCA actions and actors related to SDG 11 – Sustainable Cities and Communities. Here it is surprising to find that in spite of the UN acknowledgement that cities are where the battle for climate change will be won or lost, there is significantly a smaller number of cities listed in relation to SDG 11 versus SDG 13. Additionally, as in the case of SDGs 1 and 7, the NAZCA portal provides no information as to specific actors and actions in relation to the companies, investors and organizations. This latter problem poses a challenge because it would appear that this data as to companies, investors and organizations in relation to all three SDGs are either missing or inaccessible to the public.

The good news is that there is growing evidence that NSAs ranging from environmental NGOs, local governments and businesses are taking it upon themselves to undertake action to mitigate climate change and reduce air pollution that improves human well-being across the globe. But the UN-led climate negotiations and its existing NAZCA portal do not provide the level of information, as well as engagement needed to implement global action aimed simultaneously at addressing poverty reduction, increasing access to clean/non-polluting energy and promoting sustainable cities. Delivering progress on linked NNSAs partnerships that can jointly address SDGs 7 and 13 requires a new more inclusive partnership framework that builds on political buy-in from leaders and parliamentarians as well as engagement and delivery of concrete results from local governments, business leaders and civil society as a whole. Cities, sustainability-focused community groups and a diverse array of business sectors need to be factored in as the locus of real and deliverable action on energy and climate. The emergence of the definitive role of local, sub-regional and regional NNSAs in delivering real results and assisting national governments in the mammoth undertaking that comprises the SDGs offers promise especially when contrasted with the long-drawn out and fraught intergovernmental-driven climate change negotiations. The actual implementation of the SDGs will take place at national level and will therefore require concrete, verifiable mechanisms to track, monitor, and review partnerships. The problem is that both at the global and the national level – as the search of the NAZCA portal has evidenced – such frameworks for monitoring and review of partnership actions involving NNSAs are still woefully inadequate even in the midst of such partnerships being hailed as vital to the success of the SDGs and the PA.

2020 was etched as a crucial year for traction on climate change. The anticipated aim of COP-25 was to finalize the rulebook of the PA for carbon markets and other forms of international cooperation under Article 6 Mechanism. But, there was no consensus on key areas, and instead COP-25 enjoyed the dubious distinction of being the longest COP on record. It was finally gavelled to a close at 1:55 pm on Sunday, 15 December 2019, nearly 44 hours after its previously scheduled end, thereby eclipsing COP-17 in Durban in terms of the duration of negotiations (Carbon Market Watch 2019). It is now necessary to reflect sombrely on the fact that 26 cycles of annual global climate negotiations conferences have resulted in a voluntary ad-hoc process based on countries failing to scale up climate responsive actions while GHG emissions continue to grow with no global protocols focused on curbing SLCPs and toxic levels of energy related air pollution. The costs in terms of delayed climate responsive action for those on the frontline in the battle against climate adversity and energy related air pollution, as well as monetary costs and the escalation in global travel energy related emission are hard to justify. These delays and costs do constitute a global collective action failure that consigns millions to suffer and die as a result of unabated pollution and lack of access to clean energy.

Given the urgency for cooperative and collective climate responsive actions, it is hard to ignore the glaring fact that currently NNSAs involvement with UN intergovernmental outcomes on the inherently linked global challenges of climate change and sustainable energy are governed by two separate and segregated policy/goal silos within the broader UN sustainable development negotiations arena. The 2030 SDA with its exhaustive list of 17 SDGs and 169 targets was globally agreed upon as the foundation on which countries are anticipated to forge their own national SDGs and track any subsequent progress. The UN-led global community has put pre-eminent priority on poverty eradication as the lens through which the 17 SDGs (including SDG 7 and SDG 13) that anchor its transformative 2030 agenda need to be achieved. In crafting the universally agreed-upon 2015 SDA, all UN member state governments clearly recognized that implementation of this 'transformative' agenda would involve global 'partnerships' between various actors, both public and non-public, or state and non-state but both the UN SDG partnerships database and the NAZCA portal on NNSAs involvement with SDGs 7, 13 and 11 have major shortcomings. Existing global goal silos need to be broken down with an aim towards unleashing the innovative power of NNSAs including cities, civil society actors, as well as the private energy sector in partnerships that allow for linked action on clean air, clean energy, climate change and poverty reduction.

References

Ackerman, F. et al. (2013). Epstein–Zin utility in dice: Is risk aversion irrelevant to climate policy? *Environmental and Resource Economics* 56 (1): 73–84. https://doi.org/10.1007/s10640-013-9645-z (accessed 24 November 2021).

ADB (2018). *Decoding Article 6 of the Paris Agreement*. Manila: ADB https://www.adb.org/sites/default/files/publication/418831/article6-paris-agreement.pdf (accessed 5 May 2019).

Agarwal, A. and Narain, S. (1991). *Global Warming in an Unequal World*. New Delhi: Centre for Science and Environment.

Ashe, J. (2014). *Seizing the Moment: Our Planet*, 12–15. Nairobi: UNEP.

Ashe, J. (2016). *Personal Notes and Interview*. New York: NY.

Baker & Mckenzie (2016). The Paris Agreement. https://www.bakermckenzie.com/-/media/files/insight/publications/2016/01/the-paris-agreement/ar_global_climatechangetreaty_apr16.pdf?la=en (accessed 1 June 2021).

Barnett, J. and Adger, W.N. (2003). Climate dangers and atoll countries. *Climatic Change* 61 (3): 321–337.

Bathiany, S., Dakos, V., Scheffer, M., and Lenton, T.M. (2018). Climate models predict increasing temperature variability in poor countries. *Science Advances* (2 May 2018). https://advances.sciencemag.org/content/4/5/eaar5809 (accessed 18 May 2021).

Benedick, R. (1991). *Ozone Diplomacy*. Cambridge: Harvard University Press.

Bezos, J. (2021). 2020 Letter to Shareholders. (15 April 2021). https://www.aboutamazon.com/news/company-news/2020-letter-to-shareholders (accessed 21 April 2021).

Busby, J., Urpelainen, J., Hale, T. et al. (2015). NGOs and Climate Advocacy. *International Politics Review* 3, 84–93.

Carbon Market Watch (2017). Goodbye Kyoto: Transitioning away from offsetting after 2020. https://carbonmarketwatch.org/wp-content/uploads/2017/04/Good-bye-Kyoto_Transitioning-away-from-offsetting-after-2020_WEB_1final.pdf (accessed 22 June 2021).

Carbon Market Watch (2019). Paris treaty establishes new carbon trading mechanisms (2015). *Carbon Market Watch*. http://carbonmarketwatch.org/news-paris-treaty-establishes-new-carbon-trading-mechanisms (accessed 8 June 2021).

Chattopadhyay, D. and Parikh, J. (1994). *CO2 Emissions Reductions from the Power System in India*. Bombay: Indira Gandhi Institute for Development Research.

Cherian, A. (1998). Emissions reduction activities and the clean development mechanism: key unresolved issues. *Industry and Environment (UNEP)* 21 (1–2): 74–76.

Chestney, N. (2011). EDF Trading quits Honduras Biogas Project. *Reuters* (14 April 2011). https://uk.reuters.com/article/greenbiz-us-cdm-edf/edf-trading-quits-honduras-biogas-project-idUKTRE73D4P620110414 (accessed 28 June 2021).

Chichilinsky, G. and Sheeran, K. (2009). *Saving Kyoto*. London: New Holland Publishers.

CI-Dev Website (2021). Who we are. https://www.ci-dev.org/who-we-are (accessed 21 March 2021).

Claussen, E. and McNeilly, L. (1998). *Equity & Global Climate Change: The Complex Elements of Global Fairness*. Arlington, VA: Pew Center on Global Climate Change.

Climate Pledge website (2021). About the Pledge. https://www.theclimatepledge.com/us/en/the-pledge (accessed 21 April 2021).

CRU/IERL (1992). *Development of a Framework for the Evaluation of Policy Options to Deal with the Greenhouse Effect: Economic Evaluation of Impacts and Adaptive Measures in the European Community*. Norwich: University of East Anglia.

CSDA/FIELD/WRI (1998). *Clean Development Mechanism: Draft Working Papers*. Washington DC: CSDA.

CSE (2008). Climate Inaction at Poznan. https://www.cseindia.org/climate-inaction-3007 (accessed 21 April 2021).

Davenport, C. (2014). A climate accord based on global peer pressure. *New York Times* (14 December 2014). https://www.nytimes.com/2014/12/15/world/americas/lima-climate-deal.html (accessed 20 April 2021).

Dubash, N. (ed.) (2015). *Handbook of Climate Change and India*. London: Routledge.

Fankhauser, S. and Tol, R.J.S. (1997). The social costs of climate change: the IPCC second assessment report and beyond. In: *Mitigation and Adaptation Strategies for Climate Change*, 385–402. Amsterdam: Kluwer Publications https://www.researchgate.net/publication/221678659_The_Social_Costs_of_Climate_Change_The_IPCC_Second_Assessment_Report_and_Beyond (accessed 12 March 2021).

Fink, L. (2020). Fundamental Reshaping of Finance. https://www.blackrock.com/corporate/investor-relations/2020-larry-fink-ceo-letter (accessed 27 February 2021).

Francis, P. (2015). *Encyclical Letter – Laudato Si' of the Holy Father Francis – On Care of our Common Home*. Rome: The Vatican https://www.vatican.va/content/dam/francesco/pdf/encyclicals/documents/papa-francesco_20150524_enciclica-laudato-si_en.pdf (accessed 21 April 2021).

GAIA (2019). What is the Clean Development Mechanism? Global Anti-Incinerator Alliance (GAIA) Factsheet on Clean Development Mechanism. http://no-burn.org/downloads/GAIA_CDMFactsheet.pdf (accessed 29 May 2021).

Gelbspan, R. (1997). *The Heat is On – The High States Battle over Earth's Threatened Climate*. New York: Addison Wesley Publishing.

Gillenwater, M. and Seres, S. (2011). *The Clean Development Mechanism: A Review of the First International Offset Program*. Arlington VA: Pew Center on Global Climate Change.

Global Climate Action/NAZCA website (2021). NAZCA Portal. https://climateaction.unfccc.int/views/total-actions.html (accessed 22 April 2021).

Gordon, D.J. (2016). The politics of accountability in networked urban climate governance. *Global Environmental Politics*. 16 (2): 82–100. https://direct.mit.edu/glep/article-abstract/16/2/82/14872/The-Politics-of-Accountability-in-Networked-Urban (accessed on 22 September 2021).

Greenstone, M. (2015). Surprisingly, a voluntary climate treaty could actually work. *New York Times* (13 February 2015). https://www.nytimes.com/2015/02/15/upshot/surprisingly-a-voluntary-climate-treaty-could-actually-work.html (accessed 20 April 2021).

Grubb, M., Chapuis, T., and Duong, M.H. (1995). The economics of changing course: implications of adaptability and inertia for optimal climate policy. *Energy Policy* 23: 417–432.

Grubb, M., Wieners, C., and Yang, P. (2021). Modeling myths: on DICE and dynamic realism in integrated assessment models of climate change mitigation. *Wires Climate Change (Wiley Online)* (2 February 2021). https://onlinelibrary.wiley.com/doi/10.1002/wcc.698 (accessed 12 August 2021).

Guardian Environment Network (2011). What is the Clean Development Mechanism (CDM)? (26 July 2011). https://www.theguardian.com/environment/2011/jul/26/clean-development-mechanism (accessed 1 June 2021).

Gupta, J. (1997). *The Climate Change Convention and Developing Countries: From Conflict to Consensus*. Amsterdam: Kluwer Academic Publishers.

Gupta, J. (2010). A history of international climate change policy. *Wires Climate Change*. https://onlinelibrary.wiley.com/doi/10.1002/wcc.67 (accessed 21 April 2021).

IEA (2021a). Global Energy Review: CO_2 Emissions in 2020. (2 March 2021). https://www.iea.org/articles/global-energy-review-co2-emissions-in-2020 (accessed 20 March 2021).

IEA (2021b). *Global Energy Review 2021*. Paris: IEA https://www.iea.org/reports/global-energy-review-2021# (accessed 22 April 2021).

ISA (2019) ISA: Annual Report 2019. https://www.isolaralliance.org/uploads/docs/f661c80f9c697a7645164b48fb9003.pdf (accessed 22 April 2021).

ISA Website (2021) International Solar Alliance-Background. https://www.isolaralliance.org/about/background (accessed 22 April 2021).

IPCC (1995). *IPCC Second Assessment Climate Change 1995*. Geneva: WMO/UNEP https://archive.ipcc.ch/pdf/climate-changes-1995/ipcc-2nd-assessment/2nd-assessment-en.pdf (accessed 1 June 2021).

IPCC Summary for Policy Makers (SPM) (2018). Global Warming of 1.5°C – an IPCC special report on the impacts of global warming of 1.5°C above pre-industrial levels and related global greenhouse gas emission pathways, in the context of strengthening the global response to the threat of climate change, sustainable development, and efforts to eradicate poverty. https://www.ipcc.ch/site/assets/uploads/sites/2/2018/07/SR15_SPM_version_stand_alone_LR.pdf (accessed 7 April 2021).

Keating, D. (2017). What was the point of Macron's climate summit. *Forbes* (12 December 2017). https://www.forbes.com/sites/davekeating/2017/12/12/what-was-the-point-of-macrons-climate-summit/?sh=5d2cf95a32b7 (accessed 20 April 2021).

Keating, D. (2019). Failure in Madrid as COP25 climate summit ends in disarray. *Forbes* (15 December 2019). https://www.forbes.com/sites/davekeating/2019/12/15/failure-in-madrid-as-cop25-climate-summit-ends-in-disarray/?sh=5b5487073d1f (accessed 20 April 2021).

Koakutsu, K., Amellina, A., Rocamora, A.R., and Umemiya, C. (2016). Operationalizing the Paris Agreement Article 6 through the Joint Crediting Mechanism. https://pub.iges.or.jp/pub/operationalizing-paris-agreement-article-6 (accessed 20 January 2021).

KPMG (2019). Prepared for change: 2019 index measures 140 countries' ability to respond to change and climate challenges. https://home.kpmg/xx/en/home/media/press-releases/2019/06/2019-change-readiness-index-launches.html (accessed 16 January 2021).

Mansell, A. (2016). What's ahead for carbon markets after COP 21, Center for Energy and Climate Solutions (C2ES). http://www.c2es.org/newsroom/articles/whats-ahead-for-carbon-markets-after-cop-21 (accessed 9 January 2021).

Marcu, A. (2016). Carbon Market Provisions in the Paris Agreement (Article 6), Centre for Europe an Policy Studies (CEPS) Special Report No. 128/January 2016. https://www.ceps.eu/system/files/SR%20No%20128%20ACM%20Post%20COP21%20Analysis%20of%20Article%206.pdf (accessed 8 January 2021).

Michaelowa, A. and Pallav, P. (2007). *Additionality determination of Indian CDM projects: Can Indian CDM project developers outwit the CDM Executive Board?* London: Climate Strategies.

Ministerial Declaration on Carbon Markets (2015). https://www.mfe.govt.nz/sites/default/files/media/Ministerial-Declaration-on-Carbon-Markets.pdf (accessed 9 January 2021).

Moncel, R. et al. (2011). *Building the Climate Change Regime*. Washington, DC: World Resources Institute.

Nath, K. (1995). Statement by Hon. Kamal Nath-Minister for Environment and Forests, India to the Conference of the Parties to the Climate Change Convention. Berlin, Germany. (accessed 6 April 1995).

Nicholls, R.J. et al. (2011). Sea-level rise and its possible impacts given a 'beyond 40C world'in the twenty-first century. *Philosophical Transactions: Mathematical, Physical and Engineering Sciences* 369 (1934): 161–181. https://royalsocietypublishing.org/doi/10.1098/rsta.2010.0291 (accessed 11 December 2021).

Nordhaus, W.D. (1991). To slow or not to slow: the economics of the greenhouse effect. *Economic Journal* 101: 920–937.

Nordhaus, W.D. (1992). An optimal transition path for controlling greenhouse gases. *Science* 258 (5086): 1315–1319.

Okullo, S.J. (2020). Determining the social cost of carbon: under damage and climate sensitivity uncertainty. *Environmental Resource Econmics* **75**: 79–103. https://doi.org/10.1007/s10640-019-00389-w (accessed 20 November 2021).

One Planet Summit site (2021). https://www.oneplanetsummit.fr/en/coalitions-82?fieldset%5Blocation%5D=&fieldset%5Btheme%5D=24&fieldset%5Bactors%5D=&op=&form_build_id=form-814CbUvhqDS03fpUWb30nCrRAwRYWimE_Y3i6k8tqxU&form_id=mtes_ops_specifics_coalitions_filter_form (accessed 12 March 2021).

Parikh, J. (1995). *Gender Issues in Energy Policy*. Bombay: Indira Gandhi Institute for Development Research.

Parry, M.L. and Duncan, R. (1995). *The Economic Implications of Climate Change in Britain*. London: Earthscan.

Paris Pledge for Action/ L'Appel de Paris Website (2021). http://parispledgeforaction.org (accessed 1 January 2021).

Pearce, F. (1995). Price of life send temperatures soaring. *New Scientist*. https://www.newscientist.com/article/mg14619710-400-price-of-life-sends-temperatures-soaring/ (accessed 7 January 2021).

Raustiala, K. (2001). Nonstate actors in the global climate regime. In: *International Relations and Global Climate Change* (ed. U. Luterbacher and D.F. Sprinz) Ch. 5. Cambridge: MIT Press http://graduateinstitute.ch/files/live/sites/iheid/files/sites/admininst/shared/doc-professors/luterbacher%20chapter%205%20105.pdf (accessed 12 February 2021).

Richards, M. (2001). A review of the effectiveness of developing country participation in the climate change convention negotiations. *ODI Discussion Paper*. https://www.odi.org/publications/3618-review-effectiveness-developing-country-participation-climate-change-convention-negotiations (accessed 21 April 2021).

SDIN (2005). A CSD-celebrity. *Sustainable Development Issues Network* 5 (9).

Streck, C. (2010). Expectations and reality of the clean development mechanism: a climate finance instrument between accusation and aspirations. In: *Climate Finance: Regulatory and Funding Strategies for Climate Change and Global Development* (ed. R. Stewart, B. Kingsbury and B. Rudyk). New York: New York University Press http://www.climatefocus.com/sites/

default/files/expectations_and_reality_of_the_clean_development_mechanism_a_climate_finance_instrument_between_accusation_and_aspirations.pdf (accessed 12 February 2021).
Taalab, A. (1998). *Rising Voices against Global Warming. UN Treaty Collection (2021) Paris Climate Agreement*. Frankfurt: IZE.
UN Climate Change (2018). Achievements of the Clean Development Mechanism: 2001–2018. https://unfccc.int/sites/default/files/resource/UNFCCC_CDM_report_2018.pdf (accessed 21 March 2021).
UN Treaties (2021). https://treaties.un.org/Pages/ViewDetails.aspx?src=TREATY&mtdsg_no=XXVII-7-d&chapter=27&clang=_en (accessed 12 March 2021).
UNEP/WMO/IUC (1994). *United Nations Framework Convention on Climate Change*. Geneva: UNEP/WMO/IUCC.
UNFCCC (1997). Kyoto Protocol to the United Nations Framework Convention on Climate Change. FCCC/CP/1997/L.7/Add.1. https://unfccc.int/sites/default/files/resource/docs/cop3/l07a01.pdf#page=24 (accessed 22 April 2021).
UNFCCC (2011). Fact Sheet: The Kyoto Protocol. https://unfccc.int/files/press/backgrounders/application/pdf/fact_sheet_the_kyoto_protocol.pdf (accessed 20 March 2021).
UNFCCC (2015). Adoption of the Paris Agreement. FCCC/CP/2015/L.9/Rev.1. https://unfccc.int/resource/docs/2015/cop21/eng/l09r01.pdf (accessed 20 April 2021).
UNFCCC (2020). *Annual Report: 2019*. Bonn: UNFCCC Secretariat https://unfccc.int/sites/default/files/resource/unfccc_annual_report_2019.pdf (accessed 22 April 2021).
UNFCCC website (2021a). Status of Ratification of the Convention. https://unfccc.int/process-and-meetings/the-convention/status-of-ratification/status-of-ratification-of-the-convention (accessed 21 April 2021).
UNFCCC website (2021b). The Clean Development Mechanism. https://unfccc.int/process-and-meetings/the-kyoto-protocol/mechanisms-under-the-kyoto-protocol/the-clean-development-mechanism (accessed June 20, 2021).
UNIDO website (2021). The Montreal Protocol evolves to fight climate change. https://www.unido.org/our-focus-safeguarding-environment-implementation-multilateral-environmental-agreements-montreal-protocol/montreal-protocol-evolves-fight-climate-change (accessed 21 April 2021).
Vidal, J. (2009). Copenhagen climate summit in disarray. *The Guardian* (8 December 2009). https://www.theguardian.com/environment/2009/dec/08/copenhagen-climate-summit-disarray-danish-text (accessed 21 February 2021).
Werksman, J. (1998). The clean development mechanism: unwrapping the Kyoto surprise. *Reciel* 17 (2): 147–158.
WHO (2018). *COP-24: Special Report – Health and Climate Change*. Geneva: WHO https://apps.who.int/iris/handle/10665/276405 (accessed 27 February 2021).
Widerberg, O. and Pattberg, P. (2015). *Harnessing company climate action beyond Paris*. Stockholm, Sweden: FORES.
World Bank (2017). World Bank Group Announcements at One Planet Summit. https://www.worldbank.org/en/news/press-release/2017/12/12/world-bank-group-announcements-at-one-planet-summit (accessed 12 March 2021).
YCCD (2015). Yale Climate Change Dialogue: White Paper. 23 July 2015. http://envirocenter.yale.edu/sites/default/files/yale_climate_change_dialogue_white_paper.pdf (accessed 12 March 2021)

4

On the Frontlines for Clean Air and Climate Action

Role of Cities and India in Mitigating PM Pollution

4.1 The Urgency of Curbing Urban Air Pollution: Layering of Ill Health and Morbidity Burdens

The global trend towards urbanization is amply evident. In 1996, the historic UN Habitat report entitled 'An Urbanizing World: Global Report on Human Settlements' noted that: 'The average population of the world's 100 largest cities was over 5 million inhabitants by 1990 compared to 2.1 million in 1950 and less than 200,000 in 1800. The last few decades have also brought a world that is far more urbanized, and with a much higher proportion living in large cities and metropolitan areas. Soon after the year 2000, there will be more urban dwellers than rural dwellers worldwide' (UN Habitat 1996, p. 12). This was the first globally relevant UN policy recognition that urbanization would play an increased role in policymaking at all levels – global, regional, national and state/municipal level. What was notable was that this 1996 UN report also referenced the linkages between urbanization and 'acute respiratory infections (that include pneumonia, influenza and bronchitis) as having one of the largest disease burdens'; and it pointed out that the 'full extent of acute respiratory infections, their health impact and the risk factors associated with them remain poorly understood' (p. 137).

Today, more than 25 years later, the morbidity and health impacts of fossil fuel related air pollution in conjunction with climate related health costs, both of which are disproportionately borne by the poorest and most vulnerable amongst us lie in plain view. The policy intersection of climate vulnerabilities, exposure to toxic levels of PM pollution and associated morbidity and disease burden costs of acute respiratory illnesses merits escalated global attention. Adverse impacts of climate change and air pollution have long been known to be inequitably concentrated with devastating consequences for those with least access to climate resiliency, clean air and public health. Back in 2008, Thomas Friedman pointed that the world had a problem: 'It is getting hot, flat and crowded. That is global warming, the stunning rise of middle classes all over the world, and rapid population growth have converged in a way that could make our planet dangerously unstable. In particular, the convergence of hot, flat and crowded is tightening energy supplies, intensifying the extinction of plants and animals, deepening energy poverty, strengthening petrodicatorship, and accelerating climate change. How we address these interwoven global threads will determine a lot about the quality of life on earth in the

Air Pollution, Clean Energy and Climate Change, First Edition. Anilla Cherian.

twenty-first century' (2008, p. 5). Globally relevant assessments have pointed to the growing risks associated with climate change and fossil fuel related air pollution. But, as Chapters 2 and 3 evidenced, the real-time delivery and implementation of integrated climate responsive partnerships and actions on clean air and clean energy that are directly relevant to improving the lives of those most vulnerable within the UN SDGs context are confusing and fragmented at best.

Every four years, the Population Division of UN's Department of Economic and Social Affairs issues a report. The 2018 'World Urbanization Prospects' underscored key factors related to the urgency of linked action to curb air pollution, increase access to clean energy and address climate change within urban areas particularly those in Asia and Africa:

- 'The urban population of the world has grown rapidly from 751 million in 1950 to 4.2 billion in 2018.
- Today, 55% of the world's population lives in urban areas, a proportion that is expected to increase to 68% by 2050. Projections show that urbanization, the gradual shift in residence of the human population from rural to urban areas, combined with the overall growth of the world's population could *add another 2.5 billion people to urban areas by 2050, with close to 90% of this increase taking place in Asia and Africa*' (emphasis added, UNDESA 2018, p. 14).

In its summary for policymakers, this UNDESA Report found urbanization to be a 'positive force for economic growth, poverty reduction and human development'; and that 'cities are places where entrepreneurship and technological innovation can thrive, thanks to a diverse and well-educated labour force and a high concentration of businesses' (2018, p. 3). This significant finding of cities as networks or hubs for sustainable development where livelihoods, housing, commerce, governance and transportation etc. come together is critical to addressing synergistic action on clean energy, clean air and climate policies. The report also emphasized the changing geography of urbanization and served as a compelling signal of why the future of integrated action on the clean energy–climate–clean air nexus lies with cities in Asia and Africa: 'Africa's urban population is likely to nearly triple between 2018 and 2050, while that of Asia is likely to increase by over 50 per cent. With 1.5 billion urban dwellers in 2050, Africa will have 22 per cent of the world's urban population, while Asia, with 3.5 billion persons residing in urban areas, will have 52 per cent. *Together, they will account for nearly three-quarters of the urban population of the world*' (emphasis added, 2018, pp. 24–25).

The most recent update to UN population data reflected in 'World Population Prospects 2019' includes several findings, 2 of which have great import for the nexus between air pollution and ill health, lack of access to clean energy and climate change:

'*More than half of the projected increase in the global population up to 2050 will be concentrated in just nine countries*: the Democratic Republic of the Congo, Egypt, Ethiopia, India, Indonesia, Nigeria, Pakistan, the United Republic of Tanzania, and the United States of America. Disparate population growth rates among the world's largest countries will re-order their ranking by size: for example, *India is projected to surpass China as the world's most populous country around 2027.*

In 2018, for the first time in history, persons aged 65 years or over worldwide outnumbered children under age five. Projections indicate that by 2050 there will be more than twice as many persons above 65 as children under five. By 2050, the number of persons aged 65 years or over globally will also surpass the number of adolescents and youth aged 15 to 24 years' (emphasis adde, UNDESA 2019, p. 8). As populations age globally, the additional public health challenges that accrue from long-term exposure to toxic levels of air pollution, combined with the fact that population growth will be concentrated in series of specific countries, serve to underscore the significance of future synergistic action on clean air and clean energy in countries like India.

This chapter focuses on the role of NNSAs – cities – local/municipal actors – that are on the frontline in the struggle to cope not just with climatic impacts such as SLR and coastal zone inundation but also to respond to public health challenges associated with PM pollution. The bottom line is that adverse climatic impacts layered upon existing disease and morbidity burdens accruing from PM pollution have not been adequately addressed within the SDG and PA partnerships processes, and constitute a combined policy necessities in the most densely populated and polluted urban areas of the world. The cumulative weight of inequitable climatic and energy pollution risks being borne within some of the most populated cities in Africa and Asia cannot be emphasized enough. An early 2008 OECD study highlighted that just 10 port cities accounted for half of the total world exposure to climate related coastal flooding and inundation: Kolkata and Mumbai in India; Dhaka, Bangladesh; Guangzhou, China; Ho Chi Minh City, Vietnam; Shanghai, China; Bangkok, Thailand; Rangoon, Myanmar; Miami, United States and Hai Phong, Vietnam (Nicholls et al. 2008). The IPCC Synthesis Report (2014) had very high confidence that 'In urban areas, climate change is projected to increase risks for people, assets, economies and ecosystems, including risks from heat stress, storms and extreme precipitation, inland and coastal flooding, landslides, air pollution, drought, water scarcity, sea level rise and storm surges (very high confidence). These risks will be amplified for those lacking essential infrastructure and services or living in exposed areas' (IPCC 2014, p. 69). More recently, the Global Risks Report 2019 by WEF called attention to linkages between urbanization and vulnerabilities to climatic adversities including but not limited to coastal flooding: 'Rapidly growing cities and ongoing effects of climate change are making more people vulnerable to rising sea levels. Two-thirds of the global population is expected to live in cities by 2050 and already an estimated 800 million people live in more than 570 coastal cities vulnerable to a sea-level rise of 0.5 metres by 2050. In a vicious circle, urbanization not only concentrates people and property in areas of potential damage and disruption, it also exacerbates those risks — for example by destroying natural sources of resilience such as coastal mangroves and increasing the strain on groundwater reserves. Intensifying impacts will render an increasing amount of land uninhabitable' (WEF 2019, p. 7).

In a 2009 paper entitled 'Urban Growth and Climate Change', Kahn argued that urbanization has consequences for the social costs that climate change will impose on the world's quality of life. According to Kahn, urbanization could actually reduce GHG emissions but unfortunately 'the sheer scale of consumption growth' and 'the lack of strong incentives to economize on GHG emissions' had thwarted efforts. More than a decade later, the adoption of a natural gas ban on all new buildings under seven stories high at the end of 2023 and those over seven stories in 2027 by one of the most populous cities in the US- New York City- signals a bold shift initiated by cities in the western part of the US toward climate mitigation (Disavino 2021). However, Kahn's question as to how different cities in developed and developing nations would cope with and adapt to the increasingly severe impacts of climate change remains relevant (2008, p. 2). Today, it is clear that urbanization cannot be significantly reversed, so the issue now becomes what are municipal leaders and stakeholders going to do to address the linked crises of climate change and PM pollution exposure especially if there is a lack of global consensus for linked action. Kahn's conclusions were clear and resonant:

- Cities that 'suffer from the effects of climate change can use public policies and market incentives to reduce ex ante risk taking and reduce the costs of adaptation' including zoning laws to discourage high-density development in at-risk areas because climate change was likely to increase the frequency and severity of such events.
- 'Cities in least developed countries face two additional adaptation challenges. One is the rural to urban migration accelerated by climate change. The other is the increased risks of disease, pollution exposure, and natural disaster faced by informal urban squatters' (p.13).

It was Kahn's reference of the risks experienced by the urban poor that sadly resonate for the lives of the urban poor in cities like Mumbai and Delhi: 'Climate change also poses a set of risks to the urban poor. Heat waves, exposure to high levels of urban smog, and climate-related events such as floods and mudslides all threaten this vulnerable group. In the developing world, city governments are not providing high quality services. ... If local governments do not have the revenue to provide basic services such as clean water and sanitation for a growing urban population, then climate change–induced "environmental refugees" can help to unintentionally trigger local urban quality of life challenges' (2008, p. 14).

The fact that an astonishing majority of the world's urban population has been projected to reside in African and Asian cities necessitates an immediate acceleration of resources and partnerships that can enable these cities address multiple sustainable development challenges including curbing PM pollution and building climate resiliency. What is also useful to keep in mind in terms of the frontline for linked action on clean air, clean energy and climate change is that the UNDESA 2018 Report also flagged that future increases in the size of the world's urban population are expected to be highly concentrated in just a few countries: 'Taken together, China, India and Nigeria are projected to account for 37 per cent of the increase of nearly 2.5 billion people in the urban population by 2050'. More specifically, according to the Report, from 2018 to 2050, India has been projected to add 416 million urban dwellers, nearly doubling the size of its urban population; China projected to add 255 million urban dwellers, roughly equivalent in size to 31% of its urban population, and Nigeria projected to add 189 million urban dwellers, a doubling of its population (2018, p. 43).

As evidenced in the preceding chapters, researchers and policymakers have known for years about the negative health impacts of air pollution. In 2006, 2 million premature global deaths were attributed to air pollution annually by the WHO, but by 2014, the WHO issued a media alert that the number had risen to 7 million (WHO 2014). In the ensuing year, the 'thorny' challenge which has largely been left unaddressed is that there are massive gaps in understanding the full extent and severity of PM pollution related health impacts in cities where pollution levels currently range from 'unhealthy' to 'hazardous' for human health. The US EPA has established an Air Quality Index (AQI) for five major pollutants regulated by the US Clean Air Act. Each of the five pollutants listed has a national air quality standard set by the EPA to protect public health: Ground level ozone; Particle pollution (also known as particulate matter, including $PM_{2.5}$ and PM_{10}); Carbon monoxide; Sulphur dioxide and Nitrogen dioxide. The box below excerpted from the US EPA AQI site provides a schematic overview of how the AQI levels, values and descriptions of air quality are categorized and associated with colour coding. As referenced in Figure 4.1 the AQI is akin to a yardstick measuring 0–500 – and the higher the AQI level, the greater the health concern with values listed as: 0–50 = Good; 51–150 = Moderate; 151–200 = Unhealthy; 201–300 = Very Unhealthy; and 301–500 = Hazardous.

As referenced in Chapter 1, the need to focus on curbing air pollution was made explicit within the context of WHO's global assessment on air pollution (2016). The WHO's 2016 report found that 92% of the world's population exposed to unsafe levels of air pollution and estimated that nearly 90% of air pollution related deaths occur in LMICs. The report found that exposure to air pollution also increases the risks for acute respiratory infections. From the immediate perspective of this chapter, it is worth reiterating that the WHO report was clear that air pollution should be used as a marker of sustainable development, as sources of air pollution also produce climate-modifying pollutants (e.g. CO_2 or BC) (WHO 2016). Recognizing the major challenge of lack of adequate data and air pollution monitoring stations across regions where air pollution posed an inequitable burden, this landmark assessment based on a database of information on $PM_{2.5}$ and PM_{10} from measurements for about 3000 cities and towns worldwide found that 'modelled estimates indicate that in 2014 only about one in ten people breathe clean air, as defined by the WHO Air quality guidelines'. But what should give pause as to identifying the full scope and scale of the air pollution crises is that the report also highlighted that the assessment was based on 'conservative' estimates as it considered only those: 'Health outcomes for which there is enough

How does the AQI work?

Think of the AQI as a yardstick that runs from 0 to 500. The higher the AQI value, the greater the level of air pollution and the greater than health concer. For example, an AQI value of 50 as below represents good air quality, while an AQI value over 300 represents hazardous air quality.

For each pollutant an AQI value of 100 generally corresponds to an ambient air concentration that equals the level of the short-term national ambient air quality standard for protection of public health. AQI values at or below 100 are generally thought of as satisfactory. When AQI values are able 100, air quality is unhelathy: at first for certain sensitive groups of people, then for everyone as AQI values get higher.

The AQI is divided into six categories. Each category corresponds to a different level of health concer. Each category also has a specific color. The color makes it easy for people to quickly determine whether air quality is reaching unhealthy levels in their communities.

AQI Basics for Ozone and Particle Pollution

Daily AQI Color	Levels of Concern	Values of Index	Description of Air Quality
Green	Good	0 to 50	Air quality is satisfactory, and air pollution poses little or no risk.
Yellow	Moderate	51 to 100	Air quality is acceptale. However, there may be a risk for some people, particularly those who are unusually sensitive to air pollution.
Orange	Unhealthy for Sensitive Groups	101 to 150	Members of sensitive groups may experience health effects. The general public is less likely to be affacted.
Red	Unhealthy	151 to 200	Some members of the general public may experience health effects; members of sensitive groups may experience more serious health effects.
Purple	Very Unhealthy	201 to 300	Heatlh alert: The risk of health effects is increased for everyone.
Maroon	Hazardous	301 and higher	Heatlh warning of emergency conditions: everyone is more likely to be affected.

Figure 4.1 AQI Basics. *Source:* US Air Quality Index (2021).

epidemiological evidence to be included in the analysis, comprise acute lower respiratory, chronic obstructive pulmonary disease, stroke, ischeamic heart disease and lung cancer' and excluded many other diseases that have been associated with air pollution because 'the evidence was not considered sufficiently robust'. So in the report's own words: 'The actual impact of air pollution on health presented here is a conservative figure, as it does not include the separate impacts of health from other air pollutants such as nitrogen oxides (NOx) or ozone (O_3), and excludes health impacts where evidence is still limited (e.g. pre-term birth or low birth weight)' (2016, p. 16). To be clear however, five years ahead of the release of the 2016 WHO global assessment, UNEP and the WMO issued the first ever integrated assessment on the need to curb SLCPs such as BC and O_3. The 2011 'Integrated Assessment of Black Carbon and Tropospheric Ozone: Summary for Decision Makers' provided an overview identifying the challenge and listing the benefits of emission reductions of BC and O_3, which are referenced in greater detail in the chapter that follows.

Within the context of urbanization, research has been focused on understanding the role of spatial composition, configuration and density of urban land uses – also referred to as urban forms – energy and air pollution (Anderson et al. 1996). Parikh and Shukla focused on the role of urbanization and energy use in developing countries (1995). Cole and Neumayer (2004) examined for the first time the impacts of demographic factors on a pollutant – sulphur dioxide – other than carbon dioxide at a cross-national level. By also taking into account the urbanization rate and the average household size that were neglected by many prior cross-national econometric studies, Cole and Neumayer found evidence that for carbon dioxide emissions, population increases are matched by proportional increases in emissions, while a higher urbanization rate and lower average household size increase emissions. For sulphur dioxide emissions, they found that urbanization and average household size were not found to be significant determinants of sulphur dioxide emissions. For both pollutants, the study's results suggest that an increasing share of global emissions will be accounted for by developing countries. Liang and Gong (2020), in a first of a kind comprehensive study of 626 Chinese cities, examined the urban form effects on air quality and PM_{25} pollution and found that measures of urban form are robust predictors of air quality trends for a certain group of cities and concluded that measures to curb urban air pollution are dependent on urban planning and design. Cong et al. (2019) undertook a study that was the largest of its kind to investigate the short-term impacts of air pollution on death, conducted over a 30-year period. The study analyzed data on air pollution and mortality in 24 countries and 652 cities and found that ambient PM air pollution was linked to increased cardiovascular and respiratory death rates.

Drawing from the most recent evidence produced as part of the GBD project of the Institute of Health Metrics and Evaluation, the third annual State of Global Air report highlighted air pollution as the fifth leading risk factor for mortality worldwide: 'It is responsible for more deaths than many better-known risk factors such as malnutrition, alcohol use, and physical inactivity. Each year, more people die from air pollution–related disease than from road traffic injuries or malaria' (Health Effects Institute 2019, p. 1). Additional key findings of the 2019 State of Global Air Report are included below:

- 'Air pollution exposure is linked with increased hospitalizations, disability, and early death from respiratory diseases, heart disease, stroke, lung cancer, and diabetes, as well as communicable diseases like pneumonia.

- PM 2.5 pollution contributed to nearly 3 million early deaths in 2017. More than half of this disease burden fell on people living in China and India.
- In the least-developed countries, air pollution accounts for a higher proportion of deaths overall, with household air pollution responsible for the largest portion of these deaths. These countries may face a double burden from high exposure to both household and ambient air pollution.
- Air pollution collectively reduced life expectancy by 1 year and 8 months on average worldwide, a global impact rivaling that of smoking. This means a child born today will die 20 months sooner, on average, than would be expected in the absence of air pollution' (2019, pp. 11–16).

In the most congested and polluted cities in the world, curbing air pollution has not been responded within the context of climate change and clean energy, in part because global goal and partnership silos translate into fragmented policy agendas as the national level. There has been an exponential growth in MEAs which all countries have been actively negotiating at the global level, but keeping track of these MEAs whilst also addressing the overarching development goal of poverty reduction has not resulted in policy priority being accorded to the nexus between clean air and access to clean energy for all. But, now with the recent COVID pandemic, access to clean air is at a premium for those living with years of exposure to toxic air pollution.

The task of implementing integrated partnerships on increasing access to clean air and clean energy for all has been made even more complicated by unclear roles and involvement of municipal stakeholders within the SDGs silos. But, it is really at the municipal and local level where the climate change and clean air challenges are directly experienced. Here, it is worth referencing Homer-Dixon's 1994research finding that 'scarcities of renewable resources will increase sharply' and that environmental conflicts induced or aggravated by scarcer resources can often be 'subnational' and that: 'Poor societies will be particularly affected since they are less able to buffer themselves from environmental scarcities and the social crises they cause. These societies are, in fact, already suffering acute hardship from shortages of water, forests, and especially fertile land' (1994, pp. 5–6). Homer-Dixon's warning that fast moving, unpredictable and complex environmental problems can overwhelm states especially those already under pressure which can either lead to further fragmentation or more authoritarianism remains foreboding (1994, p. 6).

Public policy neglect of air pollution health risks in countries across the world was starkly revealed as a result of the COVID pandemic. Yet, years before the COVID pandemic hit, breathing clean air – an essential human function – was noticeably at risk in many cities across the world. In 2009, Langrish et al. assessed the beneficial cardiovascular effects of reducing PM air pollution exposure in Beijing by wearing a facemask. The study's conclusion that wearing a facemask ameliorated the adverse effects of air pollution on blood pressure and heart rate variability offers a current policy intervention that is highly pertinent in the megacities of South Asia coping with high concentrations of ambient air pollution. But the problem was that the measurement and data of PM pollution remains sparse and inconsistent in those regions where PM pollution levels are amongst the highest in the world. Additionally, research related to chronic exposure to air pollution worsening the risks of viral respiratory infections that was expressly focused on developing countries has been inadequate and public health risks have been neglected. Cui, Zhang et al. 2003 analysis of 5 regions with 100 or more Severe Acute Respiratory Syndrome (SARS) cases in China showed that case fatality rate increased with the air pollution index (API), and their partially ecologic study based on short-term exposure demonstrated that patients with SARS from regions with moderate APIs had an 84% increased risk of dying from SARS compared to those from regions with low APIs.

The COVID-19 pandemic, which began in the most populous country of the world, spared no borders, destroyed lives and wreaked socio-economic havoc across the world. It laid bare the disproportionate burdens borne by those who live at the intersection of socio-economic marginalization, existing co-morbidities and public health neglect in urban areas not just in the United States but across the world. The two principal precautionary measures to combat the spread of the COVID-19 pandemic highlighted by the WHO were (i) to regularly and thoroughly clean hands with soap and water or a 60% alcohol-based sanitizer; and (ii) to maintain 'social distancing' of at least 3 ft between individuals. But, these precautionary public health responses to COVID-19 also exposed major inequities including the lack of access to safe water in poor communities and households, and the grim realities of urban poverty where even 3 ft of social distancing remains an improbable luxury. The indisputable truth is that fighting COVID and other infectious diseases requires universal access to safe water and access to adequate public health care. The persistent lack of access to clean water across the globe means that poorer households and communities are less able to combat the spread of highly infectious diseases and more likely to succumb to diseases associated with water contamination. Box 4.1 excerpted from WHO (2019) evidences the enormous scale associated with unsafe water across the globe.

Analogous to the challenge of lack of access to clean energy sources experienced by the energy poor and climate vulnerable, the lack of access to safe water for poorer communities is not a new policy challenge. All UN member states universally agreed on SDG 6 – increasing access to safe water and sanitation in 2015. But lack of access to clean water combined with existing burdens of disease impacting upon poorer communities especially

Box 4.1 Drinking water: Key facts (Excerpt from WHO 2019).

- 785 million people lack even a basic drinking-water service, including 144 million people who are dependent on surface water.
- Globally, at least 2 billion people use a drinking water source contaminated with faeces.
- Contaminated water can transmit diseases such diarrhoea, cholera, dysentery, typhoid, and polio. Contaminated drinking water is estimated to cause 485,000 diarrhoeal deaths each year.
- By 2025, half of the world's population will be living in water-stressed areas.
- In least developed countries, 22% of health care facilities have no water service, 21% no sanitation service, and 22% no waste management service.

Source: WHO 2019.

communities of colour within individual cities was certainly brought into sharper public focus during the COVID-19 pandemic, in advanced industrialized countries. Back in 2014, a policy decision to switch the city of Flint's drinking water supplies as a cost-saving measure ended up leaching poisonous levels of lead; and to make matters worse, the treatment and testing of the polluted water was chronically ignored by government officials for 18 months. Flint, Michigan, was seen as an example of 'environmental injustice and bad decision making' by one of the largest environmental advocacy groups in the US – the Natural Resources Defense Council.

Troublingly, the COVID-19 pandemic left a wake of devastation in cities like Detroit which has one of the largest African American populations in the country. COVID-19 also revealed that indigenous communities have suffered in many of the same ways from lack of access to public health services and also lack of access to clean water. Studies of past epidemics show the same pattern repeating again as infectious diseases more easily take hold in racially and economically vulnerable groups who live in crowded conditions and work next to others and who lack access to health care. The nexus between societal factors that increase the risk of infection in impoverished communities, such as housing density and reliance on public transportation, and disproportionately high rates of disease and illness that make infection more deadly was evidenced as a result of COVID-19 (Gravlee 2020; Gay et al. 2020). In their 16 April 2020 *Lancet* article, Van Dorn et al. (2020) argued that the impacts of COVID-19 confirmed glaring morbidity disparities within urban areas: 'Confirming existing disparities, within New York City and other urban centres, African American and other communities of colour have been especially affected by the COVID-19 pandemic. Across the country, deaths due to COVID-19 are disproportionately high among African Americans compared with the population overall. In Milwaukee, WI, three quarters of all COVID-19 related deaths are African American, and in St Louis, MO, all but three people who have died as a result of COVID-19 were African American' (2020, p. 1243). When adjusted for age, the risk of death from COVID-19 has been estimated to be as much as nine times higher for African Americans than it is for whites (Bassett et al. 2020).

The challenges of dealing with COVID-19 clearly brought to the fore the role of cities in the frontline of dealing with public health and urban poverty crises, and it also afforded for the first time in global history, a real time look at how PM pollution was temporarily reduced over many of the world's most polluted cities. To contain the virus, countries across the world imposed quarantines and even curfews, prohibited large gatherings, restricted transportation and locked down businesses deemed non-essential. The costs of enforcing these preventive measures resulted in major socio-economic changes including growing food insecurity, hunger and unemployment. But the COVID-19 prevention measures also had the unintended benefit of sharp drops in ambient air pollution in several countries that took aggressive measures to slow transmission of the virus which were captured by satellite imagery. For example, NASA and European Space Agency (ESA) pollution monitoring satellites detected significant decreases in NO_2 (a gas emitted by motor vehicles, power plants and industrial facilities) over China (see Figure 4.2) with NASA noting that there is 'evidence that the change is at least partly related to the economic slowdown following the outbreak of coronavirus' (Earth Observatory NASA 2020). News media photos of ambient air pollution clearing over many of the large cities in developing countries, including India, provided evidence that lockdowns were having a similar short-term effect.

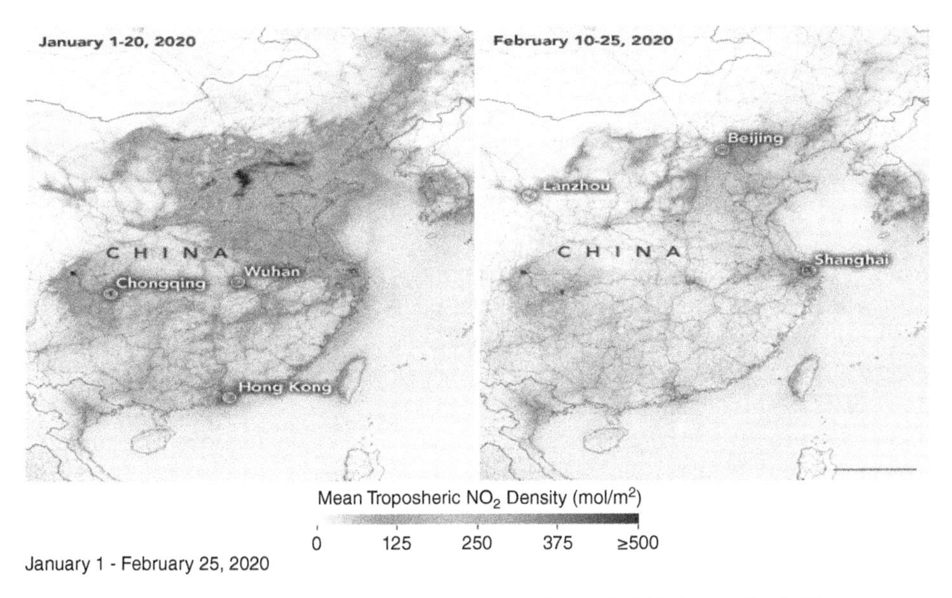

Figure 4.2 Airborne nitrogen dioxide plummets over China (NASA). *Source:* Earth Observatory NASA (2020).

Here, it is significant to highlight that the global lockdown measures put into place to curb the spread of COVID-19 were found to lead to short-term drops in coal, oil and gas production, but fossil fuel production levels continue to rise. As outlined by the 2020 Production Report which bolsters the UNEP Emission Gap report, the production gap between countries' planned fossil fuel production and PA's goals has grown. This 2020 Report jointly prepared by the Stockholm Environment Institute, UNEP and others found that: 'To follow a 1.5°C-consistent pathway, the world will need to decrease fossil fuel production by roughly 6% per year between 2020 and 2030. Countries are instead planning and projecting an average annual increase of 2%, which by 2030 would result in more than double the production consistent with the 1.5°C limit. Between 2020 and 2030, global coal, oil, and gas production would have to decline annually by 11%, 4%, and 3%, respectively, to be consistent with a 1.5°C pathway. But government plans and projections indicate an *average 2% annual increase for each fuel*. This translates to a *production gap similar to 2019*, with *countries aiming to produce 120% and 50% more fossil fuels by 2030 than would be consistent with limiting global warming to 1.5°C or 2°C, respectively*' (emphasis added, SEI et al. 2020, p. 4).

The catastrophic effects of the COVID-19 respiratory borne pandemic also revealed the extent of existing cardiovascular and respiratory ill health burdens on vulnerable and marginalized populations. Long-term exposure to air pollution has been linked to an increased risk of dying from COVID-19. The European Public Health Alliance (EPHA) – Europe's largest non-governmental public health network – warned that the coronavirus threat was greater for residents of polluted cities and that patients with chronic lung and heart conditions caused or worsened by long-term exposure to air pollution are less able to fight off lung infections and more likely to die (EPHA 2020). Additionally, air pollution linkages to hypertension, diabetes and respiratory diseases are all conditions that are linked to higher mortality rates for COVID-19. For the first time, a study published in Cardiovascular Research (2020) estimated the proportion of deaths from COVID-19 that could be attributed to the impacts of air pollution for every country in the world. The study entitled 'Regional and global contributions of air pollution to risk of death from COVID-19' estimated the fraction of COVID-19 mortality attributable to the long-term exposure to ambient $PM_{2.5}$. As the study notes: '$PM_{2.5}$, is one of the leading risk factors for cardiovascular and respiratory ill health, and responsible for many excess deaths. The global loss of life expectancy from long-term exposure to ambient air pollution exceeds that of infectious diseases, and is comparable with that of tobacco smoking'. The researchers argue that a new, though preliminary, finding of the study is that 'a significant fraction of worldwide COVID-19 mortality is attributable to anthropogenic air pollution' (Pozzer et al. 2020).

The urgency for linked action on climate mitigation, clean energy access and pollution reduction was also categorically highlighted by Shindell's key finding that: '. . . *roughly 1.4 million lives could be saved from improved air quality during the next 20 years.* As we've seen with the coronavirus lockdowns in many places, air pollution responds immediately to emissions reductions. *A rapid shift to a 2°C pathway could reduce the toll of air pollution, which leads to nearly 250,000 premature deaths per year in the US, by 40% in just a decade.* Our work shows that action now means benefits now' (emphasis added, Shindell 2020). On 13 June 2020 as direct response to the COVID-19 crisis – ***Medics for Clean Air*** – a new coalition of doctors, nurses, medical students and other health professionals in Europe called for urgent and sustainable action to tackle air pollution. The coalition of medical professionals outlined an explicit case for addressing air pollution: 'Air pollution is a global health challenge that affects us all. It increases the risk of heart disease, stroke, cancers, dementia and diabetes, causes new asthma cases in children, and damages nearly every organ in our bodies. It is a public health and climate emergency, with an unsustainable transport sector at its core' (Medics4cleanair.eu site 2021).

The question confronting decision/policymakers and public health advocates is what exactly is being done in terms of linked action at the city level to curb air pollution and increase access in developing countries in Asia and Africa where the toxic levels of air pollution are consistently much higher than Europe and the US. And why has the world's only regional air pollution treaty – LRTAP – not been replicated to cover countries and cities where exposure to toxic levels of PM pollution has been known to occur?

Chronic exposure to PM air pollution remains a massive public health concern for cities within two of the largest aggregate GHG emitters – China and India. High levels of PM pollution worsen diabetes, hypertension, coronary disease and asthma and weaken already compromised immune systems. A quantitative examination of China's economic sectors – 'Clearing the Air' – found that the building, transportation and cement production sectors caused the greatest harm. It demonstrated that 'green' taxes might not only reduce GHG emissions and health damages but even enhance China's economic growth (Ho and Nielsen 2007). The external costs of air pollution in China were previously estimated to range anywhere from 1 to 8% of GDP with discrepancies arising from uncertainties in impact modelling (Heck and Hirschberg 2011). A follow-up 2013 quantitative study of China – *Clearing the Skies* – demonstrated that China's sulphur controls achieved enormous environmental health benefits at unexpectedly low costs and that the implementation of carbon taxes could reduce not only China's carbon emissions but also its air pollution more comprehensively than current single-pollutant policies, all at little cost to economic growth. In contrast to China, cities in India have seen a worsening of air pollution levels and limited success with air pollution control measures. In 2015, New Delhi's air was found to be 'the world most toxic in part due to high concentrations of PM 2.5', especially during the winter months due to the practice of burning garbage and diesel fuel, which results in an estimated 1.5 million deaths, comprising one sixth of all national deaths annually, as a result of indoor and outdoor pollution (Harris 2015). More recently, the painful truth about the urgent scale and scope of the urban air pollution crisis extracting a tragic toll on the world's most populous and most polluted city in the world has been hard to witness.

On 30 June 2020 in the midst of the COVID-19 pandemic, the Director of Air Pollution and Health Research at Stanford University, Prunicki highlighted the relationship between PM pollution and increased incidence of diabetes: 'The small particulate matter in air pollution is about one-thirtieth the width of a human hair. It is small enough to enter the bloodstream after being inhaled and to travel to many organs. In diabetes, for example, it is thought that inflammation from small particulates increases insulin resistance. Eventually this leads to overt diabetes. In fact, in 2016, it was estimated that pollution linked diabetes cut short peoples' healthy lives by a total of 8.2 million years'. Prunicki could not have been more categorical about the inequities associated with PM pollution exposure being disproportionately experienced within the US: 'Particulate exposure is disproportionately caused by non-Hispanic whites, but disproportionately inhaled by African Americans and Hispanic minorities. African Americans are exposed to 56 per cent more pollution than they cause through their consumption. Moreover, African Americans are 75 percent more likely to live in communities adjacent to sources of pollution. For example, more than a 1 million African Americans live within a half-mile of natural gas facilities and face a cancer risk above the EPA's level of concern from toxins emitted by those facilities. There is a long history of placing pollution sources in low-income areas and communities of color and the health impacts are well documented'. And when asked how exposure to PM pollution increased the COVID risk for communities of colour, Prunicki pointed out African Americans are more vulnerable to COVID for many reasons including chronic exposure to elevated PM pollution and suffering from chronic disease burdens related to asthma and hypertension, which when coupled with health care access shortages resulted in African Americans being 'overrepresented among hospitalized COVID-19 patients and their death rates appear to be twice as high as Caucasians' (Jordan 2020).

The linkages between chronic disease burden, lack of access to health care, combined with long-term exposure to PM pollution were also flagged in a foreboding 20 October 2020 news report entitled 'COVID 19 and Pollution: Delhi Staring at a Coronavirus Disaster', $PM_{2.5}$ levels in Delhi were averaged at around 180–300 μm^3 – 12 times higher than the WHO's safe limits. The article cautioned that there were 'no studies' focused on impacts of air pollution and COVID-19 infection or recovery rates, and issued a chilling conclusion that:

'. . . doctors and epidemiologists have long warned that toxic air will only hamper India's fight against the virus. The country now has the world's second-highest caseload (7.5m and counting) and the third-highest death toll (more than 114,000) from the virus, although deaths per million of the population are relatively low. But experts say worsening air quality will likely increase these numbers. Delhi, already one of the cities that is worst-hit by the pandemic, will probably bear the brunt because its residents have been exposed to hazardous levels of pollution for years' (Pandey 2020).

Less than six months later by April 2021, the worst-case scenario of chronic exposure to hazardous levels of pollution intersecting with massive public health neglect and disease burdens of diabetes, heart and lung disease became evident in several Indian states and cities including the world's most populous and polluted cities – Delhi. News reports as well as social media posts from Delhi residents broadcasted a city where residents literally struggled to breathe. Tragically, there was a dire shortage of oxygen tanks, hospitals ran out of beds, smokestacks of crematoriums melted and funeral pyres were held round the clock in parking lots. The disease and morbidity burdens associated with chronic exposure to high levels of air pollution came into sharper focus in connection to respiratory-driven nature of the COVID-19 pandemic. The true scale of the interface between long-term exposure to $PM_{2.5}$ pollution and highly infectious respiratory diseases like COVID-19 are not fully estimated, and perhaps might never be accessible given that many of those who died in their homes in cities, particularly in India, have not been registered as COVID deaths.

In light of the human health inequities that the pandemic exposed, it is time to face up to the fact that the world's largest environmental health risk – air pollution – cannot be addressed via segregated SDGs. A few fragmented global partnerships simply do not address the massive health and morbidity costs of PM air pollution experienced in the most polluted cities. Global policy silos on clean energy, clean air and climate change impede the future development trajectory of sustainable cities. The process of waiting for incremental, textually driven, consensus agreements on climate action via the annual cycles of COP has not reduced the growing fossil fuel related air pollution crisis that casts a pall over many large cities especially those in Asia and Africa. The remaining sections focus on identifying measures that focus on NNSA actions that are aimed towards integrated action on curbing SLCPs and PM pollution.

4.2 On the Front Lines: Role of Cities as the Loci for Linked Action on Clean Air and Climate

Cities are the current and future loci for a manifold range of sustainable development actions including those that integrate clean air and access to clean energy with climate justice. Recognizing the intersections between climate inequities/injustice and urban poverty, Caniglia et al. issued a succinct warning that climate change 'portends widespread environmental injustices', and pointed to troubling linkages between growing urbanization and inequalities. Growth in urbanization corresponding to growing socio-economic inequalities gives rise to major challenges for the largest megacities in the world as Caniglia et al. highlighted: 'Urban areas, including megacities with over 10 million inhabitants . . . are widespread across the globe and their number is expected to rise in the coming decades . . . Urban centers are also bastions of inequalities, where poverty, marginalization, segregation and health insecurity are magnified. Socially disadvantaged groups are especially at risk in times of shock or change as they lack: the option to avoid, mitigate and adapt to threats posed to the system's social, environmental and cultural integrity; the capacity and freedom to exercise the options that are available to them and the ability to actively participate in obtaining these options' (2017, pp. 1–2). Drawing on the logic of cities as the frontline actors in addressing linked climate and clean energy injustice is important not just for countries like the United States (the world's second-largest aggregate GHG emitter) but most significantly for countries like India (the world's third-largest aggregate GHG emitter) where energy related air pollution, poverty reduction and climate vulnerabilities coexist on an unprecedented scale.

Sadly, the world has just witnessed the toll of chronic underlying health conditions including respiratory illnesses, combined with lack of access to clean water and public health services during the global COVID-19 pandemic. This is a watershed moment in global environmental history which offers an opportunity for responsive action that looks beyond annual climate COPs and global summits towards dynamic clean energy and climate resilient measures directly relevant to the needs of cities. Echoing unmistakable alarm in '*Sustainable Nation*', Farr argued that climate change has given humanity its 'first deadline' and pointed out that: 'Humanity has never really had a deadline before. We tend to muddle forward, hoping and praying that the future improves and praying that the future improves on the present. No eve of destruction. ... That has just changed. Human induced climate change is a disaster orders of magnitude more challenging and urgent than any that came prior. . . . How humanity responds to this urgency will determine how much of this planet will be under water, eventually. Florida? Venice? Mumbai?' (2018, p. 94). Farr's argument that the world needs to urgently ramp up progress towards equitable and sustainable urban design in order to transform urban communities to address the challenge of climate change has enormous global relevance. Farr is right in pointing out that the locus for the delivery of sustainable lives and communities lies at the municipal level, not just in terms of climate action but importantly in terms of the inherent synergies between clean air, climate and clean energy responsive measures.

According to UNDESA's Population Division report, 'The World's Cities in 2018', there has been a dramatic growth in the number of cities with 1 million inhabitants, 371 cities in 2000, to 548 cities in 2018, to a projected 706 cities by 2030. This UN report also pointed out that 'the choice of how to define a city's boundaries was consequential for assessing the size of its population' and that there are three categories:

- The 'city proper', describes a city according to an administrative boundary.
- The 'urban agglomeration', considers the extent of the contiguous urban area, or built-up area, to delineate the city's boundaries.
- The 'metropolitan area', defines boundaries according to the degree of economic and social interconnectedness of nearby areas, identified by interlinked commerce or commuting patterns (2018, pp. 1–2).

Of the 1860 cities with at least 300,000 inhabitants in 2018, 55% follow the 'urban agglomeration' statistical concept, 35% follow the 'city proper' concept and the remaining 10% refer to 'metropolitan areas', so even here there are challenges as to conclusive boundary-based categorization. Nevertheless, for the purpose of this chapter, what is significant is that '*The World's Cities in 2018*', estimated that of the world's 33 megacities – that is, cities with 10 million inhabitants or more – in 2018, 27 are located in the less developed regions or the 'global South', with China having six megacities in 2018, while India had five. Nine of the 10 cities projected to become megacities between 2018 and 2030 are located in developing countries. The number of 'megacities' – is projected to rise from 33 in 2018 to 43 in 2030, with Delhi overtaking Tokyo as the world's largest megacity. Figure 4.3 excerpted from the Report demonstrates that 3 of 10 megacities are projected to be in South Asia (Delhi, Dhaka and Mumbai) and Kinshasa is set to become the tenth largest city by 2030 (UNDESA 2018, pp. 4–5). Troublingly, the Report also found that cities of 500,000 inhabitants or more facing high risk of exposure to at least one type of natural disaster were home to 1.4 billion people in 2018 (p. 9).

It should not come as no surprise, as referenced in the preceding chapters, that ample global consensus exists that adverse climatic impacts – rising sea levels, coastal zone erosion, increased desertification and forest fires, and extreme weather events to name just a few will wreak havoc on poorer households and cities that lack resilience and coping capacities to adapt to climate change. Within the US context for example, studies have evidenced that counties that will be hit hardest by climate change are likely to be in South and Southwest regions of the United States. Muro et al. (2019) provided a very detailed 'granular look at the politics of climatic impacts' that was grounded in

Delhi will overtake Tokyo as the world's largest city by 2030

Between 2018 and 2030, the population of Delhi, India is projected to increase by more than 10 million inhabitants, whereas that of Tokyo, Japan is projected to decline by almost 900,000. The two cities are thus expected to change places on the list of the world's cities ranked by size. Projections indicate that the world's tenth largest city in 2018—Osaka, Japan—will no longer be among the ten largest in 2030. Kinshasa, Democratic Republic of the Congo will grow to rank as the tenth most populous city in the world in 2030.

The world's ten largest cities in 2018 and 2030				
City size rank	City	Population in 2018 (thousands)	City	Population in 2030 (thousands)
1	Tokyo, Japan	37468	Delhi, India	38939
2	Delhi, India	28514	Tokyo, Japan	36574
3	Shanghai, China	25582	Shanghai, China	32869
4	São Paulo, Brazil	21650	Dhaka, Bangladesh	28076
5	Ciudad de México (Mexico City), Mexico	21581	Al-Qahirah (Cairo), Egypt	25517
6	Al-Qahirah (Cairo), Egypt	20076	Mumbai (Bombay), India	24572
7	Mumbai (Bombay), India	19980	Beijing, China	24282
8	Beijing, China	19618	Ciudad de México (Mexico City), Mexico	24111
9	Dhaka, Bangladesh	19578	São Paulo, Brazil	23824
10	Kinki M.M.A. (Osaka), Japan	19281	Kinshasa, Democratic Republic of the Congo	21914

Figure 4.3 World's largest cities in 2018 and 2030. *Source:* UNDESA Population Division (2018) *The World's Cities in 2018*, p. 4.

county-based assessments conducted by the Climate Impact Lab in their article entitled, 'How the geography of climate damage could make the politics less polarizing'. As Muro et al. noted, the more localized/county-based calculations on the economic costs of future climatic impacts are based on the research collaboration of climate scientists, data computational analysts, economists amongst others from the academic institutions including University of California at Berkeley, the University of Chicago, Rutgers University and the Rhodium Group. Figure 4.4 (map) excerpted from Muro et al. (2019) is based on Climate Impact Lab's data and offers a snapshot overview of the range of climate related costs anticipated to be borne by counties across the southern part of the US. More-prosperous counties in the United States are often in the Northeast, upper Midwest and Pacific regions, where temperatures are lower and communities are less exposed to climate damage.

The policy reality is that such localized data is simply not available for the majority of cities in Asia and Africa where the costs of climate change are widely anticipated to worsen socio-economically marginalized households and communities with internal migration and displacement, as well as food and water insecurity increasingly evidenced. Here it is necessary to focus on the fact that ignoring the precautionary principle in climate change enshrined with Article 3.3 of the UNFCCC poses greater burdens for the climate vulnerable and is amplified further in smaller and poorer communities, cities and countries. Quite literally, delayed action on climate change, clean air and clean energy

Climate-related costs by 2080–2099
Share of 2012 country income

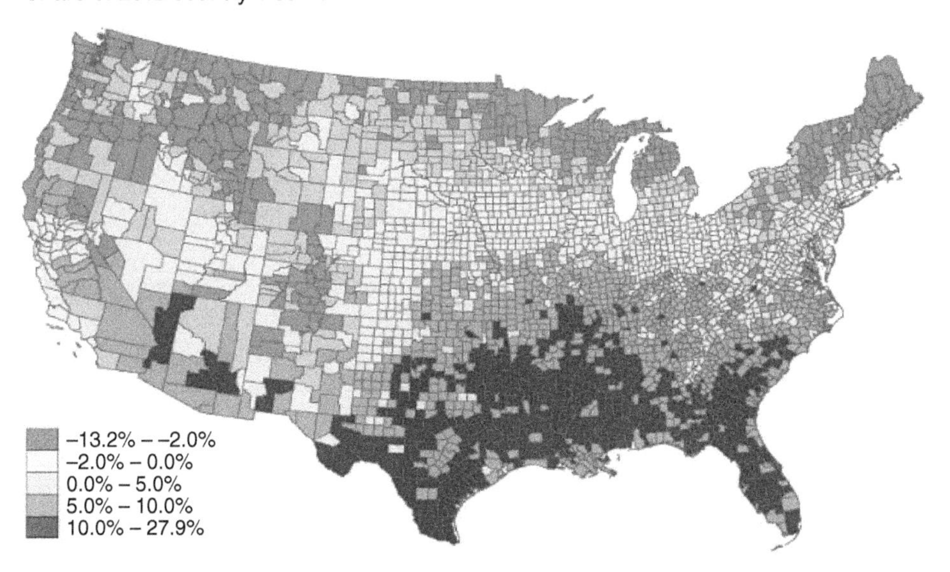

Note: Emissions projections are based on a "business-as-usual" scenario (RCP8.5), which reflects the current global trajectory
Source: Hsiang and others, 2017

B Metropolitan Policy Program at BROOKINGS

Figure 4.4 Climate related costs (excerpted from Brookings Institute). *Source:* Muro et al. (2019).

costs lives and livelihoods for countries, communities and households who have little to no safety nets that cover climatic risks and adversities. As referenced in Chapters 2 and 3, there are separate SDGs on sustainable energy (SDG 7) and climate change (SDG 13) with SDG 11 focused on the making cities inclusive, safe, resilient and sustainable. In February 2018, the Cities and Regions Talanoa Dialogue was launched at the Ninth World Urban Forum which envisaged an increased role of local and sub-national governments to undertake climate related decision-making. In March 2018, the IPCC Cities and Climate Conference in Edmonton, Canada, assessed the state of the academic and practice-based knowledge related to cities and climate change and established a global joint research agenda.

One of the most cogent and compelling arguments as to cities being the crucial loci of linked action on climate, clean energy and air as well as resilience was made at this conference. Speaking at the 27 April 2018 Forum on Urban Resilience and Adaptation, UN Climate Change Executive Secretary, Patricia Espinosa warned that the impacts of climate change are 'an incredible risk for cities'. In calling for urban action, Espinosa referenced the role of the Talanoa Dialogue – 'an initiative on behalf of the Government of Fiji—as an international conversation to determine if we're meeting the goals of the Paris Agreement, and to increase global ambition towards them'. While conversations/dialogues are the buzz words du jour of UN PA negotiations, effectively scaling up strategic partnerships with timely results looms as a challenge for poor and marginalized urban communities faced with coastal zone inundation and the rapid destruction of livelihoods and fragile infrastructure. In her speech, Ms Espinosa listed momentum for action by citing the example of Los Angeles which aims to significantly reduce its urban heat totals over 20 years by creating strong cool roofing requirements; cities throughout China embracing the Low Carbon City Initiative which aims to improve energy efficiency in their industry, construction and transportation sectors; cities such as Athens, Barcelona and Paris which have mapped their urban heat, but also their vulnerable populations so as to provide them with access to cooling centres; and making buildings more efficient by using sustainable materials (UN Climate Change 2018). But it is an understatement to point out that there is so much more to be done in cities with some of the highest urban densities in the world in terms of expanding access to sustainable transportation, establishing access to clean energy markets for the poor, investing in climate resilient infrastructure and mainstreaming climate adaptation within sustainable urban planning. The question for these city and local stakeholders is whether they can more effectively drive the engine of the PA in the direction of linked action on urban climate resilience and curbing toxic levels of energy related pollution.

There is an overwhelming amount of global scientific consensus that the adverse impacts of climate change are going to pose a clear and present danger especially those who are already socio-economically marginalized within cities. In his 2019 speech to the C-40 World Mayors Summit, UN Secretary General highlighted the enormous potential for cities as the loci for integrated action on clean air, clean energy and climate resilience. He specifically referenced that cities consume more than two-thirds of the world's energy and account for more than 70% of global carbon dioxide emissions. Echoing UNFCCC Executive Secretary Espinosa, the UN SG stated: 'Friends, cities are where the climate battle will largely be won or lost. With more than half the world's population, cities are on the frontlines of sustainable . . . and inclusive development. With air pollution a grave and growing issue, people look to you to champion better urban air quality. With environmental degradation driving migration to urban areas, people rely on you to make your cities havens for diversity, social cohesion and job creation. You are the world's first responders to the climate emergency' (UN Press Release 2019). His point about mayors and elected local municipal authorities being the first responders on shared frontline against the climate and energy pollution crisis is critical as the choices made by cities on urban infrastructure – housing, buildings, energy efficiency, clean power generation and sustainable transportation will not just impact on climate change but ameliorate public health costs of current air pollution.

For millions living in the world's most congested and largest cities in the world, the challenge of curbing PM air pollution in particular $PM_{2.5}$ pollution is extremely grave in terms of lives and livelihoods lost and destroyed. But a major problem that Hansell et al. referenced is that there are very few studies in general that have examined the health effects of very long-term (> 25 years) exposure to air pollution. In one of the few large national studies on air pollution exposure, Hansell et al. concluded that air pollution exposure has long-term effects on mortality that persist decades after exposure and that historic air pollution exposures influence current estimates of associations between air pollution and mortality (2016). Another problem is that there has been little to no tracking of the health impacts of PM pollution in cities in developing countries where chronic exposure to PM pollution levels is most dangerously and consistently hazardous.

Cities, in particular the megacities in Asia and Africa, are the unmistakable foci for linked action that is essential to reducing the toxic burden of SLCPs and air pollution. The heavy burdens and costs of exposure to PM pollution and the real-time curbing of the same hinges entirely on clean energy and clean air actions implemented in cities in Asia and Africa where PM pollution levels are the worst. The '2018 World Air Quality Report' (published by Swiss-based IQ Air) on annual regional and city air pollution levels found that 100 of the most polluted cities in the world are in Asia. Half of the world's 50 most polluted cities are in India, and 22 are in China, with Pakistani and Bangladeshi cities making up the rest. (IQ Air 2018), so focusing on the potential of regional air pollution protocols and partnerships matters. Data on the capital cities that deal with the highest level of $PM_{2.5}$ air pollution contained in 'World Air Quality Report 2020' is excerpted in Box 4.2:

From the immediate perspective of cities located in developing countries, the air pollution crisis is compounded by lack of real-time data. Real-time air quality data based on a network of monitoring sites under the rubric of the 'World Air Quality Index' (established in 2007) also reveals disproportionate levels of PM pollution across the world. The network uses over 10,000 monitoring stations across the world to deliver near real-time AQI readings (WAQI site 2021). There is, however, a big gap in that there appears to be major shortages in air quality monitoring data coverage for Africa, and here it is disconcerting that large swathes of an entire continent are shut out of data gathering. Within the WAQI site, typing in the name of a large city allows for an AQI number that reflects the PM levels the citizens of that city currently face.

Box 4.2 World capital city ranking arranged by annual average $PM_{2.5}$ concentration ($\mu g/m^3$).

1) Delhi, India (84.1)
2) Dhaka, Bangladesh (77.1)
3) Ulaanbaatar, Mongolia (46.6)
4) Kabul, Afghanistan (46.5)
5) Doha, Qatar (44.3)
6) Bishkek, Kyrgyzstan (43.5)
7) Sarajevo, Bosnia and Herzegovina (42.5)
8) Manama, Bahrain (39.7)
9) Jakarta, Indonesia (39.6)
10) Kathmandu, Nepal (39.2)
11) Hanoi, Vietnam (37.9)
12) Bamako, Mali (37.9)
13) Beijing, China (37.5)
14) Kuwait City, Kuwait (34.0)

By way of contrast, Wellington, New Zealand (6.0 $\mu g/m^3$) and Stockholm, Sweden (5.1 $\mu g/m^3$) ranked at 89 and 91 out of total of 92 capital cities.

Source: IQ Air, World Air Quality Report (2020, p. 12).

The '2018 World Air Quality Report' on annual regional and city air pollution levels found that 100 of the most polluted cities in the world are in Asia. What is also important to keep in mind is that despite air pollution being globally recognized as one of the leading environmental and public health risks, there is still no globally reliable or effective air pollution protocol or mechanism focused expressly on curbing PM pollutants including PM of 2.5 μm or less ($PM_{2.5}$) that covers the majority of cities in developing countries. For those living in urban slums amongst some of the worst traffic congestion and PM pollution, it comes as little comfort that a 2019 study published in the European Respiratory Journal using UK Biobank data on 303,887 individuals aged 40–69 years (one of the largest studies of its kind) found that exposure to ambient air pollution was associated with lower lung function and increased COPD prevalence. The research study examined particulate matter (PM_{10}), fine particulate matter ($PM_{2.5}$) and nitrogen dioxide (NO_2), which are produced by burning fossil fuels from car and other vehicle exhausts, power plants and industrial emissions (Doiron et al. 2019). Within the massive rubric of the SDGs focused on reducing poverty, increasing access to clean energy and climate change, there is a glaring data and research gap on studies focused on PM air pollution related to cities particularly those in Africa. Singh et al. (2020) in their 2020 study, 'Visibility as a proxy for air quality in East Africa', found shortage in air quality monitoring in Africa has led to scientists using visibility data for capital cities in Ethiopia, Kenya and Uganda as a substitute measurement. The study found a significant reduction in visibility since the 1970s, where Nairobi shows the greatest loss (60%), compared to Kampala (56%) and Addis Ababa (34%) due to increased PM emissions from vehicles and energy generation, and, correspondingly, PM pollution levels in the three cities are estimated to have increased by 182, 162 and 62%, respectively, since the 1970s to the current period.

As evidenced by earlier chapters, there is ample globally relevant data and knowledge particularly from global public health experts, the WHO and UNEP that air pollution contributes substantially to premature mortality and disease burden globally and has greater impacts in low-income and middle-income countries (MICs) than in HICs. But the problem is that there are huge gaps in understanding the health impacts in the countries where PM pollution has reached toxically dangerous levels ranging from 'unhealthy' to 'hazardous' for human health. The first problem as referenced previously is that there are very few studies in general examining the health effects of very long-term (> 25 years) exposure to air pollution. The second and even bigger problem is that there has been no detailed tracking of the health impacts of air pollution in some of the most populous developing countries. A key reason for these data gaps is that detailed PM pollution data and analysis focused on major urban capitals of developing countries across a period of time is virtually non-existent as air quality monitoring sites are a relatively new phenomenon. Despite broad knowledge about regional cost and disparities of air pollution, the global community has been slow to respond to the lack of adequate data measurement of PM pollution in regions where the problem is most crippling.

WHO's most recent ambient air quality database consists of urban air quality data covering more than 4000 human settlements in 108 countries from 2010 to 2016. But these settlements range in size from less than a hundred inhabitants to over 10 million and are not representative of the air pollution problem that affects the megacities in developing regions. As per Table 4.1 measurements of PM pollution are far from being homogenous with most of the measurements coming from HICs and MICs in Europe and the US. So from the start, the most significant global policy challenge is the inconsistency of air pollution data measurement and mapping. Large swathes of Africa and South-East Asia are grossly

Table 4.1 Total number of 'settlements' and countries in the WHO ambient air pollution database (2018, by region).

Region	Number of settlements	Number of countries with data	Total number of countries in region
Africa (Sub-Saharan) (LMIC)	41	10	47
Americas (HIC)	760	4	11
Americas (LMIC)	170	16	24
Eastern Mediterranean (HIC)	33	6	6
Eastern Mediterranean (LMIC)	98	9	15
Europe (HIC)	2392	31	34
Europe (LMIC)	234	12	19
South-East Asia (LMIC)	198	9	11
Western Pacific (HIC)	125	5	6
Western Pacific (LMIC)	336	6	21
Global	***4387***	***108***	***194***

AAP: Ambient air pollution database; LMIC: low- and middle-income countries; HIC: high-income countries

Source: WHO (2018, p. 2).

under-represented in terms of PM pollution data collection and measurements. This inconsistency in data assessments of PM pollution means that the full extent and scope of the air pollution in precisely those regions and countries where the problem is most acutely experienced is not even adequately monitored or referenced at this stage.

4.3 Toxic Air: Why the Future of Integrated Action on Clean Air and Clean Energy Lies with India and Indian Cities

In countries like India, where a heavy reliance of polluting solid biomass and toxic levels of PM pollution in major cities combine with climate vulnerabilities for millions, linked action on air pollution, climate change and clean energy access for the poor is absolutely vital. But to contextualize any discussion on energy access–air pollution – climate nexus within India, it is important to briefly reflect on the significance and evolution of historical versus per capita GHG emissions which was raised early on by Agarwal and Narain in their paper 'Global Warming in an Unequal World: A case of environmental colonialism' (1991). It is also useful to point out more broadly that contrary to any prevailing notions that environmental concerns were primarily acted upon only in advanced industrialized countries, India has a long history of environment legislation, some of which, for example, the Indian Forest Act of 1927 emerged even before Indian independence from British colonial rule in 1947. But, Indian politicians and policymakers historically have rarely chosen environment over development because such a binary opposition assumes parity between the two issues that simply does not exist within the national policymaking arena. In India, environmental issues have either been adapted to and framed or subsumed by national development concerns such as poverty reduction, energy sector liberalization and energy security (Khator 1992; Cherian 1997). In framing global environmental issues, priority has always been given to the crucial role that poverty eradication means for India beginning with Indian Prime Minister Indira Gandhi's historic speech at the opening session of the UN Conference on Human Environment – Stockholm Conference,

Gandhi's subsequent move towards population control via sterilization was violation of fundamental human rights, but as Gupte (2017) outlined, it was the World Bank that gave the Indian government a loan of US $66 million dollars between 1972 and 1980 for sterilization: 'Gandhi was pressed by Western democracies to implement a crash sterilization program to control India's population. The Western countries' lobby backed the sterilization program after the Emergency was imposed, even when her own advisers were unwilling to support it. The international push was so extreme that in 1965, President Lyndon B. Johnson refused to provide food aid to India – at the time threatened by famine – until it agreed to incentivize sterilization' (2017, p. 4). India has in fact long argued that poverty eradication and equity considerations need to be factored into global environmental challenges including climate change. Twenty years after Prime Minister Gandhi's Stockholm speech, Agarwal cited Indian Environment Minister Kamal Nath stating at a press conference during the 1992 Earth Summit that India did not believe it was responsible for causing global warming and that it would not agree to any global treaty that restricted or burdened its 'right to development' (Agarwal and Narain 1991). Forty years after the historic Stockholm Conference, India's Minister for External Affairs (analogue to a Foreign

Minister) to the 2012 Rio+20 Summit stated that he: '. . . represented 1.2 billion people and that our challenge was to create a robust economy and eradicate poverty while at the same time, ensure that our poor who are already deeply vulnerable to climate change disasters, are not further impacted. Permit me to point out the difference between the environmentalism of the rich and the environmentalism of the poor. The rich countries grew, developed and polluted the world. Consequently, when the environment movements came, they had the money to clean up. Our nascent growth and economy start our growth trajectory with the problems of a polluted world' (MEA 2012).

Reflecting on his own 2009–2011 tenure, former Indian Minister of Environment, Jairam Ramesh chronicled his own transformation from an 'enviro-agnostic to an enviro-believer' in his book – *Green Signals: Ecology, Growth and Democracy in India* (2015). Writing about Ramesh's legacy, Padma points out: 'When development becomes paramount in a country like India, green concerns, however justified, are ignored. When Ramesh took over, India was one half of "ChinIndia"; growing at a 8-9% and courted as a future economic powerhouse. Sections of the Indian establishment took it so seriously that they seemed to forget that vast stretches of India are lagging behind, and are closer to sub-Saharan Africa than China in development indicators such as literacy, maternal and child mortality' (2011). Ramesh's insider's view to the climate change negotiations, including a detailed account of the interactions between President Obama walking into a meeting between President Lula da Silva of Brazil, President Jacob Zuma of South Africa, Prime Minister Wen Jiabao of China and Manmohan Singh (comprising the BASIC Group – Brazil, South Africa, India and China) provides a detailed summary of how last ditch discussions within this group of leaders resulted in the Copenhagen conference narrowly avoiding a no-agreement debacle. Ramesh (an alumnus of the Indian Institute of Technology, Carnegie Mellon and Massachusetts Institute of Technology) made it clear that he did not toe the usual development versus per capita emissions policy stance and noted: 'My message was simple: historically, India has certainly not been responsible for global warming but it has the maximum vulnerabilities to it across multiple dimensions perhaps like no other country in the world. Also, we cannot forever hide behind the "per capita" argument which will always be in India's favour because of the huge denominator that is growing significantly every year. I advocated a "per capita plus" approach to demonstrate that we are very much interested in a solution because such a solution is in our national interest.

I did not want India to abandon its traditional pre-occupation with issues of equity but made clear that the time had come for India to be less argumentative and more pragmatic, less defensive and more proactive, less obstructionist and more constructive, less polemical and more substantive in its discussions and negotiations'. Ramesh went on to note that his last-minute decision to add one crucial sentence: 'All nations must take on binding commitments in an appropriate legal form' at the 2010 Cancun Climate Conference led to 'bouquets and praise abroad but brickbats at home' (Ramesh 2018).

Given the enormous human costs that accrue at the intersection of energy related pollution and climate change in India, the point that India has maximum vulnerabilities to climate change across multiple dimensions, the 'per capita plus' approach appears that much more relevant now. But, the national policy reality is that there has been a lack of linked national action on clean air and climate mitigation in the face of worsening levels of PM pollution. While action to curb SLCPs like BC and reduce PM pollution does not factor into the PA, it is precisely this policy nexus between clean air, clean energy, poverty reduction and climate mitigation that matters to vulnerable households within cities in developing countries particularly those in Africa and Asia. This nexus is particularly relevant for India which happens to the world's third-largest aggregate GHG emitter but also the world's lowest per capita emitter amongst the top ten aggregate GHG emitters, where a significant majority depend on polluting solid fuels for basic energy needs and whose cities suffer the worst levels of PM pollution.

The success of the PA is predicated entirely on voluntary yet 'ambitious' GHG mitigation actions by individual countries. But, in India, one of the two most populous countries in the world with a long and active history on involvement on UN-led sustainable development and climate change negotiations curbing SLCPs and PM pollution matters more than ever before. India's 2015 submission to the PA – its intended NDC – does not include aggregate emissions reduction targets instead, India's NDC pledges to reduce 'emissions intensity' – its emissions per unit of economic output.

India's push towards renewable energy showcased by the National Solar Mission (NSM) was the identified strategy to reduce emissions intensity and provide clean, sustainable energy for millions. India's NSM was launched in 2010 with the aim of generating 20 GW of grid-connected solar power capacity by 2022. Lauding the ambition of the NSM, a 2015 Stanford University Energy Policy and Finance Center's study entitled 'Reach for the Sun: How India's Audacious Solar Ambitions Could Make or Break Its Climate Commitments' pointed out that India had highlighted its renewable energy efforts to its 'willingness *demonstrate* to lead on climate change' and 'at the center of this renewable energy push' is the NSM. The Stanford study on the NSM arrived at three conclusions:

- 'First, for solar power to succeed as a serious climate tool in India, it will also need to deliver an array of domestic co-benefits – from improving power reliability and access to electricity to cutting air pollution and energy imports-thereby building significant political support and investment to dramatically ramp up solar deployment.
- Second, three very different segments of the solar industry – utility-scale, distributed, and off-grid solar – will be required to deliver both climate results and domestic co-benefits to India.
- Third, the Indian national and state governments, with the support of countries and institutions around the world, can advance the development of these diverse segments of solar by pursuing four building blocks of a successful solar strategy: reform the utility sector; harmonize federal and state policies; secure substantial and cost-effective financing; and foster the diffusion of technology and standards from abroad. With these building blocks in place, India can greatly accelerate the deployment of solar power and with it progress toward the world's climate imperative' (Sivaram 2015, p. 2).

On 17 June 2015, the Government announced a long-lasting, transformative impact on the lives of millions of Indians by ramping up renewable energy targets by calling for the goal of 1,00,000 MW via the NSM by 2022. According to the Government's press release on the NSM: 'The target will principally comprise of 40 GW Rooftop and 60 GW through Large and Medium Scale Grid Connected Solar Power Projects. With this ambitious target, India will become one of the largest Green Energy producers in the world, surpassing several developed countries' (Press Information Bureau 2015). But, India's NSM also served to highlight the challenging inversions of climate and clean energy equity considerations, as it was the US administration under President Obama that took the case to the World Trade Organization (WTO).

It is worth recalling that the Obama administration which actively negotiated and ratified the PA, however, also pursued a WTO ruling against the NSM. Several US environmental groups and solar companies wrote an open letter to President Obama before his 2015 visit to India, calling for a 'Power India' initiative along the lines of 'Africa's Beyond the Grid' – $1 billion investment. The NSM was seen as an important building block of US–India collaboration ahead of the 2015 UN Climate Conference which necessitated India's involvement and was important to companies like SunEdison who had just announced a US $4 billion solar equipment factory in India. An influential US environmental group, the Sierra Club argued that: 'By dropping the solar trade case, solar panels produced domestically in India will improve energy access and move the world's third largest carbon emitter away from dirty fossil fuels' (Manglik 2015). But, on 24 February 2016, the WTO ruled against India's NSM, and in favour of the US, arguing that 'certain domestic content requirements (DCRs)' commonly known as the 'buy-local' component violated the General Agreement on Tariffs and Trade (GATT)/Trade-related Investment Measures (TRIMS) rules. The 140-page ruling entitled 'India-Certain Measures Relating to Solar Cells and Solar Modules' found that '... the terms "products in general or local short supply" do not cover products at risk of becoming in short supply, and found that in any event India had not demonstrated the existence of any imminent risk of a short supply'. The Panel disagreed with India's argument that the DCR measures were justified 'on the grounds that they secure India's compliance with "laws or regulations" requiring it to take steps to promote sustainable development' (WTO 2016).

On 20 April 2016, India notified the WTO of its decision to appeal certain issues of law and legal interpretation in the panel report, and as in the case of numerous WTO cases, the matter has been disputed back and forth, with the US stating that India had failed to comply and India objecting and disagreeing with the US. As per the WTO timeline, as of 9 February 2018, the WTO's Dispute Settlement Board had deferred on the establishment of a compliance panel (WTO 2018). This WTO ruling cannot just be viewed narrowly about arbitrating the provenance of solar cells/modules. This ruling also revealed a half-truth about the WTO dispute settlement process in adjudicating on agreed, global SDGs, namely, the SDGs that were universally adopted by member states of another universal global organization – the UN. Arguably, the NSM can be viewed as principal national mechanism for addressing SDGs 7 (sustainable energy for all) and 13 (climate change), so the question is whether this WTO ruling stymies national actions that are directly related to the achievement of UN's globally agreed SDA. In the world according to GATT versus global environment, this ruling joins a long list that skirt around the thorny challenge of dealing with agreed global environmental goals and policies that are not designed to comport with, or be in conformity with trade issues. There is, however, an astonishing half-truth for this fault line between global environmental goals and trade rules according to the WTO itself: 'The WTO is only competent to deal with trade. In other words, in environmental issues its only task is to study questions that arise when environmental policies have a significant impact on trade. The WTO is not an environmental agency. Its members do not want it to intervene in national or international environmental policies or to set environmental standards. Other agencies that specialize in environmental issues are better qualified to undertake those tasks'. The WTO's admission about its non-competency with global environmental policies, and its claim about non-interventionism and non-standard setting role in national/international environmental policies belie the idea that globally agreed SDGs that were universally adopted by UN member states should not be accorded any primacy. Disregard for the existence of SDGs 7 and 13 that were universally adopted by all UN member states in 2015 is hard to explain in the context of the Panel diminishing the central role of the NSM in meeting India's national sustainable energy and climate change goals.

What is relevant to point out here is that most countries have different forms of energy subsidies, but only some are deemed unfair by WTO/GATT rules. A 2015 study of 50 US States by Timothy Meyer entitled 'How Local Discrimination Can Promote Global Public Goods' found that 44 state renewable energy programs in 23 states violate the WTO's ruling that national/local subsidies for renewable energy discriminate against foreign products. According to Meyer: 'Since 2001, California has given US $ 2 billion in such subsidies, while states ranging from Minnesota to Kansas and Mississippi have doled out hundreds of millions of dollars each. Cities, such as Austin and Los Angeles, have also gotten into the act, contributing millions to renewable energy firms. To build support for these measures, the local government might condition the subsidy on the recipient's use of components manufactured in the locality' (2015). Interestingly, a few years later, in 2019, India won a major renewable energy – solar case – trade dispute against the United States at the WTO with a dispute settlement panel pronouncing that renewable energy subsidies and mandatory local content requirements – DCRs – instituted by eight American states breached global trade rules. In its 100-page report, the three-member panel largely upheld India's claims that subsidies and DCRs in 11 renewable energy programmes in eight US states violated core global trade rules. The panel also asked the US to ensure that these states are in conformity with trade rules, but its impacts of the ruling are still being disputed (Miles 2019).

The bottom line is that solar energy strategies and plans are essential to climate change mitigation and increasing access to clean energy in the second most populated nation of the world. The India Energy Outlook 2015, a 718-page publication prepared by the IEA provided the first globally comprehensive review of India's energy data and trends. According to the IEA, 'India has made great strides in improving access to modern energy in recent years. Since 2000, India has more than halved the number of people without access to electricity and doubled rural electrification rates. Nonetheless, around 240 million people, or 20% of the population, remain without access to electricity' (2015 p. 436). But, the IEA report also found that: '*India also has the largest population in the world relying on the traditional use of solid biomass for cooking: an estimated 840 million people – more than the populations of the United States and the European Union combined.* There is a host of issues associated with the traditional use of solid biomass for cooking, including the release of *harmful indoor air pollutants that are a major cause of premature death,* as well as environmental degradation as a result of deforestation and biodiversity loss. The government has made a major effort to address these issues, primarily through the subsidised availability of LPG as an alternative cooking fuel' (emphasis added, 2015, p. 437). Six years later, IEA released the 2021 India Energy Outlook, and some of its key findings related to fossil fuel, renewable energy and in particular air pollution, which is clearly referenced up front, are listed below. Most notably, the IEA 2021 report on Indian energy found that the energy demand of the world's third-largest aggregate GHG emitter will expand more than any other nation over the next two decades, edging out the EU to make it the third-largest consumer, and also that solar power which currently makes up just 4% of the nation's power supply is set to grow 18-fold.

The size of the linked energy access and pollution mitigation challenges, particularly the scale and scope of reducing health inequities associated with polluting solid fuels, has revealed essential fault lines about the existence of global policy 'silos' on climate and clean energy action not just via the WTO ruling but also the way in which the UN SDGs are anticipated to be implemented as segregated goals. This implementation logic is the analogue of an implementation failure as it not only contravenes the selective invocation of renewable energy subsidies used by many countries but also the global importance accorded to achieving globally agreed SDGs 7 and 13 within a so-called integrated SDA with implications for countries like India. Increasing access to sustainable energy for the poor and

Box 4.3 Key findings excerpted from India Energy Outlook.

- 'India's continued industrialization and urbanization will make huge demands of its energy sector and its policy makers. Energy use on a per capita basis is well under half the global average, and there are widespread differences in energy use and the quality of service across states and between rural and urban areas. The affordability and reliability of energy supply are key concerns for India's consumers.
- Over 80% of India's energy needs are met by three fuels: coal, oil and solid biomass. Coal has underpinned the expansion of electricity generation and industry and remains the largest single fuel in the energy mix. Oil consumption and imports have grown rapidly on account of rising vehicle ownership and road transport use. Biomass, primarily fuelwood, makes up a declining share of the energy mix but it is still widely used as a cooking fuel. Despite recent success in expanding coverage of LPG in rural areas, 660 million Indians have not fully switched to modern, clean cooking fuels or technologies.
- Natural gas and modern renewable sources of energy have started to gain ground, and were least affected by the effects of the Covid-19 pandemic in 2020. The rise of solar PV in particular has been spectacular; the resource potential is huge, ambitions are high, and policy support and technology cost reductions have quickly made it the cheapest option for new power generation.
- India is the third-largest global emitter of CO2, despite low per capita CO2 emissions. The carbon intensity of its power sector in particular is well above the global average. Additionally, particulate matter emissions are a major factor in air pollution, which has emerged as one of India's most sensitive social and environmental issues: in 2019, there were well over 1 million premature deaths related to ambient and household air pollution'.

Source: IEA (2021, pp. 17–18).

mitigating climate change are essential elements of the UN's 2030 SDA which share the common connector of renewable energy. The role of solar initiatives such as the NSM for example needs to be viewed from the perspective of millions whose heavy reliance on polluting and highly toxic solid fuel energy sources results in grave health and climatic impacts and in terms of linkages across SDGs 7, 13 and 17 (GPSD).

In his 23 September 2019 speech to the UN Climate Action Summit (where the UN SG prefaced to world leaders that they not come onstage if they had only beautiful speeches but no demonstrable action), Indian Prime Minister Modi highlighted the need for a 'comprehensive' approach to climate change and development that looked education, values, lifestyle choices as well the need to incorporate behavioural change. Stating that 'need not greed has been our guiding principle', he noted that India had come to the UN Summit with a 'road map' which was based on the belief 'that an ounce of practice is worth more than a ton of preaching'. In this regard, he reiterated the national goal of 175 GW renewable energy by 2022, stating that India would move towards a goal of 450 GW of renewable energy in the future. Giving emphasis to the role of e-mobility, he also referenced that 150 million families lives had been improved via the Clean Cooking Gas connection and announced Mission Jal Jeevan (translated as life water) with a $50 billion in spending projected as well as India's leadership in the ISA.He closed by pointing out that India would unveil its donation of 193 solar panels (representing each one of the UN member states) installed on the UN building on 24 September 2019 as a dedication to Mahatma Gandhi and noted, 'The time for talking is over, the world needs to act now' (Times of India 2019). But, India's inaction in relation to the public health fiasco associated with the second calamitous and debilitating wave of COVID-19 would be starkly highlighted a year later.

Arundhati Roy in her article, 'We are witnessing a crime against humanity', described the devastation and chaos that COVID-19 unleashed in the nation's capital, Delhi: 'Hospital beds are unavailable. Doctors and medical staff are at breaking point. Friends call with stories about wards with no staff and more dead patients than live ones. People are dying in hospital corridors, on roads and in their homes. Crematoriums in Delhi have run out of firewood. The forest department has had to give special permission for the felling of city trees. Desperate people are using whatever kindling they can find. Parks and car parks are being turned into cremation grounds. It's as if there's an invisible UFO parked in our skies, sucking the air out of our lungs. An air raid of a kind we've never known. Oxygen is the new currency on India's morbid new stock exchange' (Roy 2021). And then to make matters far worse, a cyclone hit in the midst of this all bringing coastal zone inundation and flooding to several states and cities in India.

The urgency of the climate change and energy related pollution crises presents a very real and present danger for India and Indian cities balancing poverty eradication, economic growth, and access to clean air, clean energy and public health services. Pointing to the GBD studies which estimated 695,000 premature deaths in 2010 due to continued exposure to PM and ozone pollution for India, Guttikunda et al. argued that the expected growth in many of the sectors (industries, residential, transportation, power generation and construction) will result in an increase in pollution related health impacts for most Indian cities. Guttikunda et al. focused on policy interventions such as urban public transportation; emission regulations for power plants; clean technology for brick kilns; measure to curb road dust and open waste burning are all relevant (2014). In 2011, the Central Pollution

Control Board noted that the major sources of ambient PM pollution in India are coal burning for thermal power production, industry emissions, construction activity and brick kilns, transport vehicles, road dust, residential and commercial biomass burning, waste burning, agricultural stubble burning and diesel generators.

The 'Burden of Disease Attributable to Major Air Pollution Sources in India: Special Report' provided the first comprehensive analysis of the levels of fine PM air pollution ($PM_{2.5}$) in India by source at the state level and their impact on health, and was the result of an international collaboration of the Health Effects Institute (Boston, MA), Indian Institute of Technology (Bombay), HEI and the Institute for Health Metrics and Evaluation (Seattle, WA). The study found that air pollution exposure contributed to some 1.1 million deaths in India in 2015. Household burning emissions and coal combustion were found to be the single largest sources of air pollution related health impacts, with emissions from agricultural burning, anthropogenic dusts, transport, diesel and brick kilns also contributing significantly (HEI 2019). It included the following findings:

- 'Exposure to outdoor PM2.5 was the third leading risk factor contributing to mortality among the behavioral, environmental, and metabolic factors analyzed;
- PM 2.5 exposure was responsible for more than 1 million deaths in 2015, which represent nearly a quarter of the 4.2 million deaths attributable to outdoor air pollution worldwide. It also accounted for 29.6 million years of healthy life lost (i.e., DALYs).
- The number of deaths attributable to air pollution has been growing steadily in India over the past 25 years. This trend is in part attributable to increases in ambient PM2.5 levels, but also to a growing and aging population with increasing numbers of people with ailments that are affected by exposure to air pollution, such as cardiovascular disease. When this loss of life is translated into economic terms, the costs are considerable. For India alone, the estimate for lost labor output was US$55 billion and for welfare losses US$505 billion'.

But what is striking is that the Report issued the caveat that these 'estimates of burden do not include the additional effects that air pollution has on society via its impacts on climate and on the environment' (HEI 2019, p. 5).

A groundbreaking survey paper published in the Lancet Planetary Health entitled 'The impact of air pollution on deaths, disease burden, and life expectancy across the states of India: the Global Burden of Disease Study 2017' pointed out that rapidly developing countries such as India face the dual challenge of exposures from both ambient and household air pollution. This 2019 paper presented troubling evidence regarding the scope and scale of air pollution and the estimation of deaths and DALYs attributable to air pollution:

- *'India has one of the highest annual average ambient particulate matter $PM_{2.5}$exposure levels in the world.*
- *In 2017, no state in India had an annual population-weighted ambient particulate matter mean $PM_{2.5}$ less than the WHO recommended level of 10 μg/m^3, and 77% of India's population was exposed to mean $PM_{2.5}$ more than 40 μg/m^3, which is the recommended limit set by the National Ambient Air Quality Standards of India.* Although the use of solid fuels for cooking has been declining in India, 56% of India's population was still exposed to household air pollution from solid fuels in 2017. Behind these high overall air pollution

exposure levels in India, there is a marked variation between the states, with a 12 times difference for ambient particulate matter pollution and 43 times difference for household air pollution.

- *India had 18% of the global population in 2017, but had 26% of global DALYs attributable to air pollution. A substantial 8% of the total disease burden in India and 11% of premature deaths in people younger than 70 years could be attributed to air pollution.*
- *An estimated 1·24 million deaths in India in 2017 could be attributed to air pollution,* including 0·67 million to ambient particulate matter pollution and 0·48 million to household air pollution' (emphasis added, Balakrishnan et al. 2019, pp. 33 and 34).

In 2018, 14 of the 15 most polluted cities of the world were in India (Times of India 2018). A Bloomberg article entitled 'World's Dirtiest Air Is Now in India' referenced a joint data study by Greenpeace and IQAirVisual that found that seven of top 10 cities with worst air quality in 2018 are in India. According to this article: 'India, the world's fastest-growing major economy, makes up 22 of the top 30 most polluted cities, with five in China, two in Pakistan and one in Bangladesh. India racks up health-care costs and productivity losses from pollution of as much as 8.5 percent of gross domestic product, according to the World Bank' (Jamrisko 2019). A 16 March 2021 article in Forbes citing data contained in the 2020 World Air Quality Report by IQAir highlighted the fact that in 2020, 22 out of the top 30 most polluted cities in the world were from India (Shetty 2021).

Given the scope of the PM pollution crisis unfolding in Indian cities, there has been an increasing focus on the urgency of curbing air pollution but scaled-up action is now imperative from a public health and humanitarian perspective. According to A. Upadhyay, promises to fight the world's most toxic air have made it to the manifestos of major political parties for the first time in Indian elections. Major political parties such as the ruling Bharatiya Janata Party, the opposition Indian National Congress and the Aam Aadmi Party have pledged to combat the crisis by taking measures ranging from setting deadlines, introducing new emission standards to promoting electric vehicles in a bid to fight toxic air. That is a change from the 2014 elections when none of the party manifestos had any mention of clean air or pollution (Upadhyay 2019). But clearly, the enormous scale of the air pollution problem when combined with existing poverty and climate inequities translates into the need for much more urgent partnership action at city, national and even regional level.

Air pollution controls have been attempted in countries like India but need to be provided with enforcement and compliance capacities. The Air (Prevention and Control of Pollution) Act, 1981 was aimed to enable the preservation of the quality of air and control of air pollution in India. It was enacted to fulfil India's commitments at the 1972 UN environment conference and serves as an example of how global environmental outcomes can influence national agendas. The law accorded sweeping powers to state and central governments to take action to improve air quality, enforce pollution control measures, shut down errant industries and send polluters to jail. Chapter 2 of the Act set up the Central and State Boards for the Prevention and Control of Air Pollution, while Chapter 4 accorded powers to Boards including: 'Power to declare air pollution control areas; Power to give instructions for ensuring standards for emission from automobiles; Restrictions on use of certain industrial plants; Persons carrying on industry, etc., not to allow emission of air pollutants in excess of the standard laid down by State Board; Power of Board to make application to court for restraining persons from causing air pollution' (Government of India, Air Act 1981). Over the years, however, the law has seen a decline in its relevance, in spite of the growing levels of air pollution with Indian cities being amongst the most polluted in the world. Gokhale (2020) noted that India's clean air law has been largely ineffective: 'Almost zero cases have been filed under the Air Act from northern Indian states in recent years, even though they face the worst pollution every winter. Even the Supreme Court and governments have ignored the law; measures like the Graded Response Action Plan or the Odd-Even scheme rely on other laws or regulations, some of which have nothing to do with the environment. The law is commonly described as "toothless". As pollution spikes in the winter season, parliamentarians, lawyers or activists demand amendments to it or to replace it with a new law, usually to add powers of enforcement'. To improve long-term air quality, the Indian Government launched the National Clean Air Programme (NCAP) in early 2019. The NCAP requires more than 120 Indian cities to create city-level Clean Air Plans (CAPs) to plan and implement mitigation measures aimed to reduce ambient PM concentrations (Government of India, NCAP 2019). But, if the COVID-pandemic crisis response is any measure, India will require considerable political will and community stakeholder engagement within cities to combat current malaise at the national leadership level.

Major sources of India's air pollution include transportation, biomass burning for cooking, electricity generation, industry, construction, waste burning and episodic agricultural burning. Transportation constitutes one of India's leading $PM_{2.5}$ emission sources, responsible for emitting pollutants and road dust. Biomass cookstoves are the main source of indoor pollution nationally, particularly affecting women and children. While India promotes access to fuels which emit less particulate pollution like liquefied gas and increases the share of clean energy extending electricity access across the country, coal currently remains the major domestic source of India's energy supply. India is not alone in continuing to focus on coal mining projects that are largely publicly financed.

The Global Energy Monitor (GEM) develops and shares information on energy projects in support of the worldwide movement for clean energy and includes the Global Coal Mine Tracker, Global Coal Plant Tracker, Global Fossil Infrastructure Tracker, Global Steel Plant Tracker, Europe Gas Tracker, CoalWire newsletter, Global Gas Plant Tracker, Global Registry of Fossil Fuels, Latin America Energy Portal and GEM.wiki. The Global Coal Mine Tracker, which provides information on the world's major coal mines, includes information on every operating mine producing 3 mtpa or greater and every proposed mine with a capacity of 1 mtpa or greater. Some of the key findings of a June 2021 report published by GEM entitled 'Deep Trouble: Tracking Global Coal Mine Proposals' include:

- '*Coal rush from hundreds of proposed mines.* There are 432 new mine developments and expansion projects currently announced or under development worldwide, amounting to 2,277 mtpa of new capacity. Of this, 614 mtpa is under construction and 1,663 mtpa is in planning.
- *New developments breach a 1.5°C pathway.* Development of these new mines runs contrary to the IEA's new roadmap for net-zero emissions, which requires no new coal mines or mine extensions beyond 2021, and the findings of the UN and leading research organizations that coal production must decline 11% each year through 2030 to remain consistent with a pathway to 1.5°C. If all proposed coal mine capacity currently under

development is realized, coal production in 2030 will be over four times the 1.5°C-compliant pathway.

- *China, Australia, India, and Russia make up over three-fourths of new mine developments.* New capacity development is strongest in China, Australia, India, and Russia. Together, these countries represent 77% (1,750 mtpa) of global coal mine development: China has 452 mtpa of coal mine capacity under construction and another 157 mtpa in planning; Australia has 31 mtpa under construction and 435 mtpa in planning; India has 13 mtpa under construction and 363 mtpa in planning; and Russia has 59 mtpa under construction and 240 mtpa in planning.
- *Greenhouse gas emissions from new mines comparable to US emissions.* The emissions from coal mine projects now on the drawing board would total between 5,000 and 5,800 Mt of CO2 equivalent (CO2e) each year from combustion and methane leakage (for CO2e100 and CO2e20, respectively), comparable to the current annual CO2 emissions of the United States (5,100 Mt).
- *Stranded asset risk.* Coal mines and related infrastructure such as ports and railways are capital-intensive projects that cost tens of millions of dollars per mtpa mined to open. Yet the prospect of a low-carbon transition and tighter emission policies put these projects at risk of shutting down early, representing up to $91 billion USD in stranded assets from coal mines alone' (emphasis included, Tate, Shearer and Matikinca 2021, pp. 3–4).

In terms of the sheer magnitude of the air pollution crisis confronting countries like India, it is necessary to underscore the significance of groundbreaking research on air quality being conducted by the Energy Policy Institute at the University of Chicago (EPIC). The Air Quality Life Index (AQLI) is based on cutting-edge research by EPIC which has quantified the causal relationship between human exposure to air pollution and reduced life expectancy. Specifically, the AQLI is novel for a variety of reasons: (i) AQLI is grounded in pollution data at the very high concentrations that exist in Asia currently, whereas past comparable research has relied on extrapolations of associational evidence from much lower levels in the US or on extrapolations from cigarette studies, (ii) the causal nature of the AQLI's underlying research allows it to isolate the effect of air pollution from other factors that impact health; (iii) AQLI delivers estimates of the loss of life expectancy for the average person, whereas prior comparable research on the number of people who die prematurely due to air pollution, leaves unanswered how much their life was cut short or if they were more predisposed to be impacted from it (e.g. elderly or sick) and (iv) the AQLI uses highly localized satellite data, making it is possible to report life expectancy impacts at the county or similar level around the world, rather than much more aggregated levels reported in previous studies (EPIC website 2021). EPIC produced two landmark AQLI fact sheets – one focused on China and the other on India, and it is here that some of the sobering key findings of EPIC's 2020 AQLI: India Fact Sheet need to be highlighted:

- '*All of India's 1.4 billion people live in areas where the annual average particulate pollution level exceeds the WHO guideline.*
- Ninety four percent live in areas where it exceeds India's own air quality standard.
- Particulate pollution has sharply increased over time. Since 1998, average annual particulate pollution has increased 22 percent, cutting 1.3 years off the life of the average resident over those years.

- *A quarter of India's population is exposed to pollution levels not seen in any other country, with 248 million residents of northern India on track to lose more than 8 years of life expectancy if pollution levels persist.*
- *Delhi, the national capital, has the highest level of pollution in the country, with pollution 14 times greater than the WHO guideline.* Residents of Delhi stand to lose 13 years of life expectancy if pollution persists' (emphasis added, EPIC 2020, p. 1).

4.4 Looking towards a Cleaner/Greener Future

On 16 July 2019, the UN, the WHO, the UNEP and the CCAC issued an urgent call to all cities, regions and countries 'to commit to achieving air quality that is safe for its citizens, and to align its climate change and air pollution policies by 2030'. The impetus for this call was that 'the *health burden of polluting energy sources* is *now so high* that *moving to cleaner and more sustainable energy* choices for energy supply, transport and food systems *effectively pays for itself*' (emphasis added, CCAC 2019). This commitment was seen to be 'in line with the Paris climate agreement'. What is particularly noteworthy is that global call for aligning climate change and air pollution was being issued 'because climate change and air pollution are closely linked: the main driver of climate change, fossil fuel combustion, also contributes about two-thirds of outdoor air pollution – and the contribution of air pollution alone to poor health is staggering. Each year, air pollution causes 7 million premature deaths (or about 1 in every 8 deaths), costs the global economy an estimated US$ 5.11 trillion (equivalent to all goods and services produced by the entire economy of Japan produced in 2013) in welfare losses and kills 600,000 children every year. Its death toll is similar to that from tobacco smoking, making it one of the largest avoidable risks to human health. In the 15 countries that emit the most greenhouse gas emissions, the health impacts of air pollution are estimated to cost more than 4 per cent of their GDP' (CCAC 2019).

The State of Global Air 2020 Report paints the dismal reality as to grave health burdens associated with air pollution: 'Air pollution was the fourth leading risk factor for early death worldwide in 2019, surpassed only by high blood pressure, tobacco use and poor diet. According to the Report: 'Despite all that is known about the effects of air pollution on health, the findings in 2019 show that little or no progress has been made in many parts of the world. Major disparities continue to exist; air quality has improved in many high-income countries over the past several decades, while dangerous levels of air pollution persist in low- and middle-income countries. And evidence is mounting that air pollution can cause harm at much lower levels than previously thought, suggesting that even high-income countries need to continue their efforts to reduce exposure and to ensure that the progress achieved is not lost over time. Systematic and consistent efforts to track progress toward reducing air pollution and the impacts it has on human health remain essential' (HEI 2020, pp. 3–4). Curbing SLCPs and air pollution therefore should not be compartmentalized into separate global policy silos. It is time to look beyond the protracted intergovernmental negotiations arena towards new forms of city and NNSA partnership actions that can result in a climate resilient, clean energy, low and zero carbon communities that can save millions of lives and reduce disease burdens.

The role of cities as loci of integrated action on climate change and sustainability measures has gained significant traction via city-driven partnerships such as the C-40 Initiative. The following chapter focuses on the Convention on Long Range Transboundary Air Pollution (CLRTAP) and its Gothenburg Protocol and examines lessons learned that are relevant to the needs of cities in developing countries including the enhanced role of initiatives such as the C-40. What needs to be acknowledged is that intergovernmental outcomes related to sustainable energy and climate change – the SDGs 7 and 13 and the PA – have so far failed to act comprehensively and conclusively on an integrated agenda which has meant that curbing the world's largest environmental health risk – air pollution has received fragmented attention. For countries like India and for cities that are the most polluted in the world, there is even more of a need now to broaden the dialogue for partnership action on clean air, clean energy access and climate mitigation that can actively include NNSAs and municipal level actions.

Stakeholder groups including local and regional governments (LRGs), municipal and civil society stakeholders including the private sector are key partners in the overall implementation of any global sustainability agenda at the local level, but now play an even more crucial role in the delivery of integrated responses to the climate and air pollution crises. The global policy reality is that the Local Governments and Municipal Authorities (LGMA) stakeholder group has represented local and regional governments at the processes under the UNFCCC since the first COP in 1995, but the emphasis on textually driven UN member state climate negotiations has relegated LGMA group largely as additional to, but not central in climate change, clean air and clean energy outcomes within the PA negotiations.

There is an existing history of local governmental/municipal engagement on climate and sustainability which began with the First World Assembly of Cities and Local Authorities, which convened the UN Habitat II Conference in Istanbul in 1996. One of the key outcomes of the 1996 World Assembly of Cities and Local Authorities was the creation of the UN Advisory Committee on Local Authorities. Currently, ICLEI, also known as the Local Governments for Sustainability, comprises a global network of over 2500 local governments and municipal leaders that act as the LGMA focal point within UN processes and works with the Global Task Force of Local and Regional Governments (a joint policy advocacy initiative of local governmental networks) in the area of climate (Cities and Regions.org website 2021). The coalescing of LGMA stakeholder groups under the aegis of the World Assembly of Local and Regional Governments (WALRG) resulted in the adoption of an 'Outcome Document' that provides a germane framework for action by municipal leaders on the nexus of clean energy–clean air–climate and includes the following priorities:

- 'All urban development must be based on a low emissions pathway with the aim to achieve climate neutrality in the LRG's own infrastructure and operations before 2050.
- To make resilience a core part of the LRG's planning strategies and prepare for new risks and impacts, taking into account the rights and needs of vulnerable sections of the society.
- To prioritise nature-based solutions and the mainstreaming of nature in cities and regions, recognising the value of nature as fundamental to our collective economic and social wellbeing, within the context of the Post-2020 Global Biodiversity Framework and Earth's planetary boundaries.
- To pursue secure and safe access to food water, energy, sanitation for all, culture and education as well as clean air and soil.
- To enable the creation and sustainability of people centred, safe and culturally vibrant communities' (Global Task Force on Local and Regional Governments 2019).

In the context of the COVID-19 pandemic, reducing pollution and SLCPs emissions assumes even greater importance. But, for developing countries, curbing toxic PM pollution poses numerous challenges because it is directly associated with a host of other pressing development concerns – increased urbanization, economic development, poverty reduction, public health, transportation and infrastructure development – and so it requires complex socio-economic and tough disciplined political choices to be negotiated and made. Measures such as sustainable mass transit options, diesel particulate filters, stringent traffic regulations governing trucks and passenger vehicles; monitoring protocols to reduce discharge and dust at construction sites and banning of open field burning of waste are some important ways by which PM air pollution can be curbed and are discussed in chapters that follow. But there is another green/nature-based option – urban tree corridors and green spaces – which are necessary and remedial but also very difficult to implement in the midst of congested cities in India and elsewhere where sprawling and dense slums sit cheek by jowl with modern elite urban enclaves. Urban tree corridors and green spaces can be a way to remove PM pollution. A 2013 study conducted by the US Forest Service and the Davey Institute estimated the amounts of PM removed by urban trees and forests in 10 cities (Nowak et al. 2013). The study found that trees in urban cities can remove PM from the atmosphere and consequently improve air quality and human health, and that the greatest effect of trees on reducing health impacts of PM occurred in New York City due to its relatively larger population size, and the moderately high removal rate and reduction of PM concentrations based on tree coverage. But, finding the appropriate mix of responses that can tackle the PM pollution in the midst of urbanization, poverty reduction and development considerations requires citizenry advocacy, political commitment, as well as careful policy planning, implementation and partnering investments. To be clear, urban green spaces are not easy or quick remedies, especially in the context of the most polluted, densely populated cities of the world. Urban tree corridors along congested highways and green areas in the midst of dense urban metropolises are difficult to maintain in the face of pressing poverty reduction needs and development pressures. Then there are additional concerns about ensuring that these green spaces are carefully considered and use native and appropriate species that do not compete for already scarce water resources, do not displace critical human development infrastructure/needs and do not result in endemic ecosystem damage and loss. Making concerted policy choices to collectively invest in urban green space can improve human well-being. Clearing bilious, polluted skies will not be easy, but maintaining 'green' on the ground via stakeholder city-based partnerships can be promising investments in improving human health.

References

Agarwal, A. and Narain, S. (1991). *Global Warming in an Unequal World*. New Delhi: Centre for Science and Environment.

Anderson, W.P., Kanaroglou, P.S., and Miller, E.J. (1996). Urban form, energy and the environment: a review of issues, evidence and policy. *Urban Stud* 33 (1): 7–35.

Balakrishnan, K. et al. (2019). The impact of air pollution on deaths, disease burden and life expectancy across the states of India. *The Global Burden of Disease Study* 2017 3 (21):

E-26–E39. https://www.thelancet.com/journals/lanplh/article/PIIS2542-5196(18)30261-4/fulltext (accessed 16 May 2021).

Barker, M.H. et al. (2015). Urbanization and climate change. In: *Handbook of Climate Change Adaptation* (ed. W. Leal Filho). Berlin, Heidelberg: Springer.

Bassett, M.T. et al. (2020). The unequal toll of COVID-19 mortality by age in the United States: quantifying racial/ethnic disparities. *Harvard Center for Population and Development Studies*. Working Paper 3 19. https://cdn1.sph.harvard.edu/wp-content/uploads/sites/1266/2020/06/20_Bassett-Chen-Krieger_COVID-19_plus_age_working-paper_0612_Vol-19_No-3_with-cover.pdf (accessed 18 December 2021).

Bloomberg NEF (2021). Energy Transition Investment Trends. https://assets.bbhub.io/professional/sites/24/Energy-Transition-Investment-Trends_Free-Summary_Jan2021.pdf (accessed 4 April 2021).

Caniglia, B. et al. (2017). *Resilience, Environmental Justice and The City*. New York: Routledge.

CCAC (2019). UN urges governments to act on climate and air pollution doe health's sake. https://ccacoalition.org/en/news/un-urges-governments-act-climate-and-air-pollution-health%E2%80%99s-sake-0 (accessed 2 May 2021).

Cherian, A. (1997). Energy policies, liberalization and the framing of climate change policies in India. Doctoral Dissertation. https://scholarworks.umass.edu/cgi/viewcontent.cgi?article=2968&context=dissertations (accessed 1 May 2021).

Cities and Regions.org website (2021), About the Local Governments and Municipal Authorities Constituecy. https://www.global-taskforce.org (accessed 12 April 2021).

Cole, M.A. and Neumayer, E. (2004). Examining the impact of demographic factors on air pollution. *Population and Environment* 26 (1): 5–21.

Cong, L. et al. (2019). Ambient particulate air pollution and daily mortality in 652 cities. *New England Journal of Medicine* 381 (8): https://www.nejm.org/doi/10.1056/NEJMoa1817364 (accessed 30 April 2021).

Ciu, Y and Zhang, Z et al. (2003) Air pollution and case fatality of SARS in the People's Republic of China: an ecologic study. *Environmental Health*. 2(1):15. doi: https://doi.org/10.1186/1476-069X-2-15.

Disavino, S. (2021) New York City bans natural gas in new buildings. *Reuters* (15 December 2021). https://www.reuters.com/markets/us/new-york-city-set-ban-natural-gas-new-buildings-2021-12-15/ (accessed 19 December 2021)

Doiron, D. et al. (2019). Air pollution, lung function, and COPD: results of the population-based UK Biobank study. *European Respiratory Journal*. January 2019 1802140. https://erj.ersjournals.com/content/early/2019/05/01/13993003.02140-2018 (accessed 30 April 2021).

Earth Observatory NASA (2020). Airborne Nitrogen dioxide plummets over China. https://earthobservatory.nasa.gov/images/146362/airborne-nitrogen-dioxide-plummets-over-china (accessed 12 May 2021).

EPHA (2020). *Coronavirus Threat Greater for Polluted Cities*. EPHA https://epha.org/coronavirus-threat-greater-for-polluted-cities (accessed 7 May 2021).

EPIC website (2021). About the Air Quality Life Index (AQLI). https://dev-aqli-epic.pantheonsite.io/about (accessed 31 August 2021).

Farr, D. (2018). *Sustainable Nation: Urban Design Patterns for the Future*. Hoboken: Wiley Publications.

Friedman, T. (2008). *Hot, Flat and Crowded: Why We Need a Green Revolution and How It Can Renew America*. New York: Farrar, Straus and Giroux.

Gay, T. et al. (2020). Examining the relationship between institutionalized racism and COVID-19. *Sage Journals.* 19 (3): https://journals.sagepub.com/doi/full/10.1111/cico.12520 (accessed 19 December 2021).

Gehring, U. et al. (2012). Traffic-related air pollution and lung function in children: the ESCAPE project. *European Respiratory Journal.* January 2012 40: P4778. https://erj.ersjournals.com/content/40/Suppl_56/P4778?utm_source=TrendMD&utm_medium=cpc&utm_campaign=_European_Respiratory_Journal_TrendMD_0 (accessed 29 April 2021).

Global Task Force on Local and Regional Governments (2019). Outcome Document of the World Assembly on Local and Regional Governments: Durban. https://www.global-taskforce.org/sites/default/files/2019-11/WALRG%20Outcome%20Document%20Durban%2015%20November%202019_0.pdf (accessed 30 April 2021).

Gokhale, N. (2020). India's 40 year law to combat air pollution languishes as the crisis intensifies. *Mongabay* (10 November 2020). https://india.mongabay.com/2020/11/indias-40-year-old-law-to-combat-air-pollution-languishes-as-the-crisis-intensifies (accessed 12 May 2021).

Government of India (1981). *The Air (Prevention and Control of Pollution) Act, 1981.* https://legislative.gov.in/sites/default/files/A1981-14.pdf (accessed 15 May 2021).

Government of India (2019). National Clean Air Programme (NCAP) Ministry of Environment, Forest & Climate Change. http://moef.gov.in/wp-content/uploads/2019/05/NCAP_Report.pdf (accessed 15 May 2021).

Gravlee, C. (2020). Systemic racism, chronic health inequities, and COVID-19: a syndemic in the making? *American Journal of Human Biology. 32* (5): e23482. https://www.ncbi.nlm.nih.gov/pmc/articles/PMC7441277/#ajhb23482-bib-0063 (accessed 18 December 2021).

Greenstone, M. (2020). AQLI: India Fact Sheet. https://dev-aqli-epic.pantheonsite.io/wp-content/uploads/2020/07/IndiaFactSheet2020-.pdf (accessed 30 August 2021).

Gupte, P. (2017). India: 'The Emergency' and the politics of mass sterilization. *Education about Asia* 22 (3): 40–44. https://www.asianstudies.org/wp-content/uploads/india-the-emergency-and-the-politics-of-mass-sterilization.pdf (accessed 3 May 2021).

Guttikunda, S et al. (2014) Nature of air pollution, emission sources, and management in the Indian cities. *Atmospheric Environment.* Vol 95. pp. 501–510

Hansell, A. et al. (2016). Historic air pollution exposure and long-term mortality risks in England and Wales: prospective longitudinal cohort study. *Thorax* 71 (4): 330–338. https://thorax.bmj.com/content/71/4/330 (accessed 8 May 2021).

Harris, G. (2015). Delhi wakes up to an air pollution problem it cannot ignore. *New York Times* (14 February 2015).

Health Effects Institute (2019). *State of Global Air: 2019 Special Report.* Boston: Health Effects Institute https://www.stateofglobalair.org/sites/default/files/soga_2019_report.pdf (accessed 14 May 2021).

Health Effects Institute (2020). *State of Global Air 2020. Special Report.* Boston, MA: Health Effects Institute. https://www.stateofglobalair.org/downloads (accessed 18 December 2021).

Heck, T. and Hirschberg, S. (2011). China: economic impacts of air pollution. In: *Encyclopedia of Environmental Health* (ed. J.O. Nriagu), 625–640. Burlington: Elsevier.

Ho, M. and Nielsen, C. (ed.) (2007). *Clearing the Air: The Health and Economic Damages of Air Pollution in China.* Cambridge, MA: MIT Press.

Ho, M. and Nielsen, C. (ed.) (2013). *Clearer Skies Over China: Reconciling Air Quality, Climate and Economic Goals.* Cambridge, MA: MIT Press.

Homer-Dixon, T. (1994). Environmental scarcities and violent conflicts: evidence from cases. *International Security* 19 (1): 5–40.

IEA (2015). *World Energy Outlook*. Paris: IEA.

IEA (2021). *India Energy Outlook 2021*. Paris: IEA https:// iea.blob.core.windows.net/assets/1de6d91e-e23f-4e02-b1fb-51fdd6283b22/India_Energy_Outlook_2021.pdf (accessed 29 November 2021).

IPCC (2014). Climate Change 2014: Synthesis Report. https://www.ipcc.ch/site/assets/uploads/2018/05/SYR_AR5_FINAL_full_wcover.pdf (accessed 30 April 2021).

IQ Air (2018) *World Air Quality Report 2018*. http: world-air-quality-report-2018-en.pdf (accessed 24 April 2021).

IQ Air (2020) *World Air Quality Report 2020*. http: world-air-quality-report-2020-en%20(1).pdf (accessed 24 April 2021)

Jamrisko, M. (2019). The world's dirtiest air is in India. *Bloomberg* (4 March 2019). https://www.bloomberg.com/news/articles/2019-03-05/the-world-s-dirtiest-air-is-in-india-where-pollution-costs-lives (accessed 1 May 2021).

Jordan, R. (2020). Stanford researcher discusses link between air pollution and COVID-19. *Stanford News*. https://news.stanford.edu/2020/06/30/links-covid-19-air-pollution (accessed 15 May 2021).

Kahn, M. (2008). Urban growth and climate change. *Annual Review of Resource Economics* 1 (16): 1–17.

Khator, R. (1992). *Environment, Development and Politics in India*. Lanham: University Press of America.

Langrish, J. et al. (2009). Beneficial cardiovascular effects of reducing exposure to particulate air pollution with a simple facemask. *Part Fibre Toxicol* (13 March 2009). doi: https://doi.org/10.1186/1743-8977-6-8 https://pubmed.ncbi.nlm.nih.gov/19284642 (accessed 2 May 2021).

Laumbach, R.J. and Kipen, H.M. (2012). Respiratory health effects of air pollution: update on biomass smoke and traffic pollution. *Journal of Allergy and Clinical Immunology* 129 (1): 3–13.

Liang, L. and Gong, P. (2020). Urban and air pollution: a multi-city study of long-term effects of urban landscape patterns on air quality trends. *Scientific Reports* 10 (18618): https://www.nature.com/articles/s41598-020-74524-9 (accessed 30 April 2021).

Manglik, V., (2015). Businesses and environmental groups call on President Obama to create clean 'Power India' initiative. *Sierra Club* (23 January 2015). https://www.sierraclub.org/compass/2015/01/businesses-and-environmental-groups-call-president-obama-create-clean-power-india (accessed 1 May 2021).

MEA (2012). Opening remarks by hon'ble minister of state for environment and forests at the media interaction in Rio de Janeiro. (20 June 2012). https://www.mea.gov.in/in-focus-article.htm?19813/Opening+Remarks+by+Honble+Minister+of+State+for+Environment+and+Forests+at+the+Media+Interaction+in+Rio+de+Janeiro (accessed 1 May 2021).

MEA (2018). Statement by EAM At UNSG's climate change event. (26 September 2018). https://www.mea.gov.in/Speeches-Statements.htm?dtl/30429/statement+by+eam+at+unsgs+climate+change+event (accessed 1 May 2021).

Medics4cleanair.eu website (2021). Sign our Manifesto. https://medics4cleanair.eu/sign-our-manifesto (accessed 7 May 2021).

Meyer, T. (2015). How local discrimination can promote global public goods. *Boston University Law Review* 95 (2015): 1937–2001.

Miles, T. (2019). India wins solar case at WTO but impact disputed. *Reuters* (27 June 2019). https://www.reuters.com/article/us-usa-trade-india-wto/india-wins-u-s-solar-case-at-wto-but-impact-disputed-idUSKCN1TS2B0 (accessed 1 May 2021).

Muro, M., Victor, D.G., and Whiton, J. (2019). How the geography of climate change could make the politics less polarizing. *Brookings Institute*. https://www.brookings.edu/research/how-the-geography-of-climate-damage-could-make-the-politics-less-polarizing/ (accessed 9 May 2021).

Nicholls, R.J., et al. (2008). Ranking port cities with high exposure and vulnerability to climate extremes: exposure estimates. Working Paper 1, Paris: OECD.

Nowak, D.J., Hirabayashi, S., Bodine, A., and Hoehn, R. (2013). Modeled PM2.5 removal by trees in ten U.S. cities and associated health effect. *Environmental Pollution* 178 (2013): 395–402. https://www.fs.fed.us/nrs/pubs/jrnl/2013/nrs_2013_nowak_002.pdf (accessed 8 May 2021).

Padma, T.V. (2011). Jairam Ramesh's legacy is an Indian Environment Minister with an identity. *The Guardian* (13 July 2011). https://www.theguardian.com/environment/blog/2011/jul/13/jairam-ramesh-india-environment-ministry (accessed 25 April 2021).

Pandey, V. (2020). COVID 19 and pollution: Delhi staring at coronavirus disaster. *BBC*. https://www.bbc.com/news/world-asia-india 54596245#:~:text=India's%20dreaded%20pollution%20season%20has,19%20case%20numbers%20and%20deaths (accessed 8 May 2021).

Parikh, J. and Shukla, V. (1995). Urbanization, energy use and greenhouse effects in economic development. *Global Environmental Change* 5 (2): 87–103.

Pozzer, A. et al. (2020). Regional and global contributions of air pollution to risk of death from COVID-19. *Cardiovascular Research*.(26 October 2020). https://academic.oup.com/cardiovascres/advance-article/doi/10.1093/cvr/cvaa288/5940460 (accessed 1 May 2021).

Press Information Bureau (2015) *India Surging Ahead in the Field of Green Energy: 100 GW Solar Scale-up Plan*. https://www.pib.gov.in/newsite/PrintRelease.aspx?relid=122566 (accessed on 2 May 2021)

Ramesh, J. (2015). *Green Signals: Ecology, Growth and Democracy in India*. Oxford: Oxford University Press.

Ramesh, J. (2018). An Insider's View of Climate Talks at Copenhagen and Cancun. *The Wire Science*. https://science.thewire.in/environment/jairam-ramesh-an-insiders-view-of-climate-talks-at-copenhagen-and-cancun (accessed 2 May 2021).

Roy, A. (2021). We are witnessing a crime against humanity. *The Guardian* (28 April 2021). https://www.theguardian.com/news/2021/apr/28/crime-against-humanity-arundhati-roy-india-covid-catastrophe (accessed 1 May 2021).

SEI et al. (2020) *The Production Gap Report: 2020 Special Report. SEI US*. https://productiongap.org/wp-content/uploads/2020/12/PGR2020_FullRprt_web.pdf (accessed 28 April 2021)

SE4All website. (2021). Global Tracking Framework. https://www.seforall.org/news/seforall-global-tracking-framework. (accessed 2 April 2021).

Shetty, Disha (2021). 22 out of the top 30 world's most polluted cities in India. *Forbes* (16 March 2021). https://www.forbes.com/sites/dishashetty/2021/03/16/22-out-of-top-30-worlds-most-polluted-cities-in-india/?sh=5a2d31cb75ad (accessed 8 May 2021).

Shindell, D. (2020). Health and economic benefits of a 2°C climate policy testimony to the house committee on oversight and reform hearing on 'the devastating impacts of climate change on health. (5 August 2020). https://oversight.house.gov/sites/democrats.oversight.house.gov/files/Testimony%20Shindell.pdf (accessed 1 May 2021).

Singh, A., Avis, W.R., and Pope, F.D. (2020). Visibility as a proxy for air quality in East Africa. *Environmental Research Letters* 2020: https://doi.org/10.1088/1748-9326/ab8b12 (accessed 9 May 2021).

Sivaram, V. (2015). *Reach for the Sun: How India's Audacious Solar Ambitions could Make or Break its Climate Commitments*. Stanford: Center for Energy Policy and Finance Stanford Law School https://www-cdn.law.stanford.edu/wp-content/uploads/2015/12/Reach-for-the-Sun-High-Resolution-Version.pdf (accessed 7 May 2021).

Tate, R.D., Shearer, C. and Matikinca, A. (2021). Deep Trouble: Tracking Global Coal Mine Proposals. Global Energy Monitor. https://globalenergymonitor.org/wp-content/uploads/2021/05/CoalMines_2021_r4.pdf (accessed 30 July 2021).

Times of India (2018). 14 of the world's 15 most polluted cities in India. (2 May 2018). https://timesofindia.indiatimes.com/city/delhi/14-of-worlds-15-most-polluted-cities-in-india/articleshow/63993356.cms (ccessed 7 May 2021).

Times of India (2019). UN Climate Action Summit 2019: Time for talking over, world needs to act now, says PM Modi. (23 September 2019). https://www.timesnownews.com/india/article/pm-narendra-modi-speech-at-un-climate-action-summit-2019-new-york-today-on-climate-change/493766 (accessed 8 May 2021).

US Air Quality Index (2021) AQI Basics. AQI Basics | AirNow.gov (accessed 2 May 2021)

U.S. Mission India NowCast Air Quality Index (2021). https://in.usembassy.gov/embassy-consulates/new-delhi/air-quality-data (accessed 15 May 2021).

UN Climate Change (2018). Climate change poses 'incredible risk' to Cities- Top UN official. (27 April 2018). https://unfccc.int/news/climate-change-poses-incredible-risk-to-cities-un-s-top-climate-official (accessed 29 April 2021).

UN Habitat (1996). *An Urbanizing World: Global Report on Human Settlements*. Nairobi: UN https://mirror.unhabitat.org/downloads/docs/GRHS.1996.1.pdf (accessed 29 April 2021).

UN Press Release (2019). Climate Battle 'Will Largely Be Won or Lost' in Cities, Secretary-General Tells World Mayors Summit' UN. (11 October 2019). https://www.un.org/press/en/2019/sgsm19808.doc.htm (accessed 25 April 2021).

UNDESA (2018). *World Urbanization Prospects: 2018 Revision*. New York: UN https://population.un.org/wup/Publications/Files/WUP2018-Report.pdf (accessed 25 April 2021).

UNDESA (2019). *World Population Prospects 2019 Highlights*. New York: UN World Population Prospects 2019 Highlights (un.org) (accessed 2 January 2021).

UNDESA, Population Division (2018). *The World's Cities in 2018 – Data Booklet (ST/ESA/SER.A/417)*. New York: UN https://digitallibrary.un.org/record/3799524?ln=en (accessed 24 April 2021).

Upadhyay, A. (2019). World's worst air pollution finally emerges as an election issue in India. *Bloomberg News* (28 April 2019). https://www.bloomberg.com/news/articles/2019-04-28/finally-world-s-worst-air-emerges-as-an-election-issue-in-india (accessed 1 May 2021).

Van Dorn, A., Cooney, R.E., and Sabin, M.L. (2020). COVID 19 exacerbating inequalities in the US. *Lancet* 395 (10232): 1243–1244. (16 April 2020). https://www.ncbi.nlm.nih.gov/pmc/articles/PMC7162639. (accessed 12 May 2021).

WAQI site (2021). World's Air Pollution: Real-time Air Quality Index. https://waqi.info (accessed 1 June 2021).

WEF (2019). *Global Risks Report 2019*. Geneva: WEF http://www3.weforum.org/docs/WEF_Global_Risks_Report_2019.pdf (accessed 22 April 2021).

WHO (2014). 7 million premature deaths linked to air pollution. *WHO Media Centre*. https://www.who.int/mediacentre/news/releases/2014/air-pollution/en (accessed 2 May 2021).

WHO (2016). Ambient Air Pollution: A Global Assessment of Exposure and Burden of Disease. https://apps.who.int/iris/bitstream/handle/10665/250141/9789241511353-eng.pdf?sequence=1 (accessed 2 May 2021).

WHO (2018). *COP-24: Special Report- Health and Climate Change*. Geneva: WHO https://apps.who.int/iris/handle/10665/276405 (accessed 1 June 2021).

WHO (2019). *Drinking Water: Key Facts*. WHO. https://www.who.int/news-room/fact-sheets/detail/drinking-water. (accessed 22 April 2021).

WTO (2016). India – Certain Measures Relating to Solar Cells and Solar Modules. https://docs.wto.org/dol2fe/Pages/FE_Search/FE_S_S009-DP.aspx?language=E&CatalogueIdList=227076,227077&CurrentCatalogueIdIndex=0&FullTextHash=&HasEnglishRecord=True&HasFrenchRecord=True&HasSpanishRecord=True (accessed 3 May 2021).

WTO, 2018 Summary of Disputes- India- Certain Measures relating to Solar Cells and Solar Modules. https://www.wto.org/english/tratop_e/dispu_e/cases_e/ds456_e.html (accessed 3 May 2021).

5

The Urgency of Curbing SLCPs

Why Reducing BC Emissions Matters

5.1 Understanding the Relevance of Reducing PM Pollution: Context and Background

Breathing air is essential to human life, but for those who live at the intersection of a lack of access to clean energy and extreme vulnerability to adverse climatic impacts, breathing toxic PM emissions from fossil/solid fuels comes with a definitive burden of death and disease. The scope of PM air pollution and the missed opportunities to address this challenge within the UN's PA and the SDA discussed in the preceding chapters underscore the development reality that the poorest households and communities lack climate adaptive resilience, the ability to relocate to neighborhoods where the air is cleaner and access to necessary public health services. The IPCC's SPM for Working Group III highlighted that climate mitigation options are critically linked to the growing global trend towards urbanization and pointed to the importance of 'bundling' policy instruments for efficacy. This point is particularly crucial for addressing impact analysis of city-driven interlinked actions on climate mitigation, clean air and clean energy. The IPCC's scientific consensus was both 'robust' and reflective of 'high agreement' that: 'Thousands of cities are undertaking climate action plans, but their aggregate impact on urban emissions is uncertain. There has been little systematic assessment on their implementation, the extent to which emission reduction targets are being achieved, or emissions reduced' (IPCC 2014a, p. 25).

As our planet continues to urbanize at unprecedented levels, there is a concentration of disease and morbidity burdens in cities primarily in developing countries that urgently need to take stock of the fact that clean air matters now more than ever. Millions in the dense megacities of Asia and Africa are exposed to toxic levels of PM emissions as they go about their daily lives, but sadly there is a dearth of consistent data collection and analysis as to the full extent of persistent PM pollution exposure caused by SLCPs. SLCPs are powerful climate forcers with global warming potentials many times that of carbon dioxide and they are also toxic air pollutants. As referenced earlier, the four key SLCPs: BC, methane (CH_4), tropospheric ozone (O_3) and HFCs are commonly associated with refrigeration, diesel-fueled vehicles and solid fuel cooking fires. PM emissions from diesel exhaust—a major source of BC—have been linked to lung and heart disease as well as cancer resulting

Air Pollution, Clean Energy and Climate Change, First Edition. Anilla Cherian.

in more stringent diesel emissions standards in countries like the US since the 1970s in response to public health concerns.

While global and regional UN-led action to reduce BC emissions across Asia and Africa has still not been implemented, interestingly, it was the UN's principal environmental agency, UNEP that commissioned one of the earliest assessments of the climate implications of BC via the Indian Ocean Experiment which focused on the haze located principally over South Asia and the Indian Ocean. The panel's 2002 report results were discussed in an influential paper written by Ramanathan and Crutzen (scientists at the Scripps Institution of Oceanography) and others which summarized the research findings and evidenced that "brown clouds" laden with the dark particles were found to influence weather patterns over South Asia (2002). In their paper, Ramanthan, Crutzen et al. forcefully argued: 'The South Asian brown haze covers most of the Arabian Sea, Bay of Bengal, and the South Asian region. It occurs every year and extends from about November to April and possibly longer. The black carbon and other species in the haze reduce the average radiative heating of the ocean by as much as 10% and enhance the atmospheric solar radiative heating by 50 to 100%. . .*The long duration of the haze, its black carbon content, the large perturbation to the radiative energy budget of the region and its stimulated impact on the rainfall distribution, if proved correct, have significant implications to regional water budget, agriculture and health. The link between anthropogenic aerosols and reduction of monsoonal rainfall over South Asia also has been made by fifteen model studies preceding the UNEP report*' (emphasis added, 2002, p. 947). Decades later, the urgency of responding to the linkages between water scarcity, food security and public health particularly in South Asia remain a pressing need.

In their 2009 article in *Foreign Affairs* entitled, 'The other climate changers: Why black carbon and ozone also matter', Seedon Wallack and Ramanathan point out that: ' . . . little attention has been given a low-risk, cost effective, and high-reward option: reducing emissions of light-absorbing carbon particles (known as "black carbon") and of the gases that form ozone. Together, these pollutants' warming effect is around 40-70 percent of that of carbon dioxide. Limiting their presence in the atmosphere is an easier, cheaper and more politically feasible proposition than the most popular proposals for slowing climate change-and it would have a more immediate effect' (2009, p. 205). As discussed in Chapter 1, in 2011, a joint report by UNEP and WMO entitled 'Integrated Assessment of Black Carbon and Tropospheric Ozone' specifically called attention to the need to curb emissions of SLCPs such as BC and O_3 together with other major GHGs like CO_2 and CH_4 not only in order to prevent global temperatures from crossing a dangerous threshold, but also because a small number of targeted emission reduction measures "could immediately begin to protect climate, public health, water and food security and ecosystems' (UNEP/WMO 2011, p. 2). To be clear, this 2011 Report clearly pointed out that identifiable and targeted measures to curb SLCPs "complement" but do not replace CO_2 reduction measures. IPCC's 2014 AR 5 Working Group III's 'Technical Summary' report on climate mitigation referenced the linkages between solid fuel related air pollution, curbing SLCPs and health: 'Most notably, about 1.3 billion people worldwide do not have access to electricity and about 3 billion are dependent on traditional solid fuels for cooking and heating with severe adverse effects on health, ecosystems and development. Providing access to modern energy services is an important sustainable development objective. The costs of achieving nearly universal access to electricity and clean fuels for cooking and heating are projected to be between 72 to 95 billion USD per year until 2030 with minimal effects on GHG emissions (limited evidence, medium agreement). *A transition away from the use of traditional biomass and the more efficient combustion of solid fuels reduce air pollutant emissions, such as sulfur dioxide (SO2), nitrogen oxides (NOx), carbon monoxide (CO), and black carbon (BC), and thus yield large health benefits (high confidence)*' (emphasis added, IPCC 2014b, p. 63).

In 2016, Ramanathan et al. in 'The Next Front on Climate Change: How to Avoid a Dimmer, Drier World' underscored the need that curbing climatic impacts related to water, food and health insecurities required attention to be paid to PM pollution: 'Many of the activities that cause greenhouse gas emissions—burning coal for power, diesel for transport, and wood for cooking, for example—also yield ultra-small particles known as aerosols, which blanket vast areas in a haze that blocks and scatters sunlight' (2016). Echoing this point, Burney et al. 2016, 'Getting serious about the new realities of climate change', argued that reducing SLCPs should be part of the intergovernmental climate change negotiations system given that: 'Slightly less than half of current global warming is due to four categories of non-carbon dioxide pollutants: dark soot particles often called black carbon, methane gas, lower atmospheric ozone, and hydrofluorocarbons (industrial gases used as coolants). Nearly all have life spans of a few weeks to a decade, much shorter than carbon dioxide. Yet they are potent warmers. Emitting one ton of black carbon, for example, has the same immediate effect on warming as emitting 500 to 2,000 tons of carbon dioxide' (p. 51).

Recently, the second to last page of the findings of the 2021 IPCC AR6 Working Group 1's SPM highlighted the significance of climate mitigation and air quality in the midst of the COVID-19 pandemic and for the future: 'Emissions reductions in 2020 associated with measures to reduce the spread of COVID-19 led to temporary but detectible effects on air pollution (high confidence) . . . Reductions in GHG emissions also lead to air quality improvements. However, in the near term, even in scenarios with strong reduction of GHGs, as in the low and very low GHG emission scenarios (SSP1-2.6 and SSP1-1.9), these improvements are not sufficient in many polluted regions to achieve air quality guidelines specified by the World Health Organization (high confidence). Scenarios with targeted reductions of air pollutant emissions lead to more rapid improvements in air quality within years compared to reductions in GHG emissions only, but from 2040, further improvements are projected in scenarios that combine efforts to reduce air pollutants as well as GHG emissions with the magnitude of the benefit varying between regions (high confidence)' (2021, p. 40). The finding that targeted reductions of air pollutant emissions lead to rapid improvements in air quality within years cannot be underestimated within cities and regions reeling from polluted air.

It is important to also underscore that IPCC Special Report, 'Global Warming of 1.5°C' had called attention to the fact that its global scientific assessment process had 'high confidence' that ensuring the 1.5 °C temperature threshold would require reaching net zero emissions by 2050 in both CO_2 and non-CO_2 emissions: 'Modelled pathways that limit global warming to 1.5°C with no or limited overshoot involve deep reductions in emissions of methane and black carbon (35% or more of both by 2050 relative to 2010). These pathways also reduce most of the cooling aerosols, which partially offsets mitigation effects for two to three decades. Non-CO_2 emissions can be reduced as a result of broad mitigation measures in the energy sector. In addition, targeted non-CO_2 mitigation measures can reduce nitrous oxide and methane from agriculture, methane from the waste sector, some sources of black carbon, and hydrofluorocarbons. High bioenergy demand can increase emissions of nitrous oxide in some 1.5°C pathways, highlighting the importance of appropriate management approaches. *Improved air quality resulting from projected reductions in many non-CO_2 emissions provide direct and immediate population health benefits in all 1.5°C model pathways* (high confidence)' (2018, p. 12).

The policy nexus between curbing SLCPs and a transition towards cleaner, non-polluting energy is arguably critical to the future of PA and SDA given the shared priority accorded to poverty eradication by 2030. There is an ample body of globally relevant research regarding the fact that mitigating energy related GHG emissions lies at the heart of the collective global effort to resolve climate change conclusively. Continuing to ignore the linkages between increasing access to non-polluting forms of energy, reducing human exposure to PM air pollution and benefits of the same on human health and poverty reduction goes against the grain and mandate of the UN 2030 SDA and SDGs 7 and 13. The combined impacts of lack of access to clean energy and the lack of adequate air quality control and enforcement of clean air measures endangers the lives of millions. And, the question that is long overdue is whether enough is being done in response to PM air pollution being the single largest environmental health risk in the world so as to improve human life and well-being in the most polluted areas of the world. More specifically, it is time to ask why the world's most successful air pollution treaty – CLRTAP including an amended protocol aimed at curbing PM pollution – has not been actively extended to other regions, and to also ask what lessons can be learnt from the CLRTAP that could be of benefit for cities that face toget the highest levels of PM pollution.

At the outset, it is worth highlighting that regional and national actions to reduce SLCPs in particular BC and precursors of O_3 cannot be seen as a substitute or stand-in for imperative global and national action to reduce CO_2 emissions. Emission reductions of SLCPs cannot therefore replace global action towards GHG reductions, but ignoring the impacts of SLCPs such as BC that result from exposure to inefficient combustion of solid fuels has devastating consequences for human health and a range of ecosystems which impact most negatively on poorer and vulnerable lives across the world. The linkages between the energy sector and climate change for instance were referenced in the IPCC SPM:

'A single year's worth of current global emissions from the energy and industrial sectors have the largest contributions to global mean warming over the next approximately 50 to 100 years. Household fossil fuel and biofuel, biomass burning and on-road transportation are also relatively large contributors to warming over these time scales' (IPCC 2013, p. 633). As Chapters 2 and 3 pointed out, the continued existence of silos on sustainable energy for all and climate change has resulted in a lack of an integrated policy on mitigating SLCPs and increasing access to clean energy for the poor within the intergovernmental negotiations context.

As highlighted previously, in 2016, the WHO identified air pollution as the world's 'biggest environmental risk to health', and two years later, the WHO warned that the 'most direct link between climate change and ill health is air pollution' (WHO 2018). Addressing this essential linkage between climate change and ill health by curbing PM air pollution and specifically BC is an unmistakably, urgent policy imperative for cities in developing countries in Africa and Asia, and in particular for cities and countries in the South Asia region which face extremely high levels of PM pollution. Cities in India are on the frontline in the struggle to cope with intersecting challenges of climate change and staggering levels of PM pollution.

A joint report by the WHO and UNEP demonstrated that meeting the goals of the 2015 PA could save about a million lives a year worldwide by 2050 through reductions in air pollution alone. The report also estimated that in the 15 countries that emit the most GHGs, the health impacts of air pollution are estimated to cost more than 4% of their GDP. According to this report, actions to meet the Paris goals would cost around 1% of global GDP (WHO 2018). According to the 'State of Global Air 2019', nearly half of the world's population—a total of 3.6 billion people in 2017- were exposed to household air pollution and long-term exposure to outdoor and indoor air pollution contributed to nearly 5 million deaths from stroke, heart attack, diabetes, lung cancer, and chronic lung disease worldwide. The report's analysis found that while China and India together were responsible for over half of the total global attributable deaths, with both countries facing over 1.2 million early deaths from all air pollution in 2017; China has made initial progress, beginning to achieve air pollution declines; in contrast, Pakistan, Bangladesh, and India have experienced the steepest increases in air pollution levels since 2010 (HEI 2019).

The State of Global Air 2020 report issued an even more sobering finding: 'Despite all that is known about the effects of air pollution on health, the findings in 2019 show that little or no progress has been made in many parts of the world. Major disparities continue to exist; air quality has improved in many high-income countries over the past several decades, while dangerous levels of air pollution persist in low- and middle-income countries. And evidence is mounting that air pollution can cause harm at much lower levels than previously thought, suggesting that even high-income countries need to continue their efforts to reduce exposure and to ensure that the progress achieved is not lost over time. Systematic and consistent efforts to track progress toward reducing air pollution and the impacts it has on human health remain essential' (HEI 2020, pp. 3–4). The fact that reducing air pollution related morbidity and disease burdens not only offset health care but also improve productivity, and actually more than pay for the costs of transitioning to clean energy is of enormous importance and relevance in cities and countries in Asia and Africa. In terms of advanced industrialized countries, it is the health and underlying disease burdens that undergirds the rationale for the adoption of CLRTAP and its effectiveness within the countries that comprise the UNECE. But, what can be done to address the needs of residents of cities in South Asia for example especially those cities where quite literally millions of vulnerable children and the elderly are exposed to hazardous levels of air pollution consistently and over long time periods?

5.2 The Urgency of Curbing BC Emissions: Human Health and Climate Implications

The immediate focus of this section is primarily on the significance of curbing BC while recognizing that comprehensive assessments and measures to address all SLCPs are important in yielding climate, health and economic development benefits. Globally relevant multilateral institutions ranging from UNEP, WMO and the World Bank have previously and clearly recognized that reducing emissions from SLCPs not only provides climate benefits but also can lead to developmental impacts such as improved public health and agricultural benefits (UNEP/WMO 2011; World Bank 2014).

Exposure to such air borne pollutants such as BC and other SLCPs exacerbates respiratory and cardiovascular diseases, harms plants and damages buildings. For these reasons, most major urban centres try to control discharges of airborne pollutants with varying degrees of success particularly when it comes to some of the most polluted megacities in the world as evidenced by the toxic pall that clouds the skies in these cities. Unlike CO_2 and other GHGs, BC and O_3 last in the atmosphere only for a few days to a few weeks, though indirect couplings within the Earth system can prolong their impact. These pollutants are usually most potent near their area of emission or formation, where they can force local or regional perturbations to climate (IPCC 2018, p. 684). The fact that SLCPs are most potent near their areas of emission and can force local or regional changes to climate is vital to the discussion that local and regional measures in curbing SLCPs are absolutely critical.

Of the four SLCPs, the one that is of direct relevance to public health, food and water security impacts that is focused on in this chapter is BC, but that is not to discount the significance of other SLCPs. For instance, O_3 or ozone in the lower atmosphere also referred to as ground level ozone is a major pollutant, a significant GHG and reduces the effectiveness of land-based carbon sinks. In 2008, the Royal Society (2008) conducted a comprehensive assessment on ground level ozone and its report begins by clearly stating: 'Clean air is a basic requirement for human health and wellbeing. However, in many countries around the world air pollution is a serious threat to both human health and the environment . . . Ozone is a major constituent of photochemical smog. It is a powerful oxidant that damages human health and natural ecosystems, and reduces crop yields. It is also an important greenhouse gas, with a radiative forcing since 1750 third only to carbon dioxide (CO_2) and methane (CH_4)' (2008, p. 9). It is important to note that human activities do not emit ozone directly, but instead add pollutant gases such as CO, NO_x, CH_4 and other non-methane VOCs. These 'ozone precursor' gases undergo complex photochemical reactions and form ozone in the initial 10–15 km above the ground. Because of the large increase in these ozone precursors, O_3 which is toxic to humans and plants including crops threatens human life and food security.

BC or soot, on the other hand, is carbonaceous aerosol that absorbs heat in the atmosphere (leading to a 0.4 W/m^2 radiative forcing from anthropogenic fossil and biofuel emissions) and, when deposited on snow, reduces its albedo, or ability to reflect sunlight (IPCC 2018, p. 685). Chapter 1 noted that BC results from the incomplete or inefficient combustion of fossil fuels, wood and other forms of biomass, which results in the release of CO_2, carbon monoxide, VOCs, organic carbon (OC) and BC particles. Major sources of BC include fossil fuel emissions related to transportation, biofuel usage both industrial and residential and the open burning of biomass. BC emissions have been the subject of research over a period of time (Seiler and Crutzen 1980; Penner et al. 1993; Carcaillet et al. 2002; Dickerson et al. 2002; Bond and Sun (2005) Bond et al. 2007). BC is a strongly light-absorbing carbonaceous aerosol produced under conditions of incomplete fossil fuel and biomass combustion. Essentially, both the inefficient combustion of solid fuels within households hearths and cookstoves using biomass, as well as the use of diesel fuel results in very small particles consisting a mixture of graphitic and amorphous carbon. When emitted, these PM pollutants persist on average in the atmosphere for more than a week until they come into contact with a surface or are washed out of the atmosphere by precipitation. But when inhaled, $PM_{2.5}$ as evidenced in the preceding chapters has been found to contribute to morbidity and increased burden of diseases. Published assessments that have extensively reviewed and summarized the linkages between BC and climate impacts include UNEP/WMO (2011), Bond et al. (2013). Bond et al. (2013) for instance found that after CO_2, BC is the second most important anthropogenic emission in the present-day atmosphere in terms of its climate forcing.

The troubling policy truth about curbing emissions of BC is that the global community cannot claim that this is a new policy challenge or that responsive measures are somehow too esoteric or impractical to implement as evidenced by the groundbreaking 2011 UNEP/WMO report. Instead, it is possible to argue that the lack of responsive action is the result of cumulative political inability to face up to the fact that public health and disease burdens of being exposed to BC is the analogue of 'clean air apartheid' policy regime that closely parallels what the UN Special Rapporteur referred to as 'climate apartheid'. Back as 2009, as a direct result of an International Workshop on Black Carbon held on 5–6 January in London, UK, the International Council on Clean Transportation (ICCT) published a comprehensive report entitled 'A policy-relevant summary of black carbon climate science and appropriate emission control strategies'. The report noted that given the direct radiative forcing (RF)and snow albedo effects estimated by the IPCC, BC was ranked as the third most important positive climate-forcing agent after carbon dioxide and methane. But this report also cautioned that the IPCC appeared to have provided conservative guidance on BC, noting, for example, that the definition it adopted was broad and the RF estimate is at the low end of the possible range. The report provided a succinct rationale as to need to curb BC: 'Reductions of black carbon will provide substantial public health benefits and stand on their own as a strong reason to reduce emissions' (ICCT 2009, p. 5). The key findings of this 2009 report provided a clear impetus for action and are listed in Table 5.1.

BC contributes to atmospheric warming and also has direct impacts on human health so understanding its role is crucially important from a global development perspective. Hansen et al. (2000) and Jacobson (2004) found that the short atmospheric lifetime of BC meant that measures aimed at curbing BC can yield climate benefits that occur in the immediate near term. Hansen and Nazarenko (2004) discussed contributions of soot forcing from BC to global warming of the past century, including the trend toward early springs in the Northern Hemisphere, thinning Arctic sea ice and melting land ice and permafrost. They noted that '. . . melting ice and sea level rise define the level of dangerous anthropogenic interference with the climate system, then reducing soot emissions, thus restoring snow albedos to pristine high values, would have the double benefit of reducing global warming and raising the global temperature level at which dangerous anthropogenic interference occurs' (p. 423). The negative impacts of BC on the thinning of glaciers that provide water supplies for millions should not be devalued nor can the serious impacts of rising surface temperatures on human health. Xu et al. (2009) found that BC – soot aerosols – deposited on Tibetan glaciers have been a significant contributing factor to observed rapid glacier retreat and argued that curbing BC emissions are required '. . . to avoid demise of Himalayan glaciers and retain the benefits of glaciers for seasonal fresh water supplies' (p. 22114). In 2010, Menon et al. in 'Black carbon aerosols and the third polar ice cap' highlighted that the thinning of glaciers over the Himalayas raised concerns for the future of

Table 5.1 Key findings about BC.

Black carbon is a solid particle emitted during incomplete combustion. All particle emissions from a combustion source are broadly referred to as particulate matter (PM) and usually delineated by sizes less than 10 µm (PM_{10}) or less than 2.5 µm ($PM_{2.5}$). Black carbon is the solid fraction of $PM_{2.5}$ that strongly absorbs light and converts that energy to heat. When emitted into the atmosphere and deposited on ice or snow, black carbon causes global temperature change, melting of snow and ice and changes in precipitation patterns.
Fossil fuel combustion in transport, solid biofuel combustion in residential heating and cooking and open biomass burning from forest fires and controlled agricultural fires are the source of about 85% of global black carbon emissions. Maximum feasible reductions in 2030 can capture 2.8 Tg/yr of black carbon, a reduction of 60% from business as usual. Co-emitted pollutants and the location of emission activity will determine the net impact of control strategies on the climate.
Public health protection is already a strong argument for actions that control black carbon. Exposure to PM is responsible for hundreds of thousands of global deaths each year. Actions that reduce PM such as new requirements for exhaust after treatment with lower sulphur fuels, fuel switching and reductions in fuel consumption can reduce a substantial fraction of black carbon emissions. Regardless of the climate protection benefits, there is a strong case for these actions to protect public health.
The climate impacts of black carbon reinforce the public health need for actions to control PM emissions. According to the IPCC, black carbon is the third largest contributor to the positive radiative forcing that causes climate change. One kilogram is about 460 times more potent than an equivalent amount of carbon dioxide over a 100-year time horizon and 1600 times more potent over a 20-year horizon based on unofficial IPCC estimates. IPCC estimates of radiative forcing are conservative compared to others in the published literature.
Controls on black carbon can produce rapid regional and global climate benefits. Like all aerosol particles, black carbon washes out of the atmosphere within a few thousand kilometers from its source, so it produces essentially short-lived radiative forcing. This forcing produces strong regional climate impacts that extend beyond the forcing region and approach a global scale. In the aggregate these regional impacts are a global problem. A climate change mitigation strategy that incorporates short-lived forcing agents like black carbon can more rapidly reduce the positive radiative forcing that causes climate change, especially when rapid action is needed to avert tipping points for large-scale impacts like the loss of Arctic summer sea ice, the Himalayan Tibetan glaciers and the Greenland ice sheet.
Black carbon reductions supplement but do not replace actions to control carbon dioxide and other greenhouse gases. A focus of climate change mitigation is to reduce all positive radiative forcing, and carbon dioxide is the largest positive forcing agent, so any delay in CO_2 emission reductions extends its climate impacts. Actions that reduce black carbon and carbon dioxide emissions in parallel will more effectively reduce total positive radiative forcing.
Controls on black carbon will reduce both positive and negative radiative forcing, so decisions to act on a climate basis alone should focus on the net effect. Black carbon is emitted with other pollutants that reflect light and offset its positive forcing. These include primary and secondary organic carbon, sulphates, and nitrates produced in amounts that vary with the combustion and fuel type of each source. The net effect of sources is modified by the transport and deposition of its black carbon emissions onto ice and snow, so major sources that produce negative forcing in the atmosphere can still be net positive forcers if they deposit sufficient amounts into the Arctic or atop mountain glaciers.
The highest priority targets strictly from a climate mitigation perspective are sources that cause net positive radiative forcing such as combustion of fossil fuels low in sulphur and deposition of black carbon on ice and snow surfaces. On-road heavy-duty diesel vehicles, off-road agricultural and construction equipment, residential coal combustion and industrial brick kilns are generally net positive forcers. Open agricultural burning, residential biofuel burning and commercial shipping may be negative forcers, but this can be offset locally.

Source: ICCT (2009, p. 4).

river systems that support millions of lives. As noted by Menon et al.: 'BC aerosols, released from incomplete combustion, have been increasingly implicated as causing large changes in the hydrology and radiative forcing over Asia and its deposition on snow is thought to increase snow melt. In India BC emissions from biofuel combustion is highly prevalent and compared to other regions, BC aerosol amounts are high' (2010, p. 4559). The researchers from Lawrence Berkeley National Laboratory, Berkeley, Columbia University/NASA, Indian Institute for Tropical Meteorology, Climate Analysis Section and Lawrence Livermore National Laboratory estimated the impact of BC aerosols on snow cover and precipitation from 1990 to 2010 over the Indian subcontinental region using two different BC emission inventories and found that: 'Indian BC emissions from coal and biofuel are large and transport is expected to expand rapidly in coming years . . . over the Himalayas, from 1990 to 2000, simulated snow/ice cover decreases by ~0.9% due to aerosols' (p. 4559).

Understanding the significance of BC on regional rainfall patterns including monsoons which in turn impact food and water security that are already under siege in countries that include cities which are among the most polluted and inequitable is essential in preventing what increasingly looks like an apocalyptic future for millions. The layering of toxic levels of PM air pollution atop of temperature increases should give especially serious pause for the future of human well-being in many of the megacities in South Asia. The correlation between heat and human morbidity is undeniably real. More than a decade ago, Kovats and Hajat highlighted that prevention of deaths caused by extreme high temperatures (heat waves) was an 'issue of public health concern', and pointed: 'The risk of heat-related mortality increases with natural aging, but persons with particular social and/or physical vulnerability are also at risk. Important differences in vulnerability exist between populations, depending on climate, culture, infrastructure (housing), and other factors . . . Climate change will increase the frequency and the intensity of heat waves, and a range of measures, including improvements to housing, management of chronic diseases, and institutional care of the elderly and the vulnerable, will need to be developed to reduce health impacts' (2008, p. 41). The point that increased incidence of heat waves as a result of climate change impacts disproportionately and negatively on already vulnerable, health-compromised individuals and communities remains a far-reaching public health challenge in cities and countries that are attempting to bridge the gap between poverty reduction and health, water, food and energy insecurities. It is also important to underscore here that the assumption that somehow all humans can adapt to any threshold of climate related warming – heat stress – is not borne out by scientific research.

In a 2010 research article, Sherwood and Huber noted that heat stress imposes an upper limit to human adaptation and proposed that human survival has a temperature threshold – the wet-bulb temperature (T_W) of 35 °C – defined as the combined measure of temperature or dry-bulb temperature (T) and humidity (Q) that is always less than or equal to T. In essence, high values of T_W connote hot and humid conditions. Simply put, a wet-bulb temperature is taken by wrapping the bulb of the thermometer in a wet cloth which serves as proxy/stand-in for human skin. When water from the wet cloth encased bulb evaporates, the thermometer is cooled, and the wet-bulb temperature is lower than the air temperature, but in high humidity conditions, the water does not evaporate, indicating that human physiological limits to balance heat and humidity are reached. Sherwood and Huber pointed out that: 'Peak heat stress, quantified by the wet-bulb temperature T_W, is surprisingly similar across diverse climates today. T_W never exceeds 31 °C. *Any exceedance of 35 °C for extended periods should induce hyperthermia in humans and other mammals, as dissipation of metabolic heat becomes impossible*' (emphasis added, 2010, p. 9552).

The public health challenge is that rising surface air temperatures will be differentially experienced at a regional and national level with communities and households that are already vulnerable and marginalized coming under threat of this new stressor. Pal and Eltahir (2016) and Im et al. (2017) have discussed how future temperature increases and projected deadly heat waves threaten densely populated regions in South Asia. The serious risks posed to human survival as a result of warming and humid temperatures in South Asia were specifically highlighted by Im et al.: 'Previous work has shown that a wet-bulb temperature of 35°C can be considered an upper limit on human survivability. On the basis of an ensemble of high-resolution climate change simulations, we project that extremes of wet-bulb temperature in South Asia are likely to approach and, in a few locations, exceed this critical threshold by the late 21st century under the business-as-usual scenario of future greenhouse gas emissions. *The most intense hazard from extreme future heat waves is concentrated around densely populated agricultural regions of the Ganges and Indus river basins. Climate change, without mitigation, presents a serious and unique risk in South Asia, a region inhabited by about one-fifth of the global human population, due to an unprecedented combination of severe natural hazard and acute vulnerability*' (emphasis added, 2017, p. 1).

In a 2020 research article, entitled 'The emergence of heat and humidity too severe for human tolerance', Raymond et al. point to a grim new reality as human bodies enabled to efficiently shed heat are now confronted with dangerously high humid heat – a T_W of 35°C – that marks the upper physiological limit for human survival, and highlight the fact that the incidence of this temperature threshold being evidenced in certain parts of the world has been substantially underestimated. With even lower values having grave health and productivity impacts, the researchers funded by the National Oceanic and Atmospheric Administration (NASA) note that it is time to take stock that: 'Climate models project the first 35°C T_W, occurrences by the mid 21st century. However, a comprehensive evaluation of weather station data shows that some coastal subtropical locations have already reported a T_W of 35°C and that extreme human heat overall has more than doubled in frequency since 1979'. As Raymond et al. go on to note, areas where the highest values have been recorded, namely, the southern Persian Gulf shoreline and northern South Asia are areas where millions of people live, and exposure to elevated temperatures threatens the range of natural variability to which their bodies can cope. Figure 5.1 excerpted from Raymond et al. (2020) evidences the really worrisome public health threat that awaits developing countries particularly those in South Asia. Researchers found that temperature thresholds of >31°C hotspots in the weather station record emerged through surveying the globally highest 99.9th wet-bulb temperature percentiles and included eastern coastal India, Pakistan and northwestern India and the shores of the Red Sea, Gulf of California and southern Gulf of Mexico (2020). There is no way to parse the fact that the combination of the exposure to PM pollution and rising surface temperature in cities of South Asia represents a major humanitarian and public health challenge in the coming decades.

One of the most influential networks/coalitions focused on the nexus between curbing SLCPs and climate action is the CCAC. Figures 5.2 and 5.3 excerpted from the CCAC

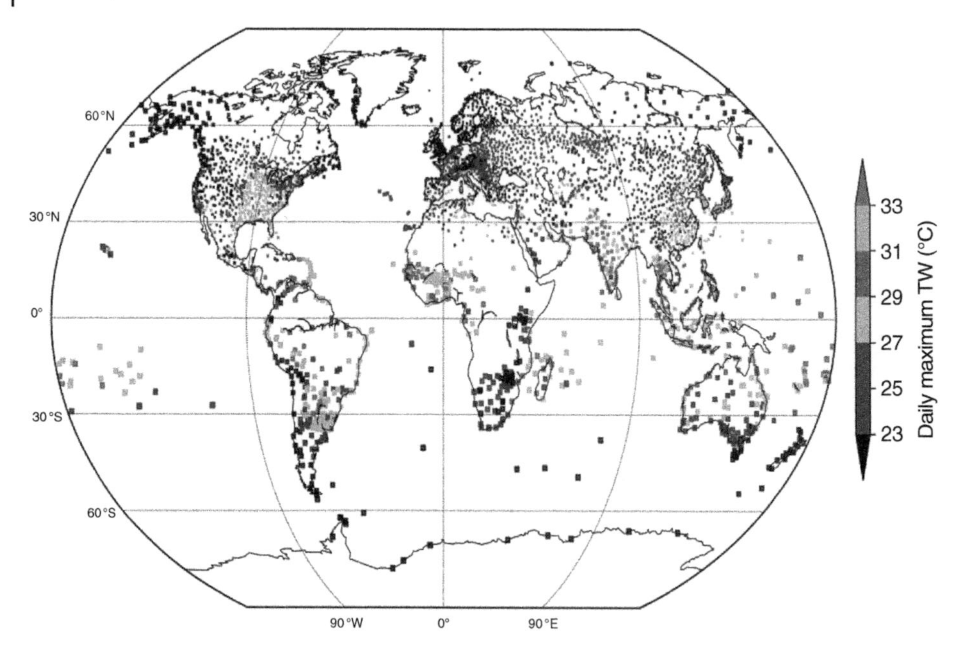

Figure 5.1 Observed global extreme humid heat. *Source:* Raymond et al. (2020). Science Advances. https://advances.sciencemag.org/content/6/19/eaaw1838/tab-figures-data accessed 23 July 2021. Distributed under a Creative Commons Attribution Non-Commercial License 4.0 (CC BY-NC).

Key figures

460–1500x	4–12 days	6.6 million tonnes	51%
Black carbon has a warming impact on climate 460–1500 times stronger than CO_2 per unit of mass	The average atmospheric lifetime of black carbon particles is 4–12 days	About 6.6 million tonnes of black carbon were emitted in 2015	Household cooking and heating account for 51% of global black carbon emissions

Primary sources of black carbon emissions

Black carbon emissions have been decreasing over the past decades in many developed countries due to stricter air quality regulations. By contrast, emissions are increasing rapidly in many developing countries where air quality is not regulated. As the result of open biomass burning and residential solid fuel combustion, Asia, Africa and Latin America contribute approximately 88% of global black carbon emissions.

Figure 5.2 BC – key figures and primary sources (excerpted from CCAC website). *Source:* CCAC website (2021a).

provide a snapshot overview of the significance of BC in terms of sources, impacts and climate change, with a key take way that 51% of BC particulate emissions arise from household energy use, with another 26% from transportation. These figures also provide an overview of the key figures and increases in primary sources of BC emissions in Asia, Africa

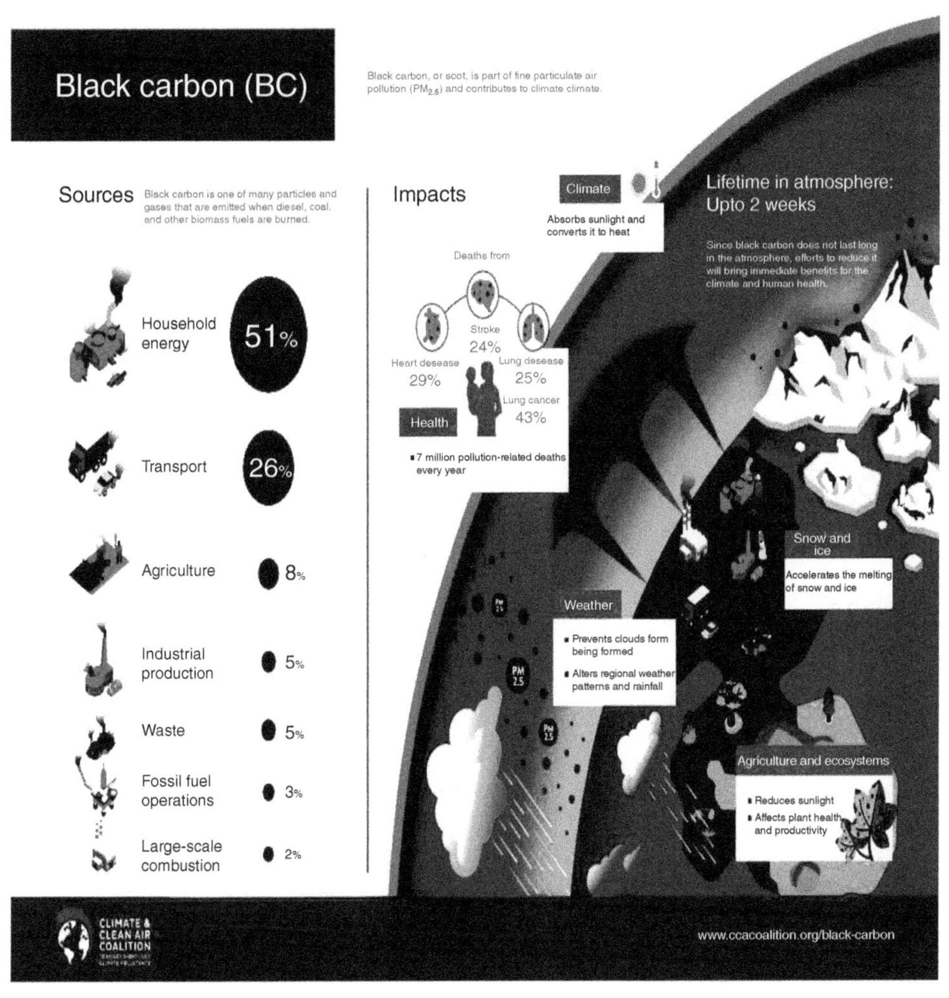

Figure 5.3 Role of BC – sources and impacts (human health and climate). *Source:* CCAC (2021a).

and Latin America which taken together contribute approximately 88% of global BC emissions. Globally and regionally verifiable comprehensive data tracking of BC.

As per Figure 5.2, BC's short atmospheric lifetime offers an opportunity for regional and city-based targeted strategies to reduce BC emissions which can provide win-win solutions in climate and health benefits within a relatively short period of time. From both environmental and human health perspectives, BC which is a key component of PM pollution has also been shown to affect the health of ecosystems by depositing on plants and increasing their temperature, dimming sunlight that reaches the earth, altering rainfall patterns such as monsoons, all of which in turn have grave consequences for the destruction of human lives, ecosystems and food security. With BC emissions emanating largely in Asia and Africa as a result of open fires and solid fuel combustion and lack of enforcement of fossil fuel/diesel emitting standards, regional costs on human health, ecosystems and agriculture are increasingly going to be dire.

To be clear, clean air and climate mitigation efforts focused on curbing SLCPs have been sponsored by entities such as CCAC. According to the CCAC's 2014-2015 Annual Report, the coalition to reduce SLCPs was founded in February 2012 by UNEP and six countries – Bangladesh, Canada, Ghana, Mexico, Sweden and the United States – and is 'the first global effort to treat these pollutants as a collective challenge'. As per this report, C CAC had a total of 109 partners comprising 49 State, 16 Intergovernmental Organizations (IGOs) and 44 NGOs as of September 2015 in its functioning as a voluntary partnership of entities committed to protecting the climate and improving air quality through actions to curb SLCPs (CCAC 2015). A review of the 2015 report reveals the following list of 49 countries as CCAC partners: 'Australia, Bangladesh, Benin, Cambodia, Canada, Central African Republic, Chad, Chile, Colombia, Cote d'Ivoire, Denmark, Dominican Republic, Ethiopia, European Commission, Finland, France, Germany, Ghana, Guinea (Republic of), Iraq, Ireland, Israel, Italy, Japan, Jordan, Kenya, Korea (Republic of), Laos, Liberia, Maldives, Mali, Mexico, Mongolia, Morocco (Kingdom of), Netherlands, New Zealand, Nigeria, Norway, Paraguay, Peru, Philippines, Poland, Russian Federation, Sweden, Switzerland, Togo, United Kingdom, United States and Uruguay'. Notably absent from the list are both China and India. Additionally, 16 IGOs are listed including amongst them UNDP, UNEP, UNIDO, WHO, WMO and UNECE (under whose purview the CLRTAP operates), notably absent are the UN Economic Commissions for other regions of the world where PM pollution is most rampant (CCAC 2015, p. 4). Additionally, what is critically important to keep in mind is that CCAC derives its work and mission from the 2011 UNEP/WMO Assessment Report which identified the following major challenges and benefits of reducing BC and O_3 – all of which have great significance particularly in Asia which is home to some of the most populous cities coping with toxic levels of air pollution.

The 2011 Integrated Assessment Report highlighted that BC exists as particles in the atmosphere and is a major component of soot. The black in BC refers to the fact that these particles absorb visible light, and this absorption leads to a disturbance of the planetary radiation balance. Since BC is an aerosol emitted directly at the source from incomplete combustion processes such as fossil fuel and biomass burning, much of the atmospheric BC is of anthropogenic origin. Given that BC contributes to anthropogenic climate forcing, accurate long-term historical emission inventories are crucial to distinguish anthropogenic influence against the natural variability of climate. But, BC emissions estimations for a range of countries and regions and the lack of longitudinal studies of the impacts of BC emissions on human populations in cities in developing countries remain a data analysis problem. This data challenge is particularly daunting for developing countries and cities in Africa and Asia where the problem of BC emissions is the greatest, but consistent and time bound BC data collection has not been consistently maintained. A number of BC emission inventories have been developed (Novakok and Hansen 2004; Bond et al. 2007; Klimont et al. 2016). The issue of regional disparities in BC emissions was highlighted by Klimont et al.'s finding that: ' . . . European, North American, and Pacific contributions to global emissions of BC dropped from nearly 30 % in 1990 to well below 15 % in 2010, while Asia's contribution grew from just over 50 % to nearly two-thirds of the global total in 2010. For all PM species considered, Asian sources represented over 60 % of the global anthropogenic total, and residential combustion was the most important sector, contributing about 60 % for BC and OC, 45 % for $PM_{2.5}$, and less than 40 % for PM_{10}, where large combustion sources and industrial processes are equally important. Global anthropogenic emissions of BC were estimated at about 6.6 and 7.2 Tg in 2000 and 2010, respectively, and represent about 15 % of $PM_{2.5}$ but for some sources reach nearly 50 %, i.e. for the transport sector' (2016, p. 8681). Here, Klimont et al.'s point that their BC estimation was higher than previously published owing primarily to the inclusion of new sources was echoed by Sun et al. (2019) paper 'Constraining a Historical Black Carbon Emission Inventory of the United States for 1960–2000'. Sun et al. highlighted the fact that discrepancies exist between inventories which underestimated BC emissions and used new findings to update the US BC inventory from 1960 to 2000. Notably, Sun et al. determined that the emissions from several key sources, including pre-1980 residential boilers and heating stoves, specific off-road engines and heavy-duty diesel and light-duty gasoline-powered vehicles assembled prior to 1970, should be increased significantly. Compared to earlier inventories, between 1960 and 1980, Sun et al. found that updated US emissions are 80% higher than previous estimates, totaling approximately 690 gigagrams per year in 1960 and 620 gigagrams per year a decade later. What is worth underscoring is that the results of this study pointed out that earlier BC emission estimates were underreported, and that inventories of other combustion by-products need to be re-evaluated. Both aspects have major implications for the developing countries, where estimated inventories of BC have been either underreported and/or are not consistently recorded over a temporal span.

The 2011 UNEP/WMO report pointed out that BC was not a GHG, but BC particles have a 'strong warming effect in the atmosphere, darken snow when it is deposited and influence cloud formation'. The regional and more localized impacts of BC on rainfall and cloud patterns and the melting of snow and ice in turn were found to have major implications for agriculture, food security and ecosystem viability. More specifically, this 2011 assessment highlighted that the short term yet potent climate pollutant impacts of BC arise from the fact that: 'The contribution to warming of 1 gramme of BC seen over a period of 100 years has been estimated to be anything from 100 to 2000 times higher than that of 1 gramme of C02. An important aspect of BC particles is that their lifetime in the atmosphere is short, days to weeks, and so emission reductions have an immediate benefit for climate and health' (2011, p. 6). Box 5.1 provides a summary of key challenges and benefits of reducing BC and O_3 excerpted from the historic UNEP/WMO 2011 report.

Figures 5.4–5.7 excerpted from the European Commission's Emissions Database for Global Atmospheric Research (EDGAR) provide data on both CO_2 emissions (global/India with most recent data year 2019) and BC emissions (global/India/with the most recent data year 2015). The figures for BC emissions worldwide when contrasted to those for India demonstrate the significance of BC emissions related to sectors such as agriculture, power and transport in India and consequently the urgency of the need to curb BC at the national level. This takes on particular urgency given that some of the world's most polluted cities coping with PM pollution are in India as evidenced by the preceding chapter. Efforts to control emissions from diesel-based transportation in developing countries like India are crucially important. So what are some of the measures and regional protocols by which PM pollution has been addressed in other parts of the world? A snapshot answer to this question can be found in a brief overview of the CLRTAP and its Gothenburg Protocol, as well

Box 5.1 Challenges and benefits of reducing emissions BC and O_3.

The challenges of BC and O_3

- **BC and O_3 in the lower atmosphere are harmful air pollutants that have substantial regional and global climate impacts,** including disruption of tropical rainfall and regional circulation models (Asian monsoons) affecting the livelihoods of millions.
- **BC's darkening of snow and ice surfaces increases their absorption of sunlight, which along with atmospheric heating exacerbates melting of snow and ice around the world,** including in the Artic, the Himalayas and other glaciated regions. This affects the water cycle and increases risks of flooding.
- **BC and O_3, both lead to adverse impacts on human health leading to premature deaths.**
- **O_3 is also an important air pollutant responsible for reducing crop yields and thus affects food security.**

Benefits of reducing emissions of BC and O_3

- **Reducing BC and O_3 now will slow the rate of climate change within the first half of this century.**
- **Small number of emission reduction measures targeting BC and ozone precursors could immediately begin to protect climate, public health, water and food security and ecosystems.**
- **Full implementation of the identified measures would reduce further global warming by 0.5°C (within a range of 0.2–0.7°C).** If the measures were to be implemented by 2030, they could halve the potential increase in global temperature projected for 2050. The rate of regional temperature increase would also be reduced.
- **Full implementation of the identified measures would have substantial benefits in the Arctic, the Himalayas and other glaciated and snow covered regions.**
- **Full implementation of the identified measures could avoid 2.4 million premature deaths (within a range of 0.7–4.6 million) and the loss of 52 million tonnes (within a range of 30–140 million tonnes), 1–4% of the global production of maize, rice, soy bean and wheat each year.**

Source: (emphasis included) UNEP/WMO (2011, pp. 2–3).

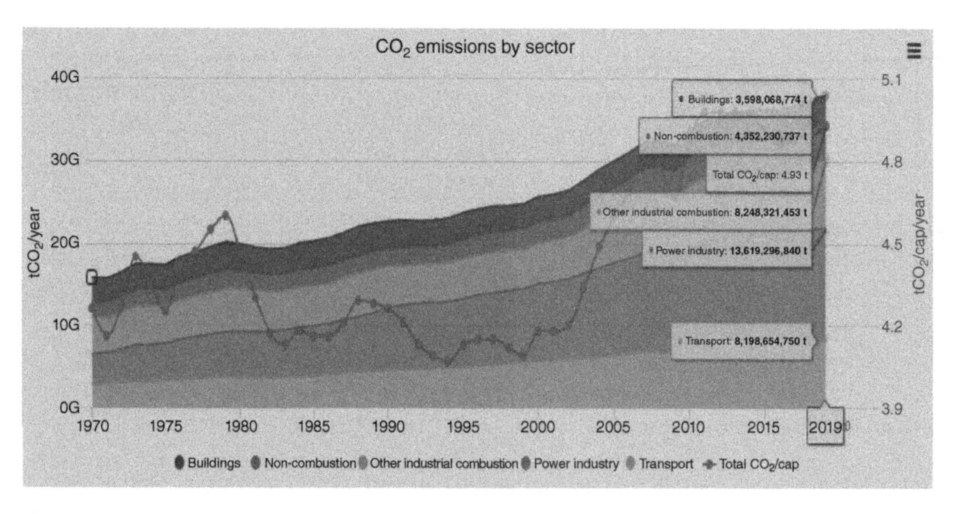

Figure 5.4 CO_2 emissions by sector: world.

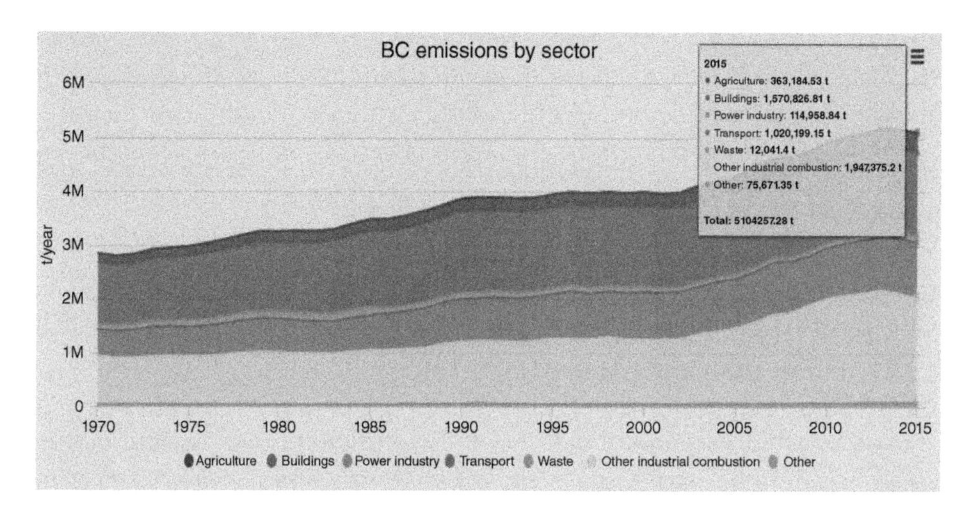

Figure 5.5 BC emissions by sector: world. *Source:* European Commission-Emissions Database for Global Atmospheric Research (EDGAR). https://edgar.jrc.ec.europa.eu/country_profile (accessed 7 July 2021).

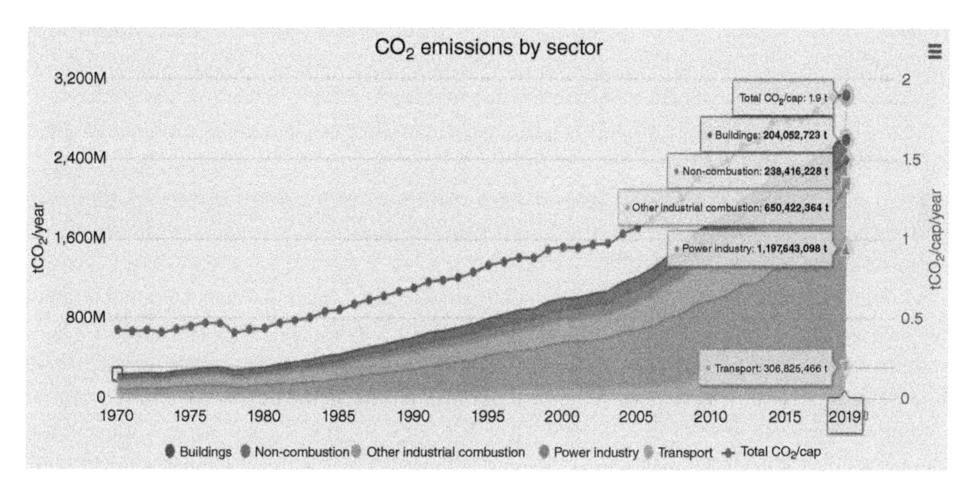

Figure 5.6 CO_2 emissions by sector: India.

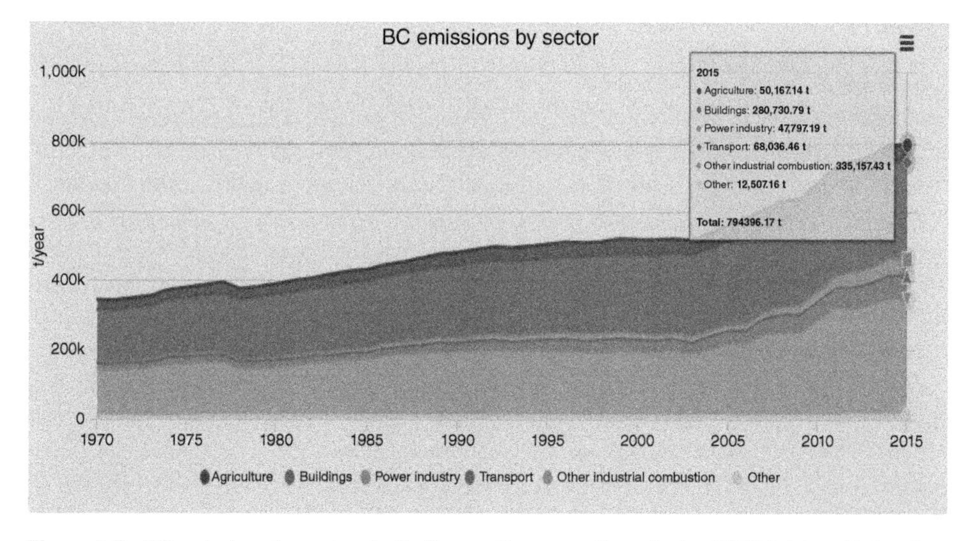

Figure 5.7 BC emissions by sector: India. *Source:* European Commission-EDGAR, https://edgar.jrc.ec.europa.eu/country_profile/IND (accessed 7 July 2021).

as the measures identified by CCAC which are derived directly from the UNEP/WMO 2011 Assessment Report and discussed in the section that follows.

5.3 The Most Successful Regional Air Pollution Treaty Which Other Regions Could Benefit From?: Brief Overview of CLRTAP and the Gothenburg Protocol

As far back as in the 1960s, scientists in Western Europe and North America studying the causes of smog and acid rain discovered that air pollutants emitted often miles away were wreaking havoc on local ecosystems and public health. As per the EPA, acid rain forms mainly through reactions with sulphur dioxide (SO_2) and nitrogen oxide (NO_2) found in fossil fuel emissions and takes form as acidic rain, snow or dust which can travel hundreds of miles in the air before falling to the Earth's surface. Acid rain was found to negatively impact aquatic and terrestrial systems including stripping nutrients from soils, destroying water quality in lakes and streams and weathering buildings. In its analysis of the legacy of acid rain research in the US, the US EPA pointed out that in the EPA's 50 years of research, 'one of the most significant environmental challenges the nation faced was the problem of acid rain' but while 'evidence of acid rain's harmful effects emerged centuries ago, it wasn't until the early 1980s that it was recognized as a major threat'. More specifically, EPA analysis highlighted that US efforts like the Acid Rain Program have had great success in reducing annual ambient concentrations of SO_2 which 'decreased 93 percent between 1980 and 2018'; while data from EPA's Long-Term Monitoring program evidenced 'improvements in the acidification of lakes and stream', and 'better air quality which led to a decrease in adult mortality' and prevented 'an estimated 230,000 premature deaths in 2020 alone' (EPA 2020). There is arguably an urgent need for comparative analysis that evaluates acid rain abatement programmes in developing countries, and this is again an immediate challenge of data and policy analysis that countries and regions of the world reeling from the highest levels of smog and acid rain could benefit from.

Air pollution has been well documented to have societal (gender and socio-economic) generational public health inequities as evidenced in the previous chapters. But the problem of controlling or curbing air pollution is that although it is directly related to local and national causal factors, its impacts on human health, ecosystems, agriculture, food security and altered weather patterns cannot be confined to national and local borders. It is precisely the spillover and intersecting effects of air pollution that were recognized early on and acted upon by developed countries within the North American and European regions. Recognizing the significant impacts of transboundary air pollution, by which pollutants generated and released at one location can travel great distances and affect air quality across national boundaries, in 1991, the United States and Canada entered in an agreement which led to reduction in acid rain in the 1990s, and was expanded in 2000 to reduce transboundary smog emissions under the Ozone Annex.

As it turns out, within the UN context, there is a comprehensive and highly successful air pollution treaty that is implemented on a regional basis but the benefits of the lessons learned in the data gathering and policy response related to this treaty have not been extended into by other regions in the world where tragically the problem of air pollution is the worst. In 1979, 32 countries in the pan-European region signed on to the CLRTAP – the world's first and to date, only transboundary regional air pollution treaty – which is implemented by the UNECE. By way of background, the UNECE was set up by the UN Economic and Social Council in 1947 and is one of the five regional commissions of the UN which includes: the Economic Commission for Africa (ECA), the Economic and Social Commission for Asia and the Pacific, the Economic Commission for Latin America and the Caribbean (ECLAC) and the Economic and Social Commission for Western Asia (ESCWA). As per the CLRTAP, 'long-range transboundary air pollution' is defined as 'the release, directly or indirectly due to human activity, of substances into the air which have adverse effects on human health or the environment in another country and for which the contribution of individual emission sources or groups of sources cannot be distinguished' (Europa-Lex site, Summary of LRTAP 2021).

In its 2019 overview of the CLRTAP UNECE highlighted the 'unparalleled' achievements of CLRTAP: 'Since its inception 40 years ago, the Convention on Long-range Transboundary Air Pollution has developed into a successful regional framework for controlling and reducing the damage to human health and the environment caused by transboundary air pollution. *The achievements until today are unparalleled:* Air pollutant emissions and economic growth have been decoupled. Emissions of certain air pollutants have been reduced by 40 to 80 per cent. Forest soils and lakes have recovered from acidification. 600,000 premature deaths have been avoided annually. With its now 51 Parties, its eight protocols setting emission reduction commitments for various pollutants, its solid science-policy interface, its effective compliance mechanism, its capacity-building programme, and its outreach beyond the UNECE region, the Convention is today more relevant than ever' (UNECE 2019).

The CLRTAP entered into force in 1983 and according to the UNECE itself has been instrumental in establishing a framework on the general principles of international cooperation for air pollution abatement and setting up an institutional framework which has since brought together new forms of research and policy. From 1983 to the present, the Convention has expanded in scope and now includes eight protocols, which has meant that the number of polluting substances covered by the Convention and its protocols has been expanded to include notably ground-level ozone, persistent organic pollutants, heavy metals and PM. The UNECE website is unequivocal as to CLRTAP's role: 'The Convention has substantially contributed to the development of international environmental law and has created the essential framework for controlling and reducing the damage to human health and the environment caused by transboundary air pollution. It is a successful example of what can be achieved through intergovernmental cooperation' (UNECE 2019). The ratification status of CLRTAP is contained in Table 5.2 and accessed from UN treaty collection depository.

The unabashed praise heaped upon the CLRTAP should make it the policy envy of the other regions of the world that grapple with toxic levels of air pollution. The fact that better air quality control measures and reduction of fossil fuel related PM pollution contributes directly to improving disease and death burdens should be enough of an incentive for regional action along the lines of the CLRTAP. The brief discussion provided here does not purport to provide an extensive review of CLRTAP over the course of its 40 year implementation trajectory or an exhaustive list of lessons learned from the CLRTAP and its Gothenburg protocol on curbing PM pollution for other regions/sub-regions reeling from toxic levels of

Table 5.2 Convention on long-range transboundary air pollution: status of ratification (as of 18 July 2021).

Participant	Signature	Ratification, Acceptance (A), Approval (AA), Accession (a), Succession (d)
Albania		2 December 2005 a
Armenia		21 February 1997 a
Austria	13 November 1979	16 December 1982
Azerbaijan		3 July 2002 a
Belarus	14 November 1979	13 June 1980
Belgium	13 November 1979	15 July 1982
Bosnia and Herzegovina		1 September 1993 d
Bulgaria	14 November 1979	9 June 1981
Canada	13 November 1979	15 December 1981
Croatia[2]		21 September 1992 d
Cyprus		20 November 1991 a
Czech Republic		30 September 1993 d
Denmark	14 November 1979	18 June 1982
Estonia		7 March 2000 a
European Union	14 November 1979	15 July 1982 AA
Finland	13 November 1979	15 April 1981
France	13 November 1979	3 November 1981 AA
Georgia		11 February 1999 a
Germany	13 November 1979	15 July 1982
Greece	14 November 1979	30 August 1983
Holy See	14 November 1979	
Hungary	13 November 1979	22 September 1980
Iceland	13 November 1979	5 May 1983
Ireland	13 November 1979	15 July 1982
Italy	14 November 1979	15 July 1982
Kazakhstan		11 January 2001 a
Kyrgyzstan		25 May 2000 a
Latvia		15 July 1994 a
Liechtenstein	14 November 1979	22 November 1983
Lithuania		25 January 1994 a
Luxembourg	13 November 1979	15 July 1982
Malta		14 March 1997 a
Monaco		27 August 1999 a
Montenegro		23 October 2006 d
Netherlands	13 November 1979	15 July 1982 A

Table 5.2 (Continued)

Participant	Signature	Ratification, Acceptance (A), Approval (AA), Accession (a), Succession (d)
North Macedonia		30 December 1997 d
Norway	13 November 1979	13 February 1981
Poland	13 November 1979	19 July 1985
Portugal	14 November 1979	29 September 1980
Republic of Moldova		9 June 1995 a
Romania	14 November 1979	27 February 1991
Russian Federation	13 November 1979	22 May 1980
San Marino	14 November 1979	
Serbia		12 March 2001 d
Slovakia		28 May 1993 d
Slovenia		6 July 1992 d
Spain	14 November 1979	15 June 1982
Sweden	13 November 1979	12 February 1981
Switzerland	13 November 1979	6 May 1983
Turkey	13 November 1979	18 April 1983
Ukraine	14 November 1979	5 June 1980
United Kingdom of Great Britain and Northern Ireland	13 November 1979	15 July 1982
United States of America	13 November 1979	30 November 1981 A

Source: UN Treaty Collection website (2021).

air pollution. Box 5.2 provides an overview of the LRTAP and the eight separate protocols that have been developed, negotiated upon and are under implementation. The information contained is derived directly from the Convention site maintained by Europa Lex which provides summaries of all key relevant legislation covering EU countries.

Of the preceding eight protocols, the 1999 Gothenburg Protocol which was amended in 2012 to include national emission reductions to be achieved by 2020 and beyond is of particular relevance. The amendments to the Protocol entered into force on 7 October 2019. The amended Gothenburg Protocol is the first and only binding agreement to target emission reductions for $PM_{2.5}$ which was recognized via the Protocol as an SLCP. The amended Gothenburg Protocol therefore represents the first regional and legally bound treaty committing countries within the UNECE to reduce $PM_{2.5}$ and thereby synergistically address air pollution and climate change policies. What is noteworthy is that the parties to the Gothenburg Protocol opened negotiations in 2007 with a view to improving further the protection of human health and the environment, including through the establishment of new emission reduction obligations for selected air pollutants to be achieved by the

Box 5.2 Eight protocols of the LRTAP (including the Gothenburg Protocol).

- The 1984 protocol on long-term financing of the cooperative programme for monitoring and evaluation of the long-range transmission of air pollution in Europe (EMEP Protocol): an instrument for the international cost sharing of a monitoring programme which forms the backbone for the review and assessment of European air pollution in the light of agreements on emission reduction.
- The 1985 protocol on the reduction of sulphur emissions or their transboundary fluxes (Helsinki Protocol) by at least 30% compared with 1980 levels.
- The 1988 protocol concerning the control of emissions of nitrogen oxides (NO_x) or their transboundary fluxes (Sofia Protocol): a first step requires the freezing of emissions of NO_x or their transboundary fluxes at 1987 levels; a second step requires the application of an effects-based approach to further reduce emissions of nitrogen compounds, including ammonia (NH_3), and of volatile organic compounds (VOCs), in view of their contribution to photochemical pollution, acidification and eutrophication and of their effects on human health, the environment and materials, by addressing all significant emission sources.
- The 1991 protocol on the control of emissions of VOCs or their transboundary fluxes: these compounds are responsible for the formation of ground-level ozone and parties have to opt for 1 of 3 emission-reduction targets, to be reached by 1999:
  - a 30% reduction in VOCs, using a year between 1984 and 1990 as a basis;
  - a 30% reduction in emissions of VOCs within the tropospheric ozone management area specified in Annex I to the protocol and ensuring that total national emissions do not exceed 1988 levels; or
  - where emissions in 1988 did not exceed certain specified levels, parties may opt for a stabilization at that level of emission.
- The 1994 protocol on further reduction of sulphur emissions (Oslo Protocol): this protocol builds on the 1985 Helsinki Protocol and sets emission ceilings until 2010 and beyond. Parties are required to take the most effective measures for the reduction of sulphur emissions, including:
  - increasing energy efficiency;
  - using renewable energy;
  - reducing the sulphur content of fuels; and
  - applying best available control technologies (BATs). The protocol also encourages the application of economic instruments for the adoption of cost-effective approaches to the reduction of sulphur emissions.
- The 1998 protocol on heavy metals (Aarhus Protocol): targets three metals – cadmium, lead and mercury. Parties will have to reduce their emissions for these below their levels in 1990 (or an alternative year between 1985 and 1995).
  - The protocol aims to cut emissions from industrial sources, combustion processes and waste incineration. It lays down stringent limit values for emissions from stationary sources and suggests BATs for these sources, such as special filters or scrubbers for combustion sources or mercury-free processes. The protocol requires parties to phase out leaded petrol. It also introduces measures to lower heavy-metal emissions from other products, such as mercury in batteries, and proposes

the introduction of management measures for other mercury-containing products, such as electrical components, measuring devices, fluorescent lamps, dental amalgam, pesticides and paint.
- The protocol was amended in 2012 to introduce more stringent emission limit values (ELVs) for emissions of particulate matter and of cadmium, lead and mercury applicable for certain combustion and other industrial emission sources that release them into the atmosphere. The emission source categories for the three heavy metals were also extended to the production of silico and ferromanganese alloys, thus expanding the scope of industrial activities for which emission limits are established.

- The 1998 protocol is on persistent organic pollutants (POPs) whose ultimate objective is to eliminate any discharges, emissions and losses of such pollutants. The protocol bans the production and use of some products outright, while others were scheduled for elimination at a later stage. It includes provisions for dealing with the wastes of products that are banned and obliges parties to reduce their emissions of dioxins, furans, polycyclic aromatic hydrocarbons and hexachlorobenzene (HCB) below their levels in 1990 (or an alternative year between 1985 and 1995). For the incineration of municipal, hazardous and medical waste, it lays down specific limit values.
  - It initially focused on a list of 16 substances that had been singled out according to agreed risk criteria. The substances comprised 11 pesticides, 2 industrial chemicals and 3 by-products/contaminants.
  - The protocol was amended in 2009 to include seven new substances: hexachlorobutadiene, octabromodiphenyl ether, pentachlorobenzene, pentabromodiphenyl ether, perfluorooctane sulphonates, polychlorinated naphthalenes and short-chain chlorinated paraffins. The parties revised obligations for the compounds DDT, heptachlor, HCB and PCBs, as well as for ELVs from waste incineration. To facilitate the protocol's ratification by countries with economies in transition, the parties introduced flexibility for these countries regarding the time frames for the application of ELVs and BATs.
- The 1999 protocol to abate acidification, eutrophication and ground-level ozone (Gothenburg Protocol): it sets national emission ceilings for 2010 up to 2020 for four pollutants: sulphur dioxide (SO_2), NO_x, VOCs and NH_3. It also sets tight limit values for specific emission sources (e.g. combustion plant, electricity production, dry cleaning, cars and lorries) and requires BATs to be used to keep emissions down. VOC emissions from such products as paints or aerosols also have to be cut, and farmers are obliged to take specific measures to control NH_3 emissions.
  - The protocol was amended in 2012 to include national emission-reduction commitments to be achieved by 2020 and beyond (these amendments were ratified by the EU in Council Decision (EU 2017/1757). Several of the protocol's technical annexes were revised with updated sets of ELVs for both key stationary sources and mobile sources.
  - The revised protocol is the first binding agreement to include emission-reduction commitments for fine particulate matter. The amended protocol also specifically includes the short-lived climate pollutant black carbon (soot) as a component of particular matter. Reducing particulate matter (including black carbon) through the implementation of the protocol will reduce air pollution, while at the same time facilitate climate co-benefits.

Source: Europa-Lex site, Summary of LRTAP (2021).

year 2020 and the updating of ELVs addressing emissions of air pollutants at source (Europa Lex 2017). In other words, despite the failure to 'seal the deal' on climate change at the 2009 Copenhagen COP and despite the complete absence of any references to SLCPs and their established linkages to air pollution and public health burdens within the 2015 PA, a select group of countries committed to reducing PM pollution via curbing SLCPs at source.

The Gothenburg Protocol contains a series of specific tables which list emissions ceilings for a range of SLCPs. Table 1 reference emission ceilings for SO_2, NO_X, ammonia and VOCs from 2010 to 2020 expressed in thousands of metric tonnes for those parties that ratified the Gothenburg Protocol prior to 2010; and Tables 2–6 reference emission reduction commitments for SO_2, NO_X, NH_4, VOC and $PM_{2.5}$ for 2020 and beyond. These commitments are expressed as a percentage reduction from 2005 emission levels. Table 5.3 excerpted next provides the details on the $PM_{2.5}$ reduction commitments and should be taken note of by countries in regions of the world where the challenge of $PM_{2.5}$ is most acutely experienced. The significance of the public health and cost benefits of having a protocol that regulates PM pollution for the first time merits special emphasis. An article on the amended Gothenburg Protocol by the International Institute of Sustainable Development put it most succinctly: 'Implementing the amended Protocol's emission reduction measures are

Table 5.3 Emission reduction commitments for $PM_{2.5}$ for 2020 and beyond (excerpted from the 2012 amended Gothenburg Protocol).

	Convention party	Emission levels 2005 in thousands of tonnes of $PM_{2.5}$	Reduction from 2005 (%)
1	Austria	22	20
2	Belarus	46	10
3	Belgium	24	20
4	Bulgaria	44	20
5	Canada[a]		
6	Croatia	13	18
7	Cyprus	2,9	46
8	Czech Republic	22	17
9	Denmark	25	33
10	Estonia	20	15
11	Finland	36	30
12	France	304	27
13	Germany	121	26
14	Greece	56	35
15	Hungary	31	13
16	Ireland	11	18
17	Italy	166	10
18	Latvia	27	16
19	Lithuania	8,7	20

Table 5.3 (Continued)

	Convention party	Emission levels 2005 in thousands of tonnes of $PM_{2.5}$	Reduction from 2005 (%)
20	Luxembourg	3,1	15
21	Malta	1,3	25
22	Netherlands	21	37
23	Norway	52	30
24	Poland	133	16
25	Portugal	65	15
26	Romania	106	28
27	Slovakia	37	36
28	Slovenia	14	25
29	Spain	93	15
30	Sweden	29	19
31	Switzerland	11	26
32	United Kingdom of Great Britain and Northern Ireland	81	30
33	United States of America[b]		
34	European Union	1504	22

[a] Upon ratification, acceptance or approval of or accession to the present Protocol, Canada shall provide: (a) a value for total estimated PM emission levels for 2005, either national or for its pollutant emission management areas (PEMA), if it has submitted one; and (b) an indicative value for a reduction of total emission levels of PM for 2020 from 2005 levels, either at the national level or for its PEMA. Item (a) will be included in the table, and item (b) will be included in a footnote to the table. The PEMA, if submitted, will be offered as an adjustment to annex III to the Protocol.
[b] Upon ratification, acceptance or approval of or accession to the amendment adding this table to the present Protocol, the United States of America shall provide: (a) a value for total estimated $PM_{2.5}$ emission levels for 2005, either national or for a PEMA; and (b) an indicative value for a reduction of total $PM_{2.5}$ emission levels for 2020 from identified 2005 levels. Item (a) will be included in the table and item (b) will be included in a footnote to the table.

estimated to cost less than 0.01% of gross domestic product (GDP) for the EU, making it a cost-effective policy solution, given that healthcare and lost workday costs due to air pollution are estimated at between 2.5% and 7% of GDP per year in Western Europe and above 20% of GDP per year for countries in the pan-European region' (IISD 2019). What is useful to note is that Canada and the US also manage transboundary pollution bilaterally under the 1991 Canada–US Air Quality Agreement so the amended Protocol which is detailed in scope has specific clauses and limits on values that pertain separately to Canada and US. Also, of note to keep in mind is that the amended Gothenburg Protocol is a multi-pollutant protocol which is meant to eventually replace the older protocols. In other words, as more Parties ratify the amended Protocol, their obligations under the following existing Protocols – SO_2 (1985 Helsinki and 1994 Oslo Protocols), NO_X (1988 Sofia Protocol) and VOC (1991 Geneva Protocol) – are anticipated to become obsolete.

The UNECE in its 2019 report, 'Protecting the Air We Breathe', which reviewed the 40 years of progress and key milestones achieved within the CLRTAP was categorical about its achievements: 'The result of the collective effort over the last 40 years has been remarkable: emissions of a series of harmful substances have been reduced by 40 to 80 per cent since 1990 in the region. In particular, the decrease in sulphur emissions has led to healthier forest soils. Particulate matter concentrations at European measurement sites declined by around a third between 2000 and 2012. National average annual concentrations of fine particulate matter (PM2.5) fell by 33 per cent between 2000 and 2012 in the United States of America, and by 4 per cent in Canada. The number of days exceeding the World Health Organization's guideline level for ozone concentrations is now about 20 per cent lower than in 1990. The decoupling of economic growth and air pollution trends has prevented 600,000 premature deaths annually in Europe and North America. The average life expectancy has increased by 12 months, thanks to emission reductions' (2019, p. 3). Tangible achievements like the ones listed only make it that much harder for countries in other regions of the world to ignore the benefits of undertaking emissions reductions of SLCPs that offer win-win benefits for public health and climate mitigation.

One of the major lessons learned from the implementation of the CLRTAP and the evolution of its eight protocols is the significance of the role of scientific consensus and a data-driven policy process. The co-operative programme for monitoring and evaluation of the long-range transmission of air pollutants in Europe (termed the 'European Monitoring and Evaluation Programme' EMEP) is a scientifically based and policy-driven programme under the CLRTAP for international co-operation to solve transboundary air pollution problems. Figure 5.8 provides a schematic of the level of detail in terms of scientific assessment and modelling undertaken within the CLRTAP. Here too the UNECE 2019 report of the 40 years of progress is clear: 'The solid scientific underpinning of the Convention has been important for its success. The scientific network under the Convention has successfully developed a common knowledge providing for joint monitoring, modelling and effects-based programmes. The Convention has also served as a platform for scientists and policymakers to exchange information, leading to innovative approaches and mutual trust and learning' (2019, p. 3). But it is really the concluding paragraphs of the UNECE 2019 Report that outlines the importance of the 2012 amended Gothenburg Protocol as a key milestone in the development of the Convention which resulted in the first ever binding agreement to include emission reduction commitments for PM pollution: 'The Parties also broke new ground for the first time in international air pollution policy by specifically including the short-lived climate pollutant black carbon (or soot) as a component of particular matter. Reducing particulate matter (including black carbon) through the implementation of the Gothenburg Protocol is thus a major step in reducing air pollution, while at the same time facilitating climate and health co-benefits'. As the 2019 Report concluded: 'The success of the protocols has not only lain in setting emissions reduction targets that clearly indicate the desired results, but also in specifying how those reductions should be achieved: by applying uniform minimum technical emission standards based on the best available techniques and energy-efficiency requirements. These technology-based requirements help to ensure a level playing field for all of the Parties' (p. 18).

The success of CLTRAP needs to be further contextualized by the lack of substantive progress made towards regional and sub-regional protocols and measures in other regions

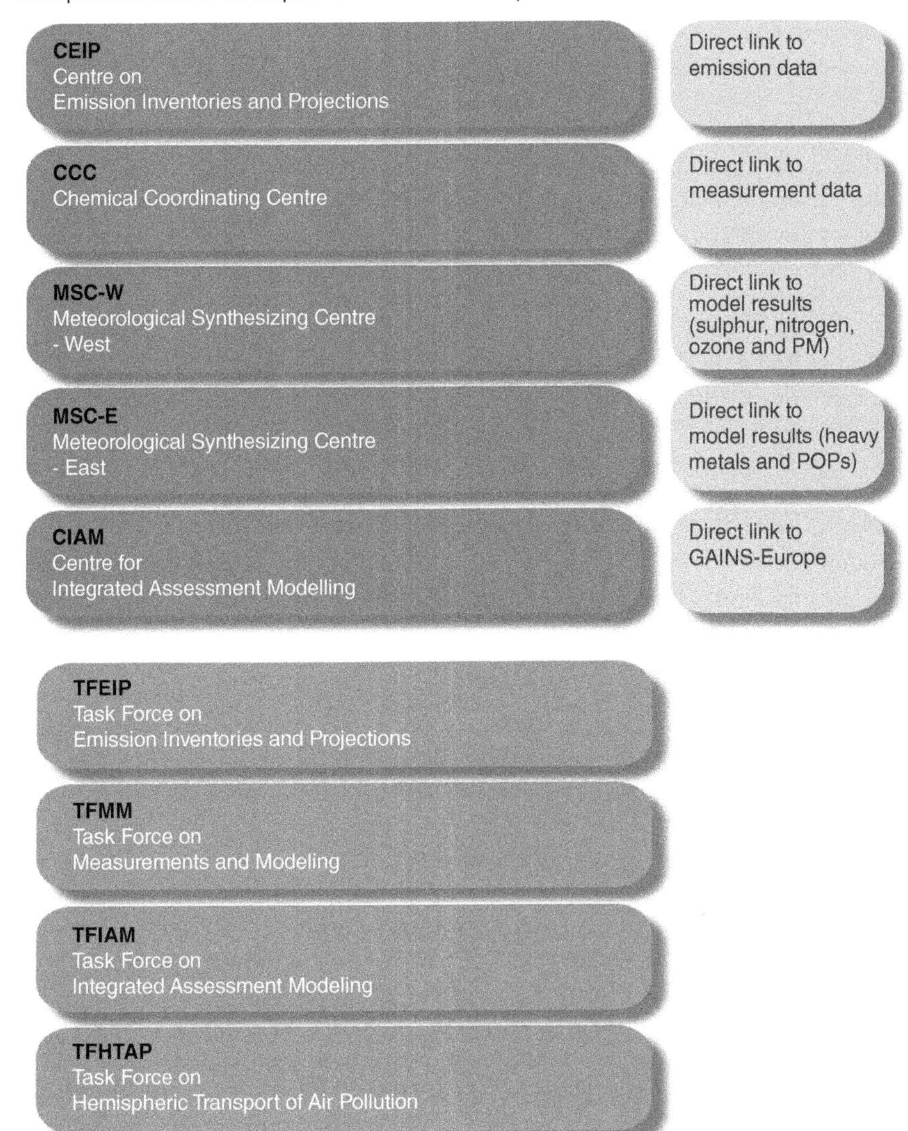

Figure 5.8 Science policy interface in CLRTAP: schematic overview of EMEP centres and task forces. *Source:* EMEP websites (2021). https://www.emep.int/ (accessed 1 August 2021).

particularly South Asia where PM pollution looms large. Surprisingly back in 1998, Bangladesh, Bhutan, India, Iran, Maldives, Nepal, Pakistan and Sri Lanka adopted the Malé Declaration which emphasized the need for countries to initiate research and programmes on air pollution within respective national frameworks. The Governing Council of South Asia Co-operative Environment Programme (SACEP) in partnership with UNEP and with the financial support from the Swedish International Development Agency implemented five phases related to the Declaration. According to study, Phase IV (2010–2014) was aimed at assisting member countries enhance their regional cooperation, monitoring, impact assessment and strengthen pre-existing initiatives, while Phase V (2014–2016) was aimed at promoting policy measures to control emissions of air pollution including SLCPs in South Asia and to ensure the sustainability and ownership of the Malé Declaration in the region. But a comprehensive 2019 study which included an evaluation of existing national air pollution measures, monitoring networks and stakeholders surveys for four of the Declaration signatories – Bangladesh, India, Nepal and Pakistan – conducted by The Energy Research Institute (TERI) with funding from the UK outlined the gaps summarized as follows:

- 'Limited knowledge of sectorial and geographical (including trans-boundary) contribution to ambient air quality, resulting in actions based on perception rather than actual contributions. The air quality management plans developed in the cities of the four countries are generally not based on scientific source apportionment studies and hence do not result in optimal, cost effective reduction in pollutant concentrations.
- Lack of intra country and inter country knowledge exchange coordination and collaboration of different institutions to improve air quality. There are examples of successful strategies which have led to some reduction of air pollution in the region, however, there is no mechanism in place for knowledge transfer and exchange for faster learning and improvement.
- Lack of regional scale inter-governmental assessments and planning. There are some studies to understand urban sources, but regional sources have not yet been fully understood. City level studies have suggested background contributions due to trans-boundary movement of pollution and hence regional scale studies and controls are required for effective control.
- Limitations of capacities in the regulatory bodies: financial, manpower, skills. Capacities in the regulatory bodies like pollution control authorities are limited in comparison to the scale and extent of the problem. Strengthening of capacities is a pre-requisite for adequate enforcement of environmental standards and laws in the four countries.
- Lack of enforcement mainly due to insufficient technical or resource capacity. Due to limitations of capacities, the existing laws are not being adequately enforced which lead to unaccounted emissions and hence deterioration of air quality.
- Limited industrial emission standards (especially for gaseous pollutants). There are several categories of industries where PM is still being controlled, but there are no standards for gaseous pollutants, which remain unattended and react in atmosphere to form secondary pollutants (also particles).
- Absence of controls on fugitive sources of ammonia and organic compounds. There are presently very limited regulation on releases of ammonia and organic compounds. These pollutants are themselves toxic and also produce secondary pollutants and hence also need to be regulated. Lack of integrated assessment of policies, action plan and projects to develop regional action plan. Sectoral/city level policies are decided in silos and integrated assessments are lacking for effective control. Regional scale integrated assessment can lead to improved efficiencies and reduced pollution levels.

- Air pollution is the least priority issue of the political parties of the region, this leads to lesser allocation of budget towards addressing issues related to air pollution. Enhanced awareness, sensitisation and capacity building can lead to more demand for better air quality and in turn can lead to more action and controls.
- Limited public awareness on air pollution issues due to inadequate information dissemination and absence of air quality forecasting framework. Except in India to some extent, there is almost no framework for forecasting and dissemination of air quality information in the region. This is essential for emergency response planning and short term control' (2019, pp. vi–vii).

The TERI report touched upon the morbidity, ill health reduced productivity of labour as well impacts on economic growth and tourism due to air pollution and estimated that the total cost of air pollution in India to be 4.5–7.7% of GDP, with costs expected to more than double by 2060, while in Pakistan, the direct cost of health damage linked with outdoor air pollution is estimated to be 6% of GDP (2019, p. 2). Curbing air pollution for cities and countries in developing countries that are not part of the CLRTAP should be understood within the broader impetus of public health including burden of disease and morbidity costs. The morbidity, health care and labour costs of air pollution are an analogue of an unclean air and unclean energy future as the threats of climate change including increased water and food insecurity dovetail with the disease burdens of PM pollution left unchecked.

5.4 Curbing the Toxic Pall Over Cities and Regions: Measures to be Considered

Integrated action to address climate change, reduce PM pollution and SLCPs and increase access to clean energy matters most to cities in countries where air pollution is on the increase. Clean air was not always the norm in cities like London for example where residents paid the price for solid fuel air pollution. Clean air was not always the norm in cities like London for example where residents paid the price for solid fuel air pollution. From Friday, 5 December to Tuesday, 9 December in 1952, London was enveloped by the Great Smog – an event that has been captured in cinematic and historic recounting. According to the United Kingdom's Meteorological Office, the weather in the preceding weeks had been very cold, and so to keep warm, Londoners burnt large quantities of coal in their homes. Under normal conditions, the smoke would have risen into the atmosphere and dispersed. But, an anticyclone was hanging over the region, and it caused an inversion, with the air close to the ground cooler than the air higher above it. So when the warm smoke came out of countless chimneys including factories and homes, it was trapped. The result was: 'A fog so thick and polluted it left thousands dead and wreaked havoc . . . The smoke-like pollution was so toxic it was even reported to have choked cows to death in the fields' (UK Met Office website 2021). Sulphur particles mixed with fumes from burning coal and made the yellow fog smell like rotten eggs. The deaths and devastation caused as result of the Great Smog propelled the UK to pass the Clean Air Act of 1956. Reduction of dark smoke pollution formed the basis of this 1956 UK Clean Air Act which was expressly adopted to abate 'the pollution of the air' and specified that 'dark smoke shall not be emitted from a chimney of any building, and if, on any day, dark smoke is so emitted, the occupier of the building shall be guilty of an offence'. What was striking about the 1956 Clean Air Act was that it specifically referenced the need for 'density meters': '... providing and installing apparatus for the purpose of indicating or recording (or indicating and recording) the density or darkness of smoke emitted from any furnace in any building or any furnace of any boiler or industrial plant not being a furnace in a building, or of facilitating the observation of smoke so emitted with a view to ascertaining its density or darkness', as well as a series of requirements that grit and dust from furnaces 'shall be minimized', that new furnaces 'arrest grit and dust' and that the grit and dust from furnaces 'be measured' (UK legislation website 1956).

Interestingly, it was only in 2016 that conclusive scientific proof as to exact cause of sulphate production in the 1952 London Fog was provided via atmospheric measurements in two Chinese megacities and complementary laboratory experiments. In their landmark study entitled, 'Persistent Sulfate Formation from London Fog to Chinese Haze', Wang et al. stated: 'The aqueous oxidation of SO_2 by NO_2 is key to efficient sulfate formation but is only feasible under two atmospheric conditions: on fine aerosols with high relative humidity and NH_3 neutralization or under cloud conditions. Under polluted environments, this SO_2 oxidation process leads to large sulfate production rates and promotes formation of nitrate and organic matter on aqueous particles, exacerbating severe haze development ... In addition to explaining the polluted episodes currently occurring in China and during the 1952 London Fog, this sulfate production mechanism is widespread, and our results suggest a way to tackle this growing problem in China and much of the developing world' (2016, p. 13630). Wang et al. highlighted that while current efforts have been focused primarily on minimizing SO_2 emissions, significant haze reduction may only be achievable by disrupting the sulphate formation process. They noted that in light of large contributions to urban NO_x, VOC and NH_3 levels from transportation, regulatory actions in minimizing traffic emissions may represent the critical step in mitigating severe haze in China.

In this regard it is also useful to briefly reflect upon the case of China where according to the 2020 World Air Quality Report, 86% of Chinese cities experienced cleaner air than 2019 and annual $PM_{2.5}$ exposure by population fell 11%. China still has 42 of the 100 most polluted cities globally but in terms of combating air pollution, there are according to UNEP, lessons to be learned from China. In its 20 year review of air pollution control, a 2019 UNEP report found that compared with 20 years earlier, the GDP, population and vehicles of Beijing sharply increased by 1078, 74 and 335%, respectively, at the end of 2017. The economic and urban growth 'resulted in the deterioration of the city's environment, especially air quality'. UNEP noted that the combined coal-vehicle pollution meant that 'heavy-pollution episodes occur regularly, with negative effects on public health'. To tackle severe air pollution, Beijing launched comprehensive air pollution control programmes in phases since 1998. In other words, there are no quick short-term fixes, but with ' . . . constant efforts in air pollution control, emission intensity decreased year by year and air quality has improved significantly. On-ground observation data shows that the annual average concentrations of SO_2, NO_2 and PM_{10} decreased by 93.3%, 37.8% and 55.3% respectively' (UNEP 2019, p. 6). Some key measures that the city of Beijing did over a 20 year period to curb air pollution are referenced by in Table 5.4. These measures undertaken by Beijing and are of relevance to other polluted cities in Asia.

Table 5.4 Review of 20 years of air pollution control in Beijing: key action areas and specific measures.

Key action areas	Specific measures
An effective air quality management system	A comprehensive and increasingly effective air quality management system has gradually taken shape over 20 years' practice. The system is characterized by (a) complete legislation and enforcement mechanism, (b) systematic planning, (c) powerful local standards, (d) strong monitoring capacity and (e) high public environmental awareness.
Economic incentives and financial support	In the past 20 years, Beijing has gradually established a number of local environmental economic policies, including subsidies, fees, pricing and other financial practices, to provide economic incentives for the effective implementation of various measures. Meanwhile, the spending on air pollution control has also been increased, especially after 2013 which manifest great ambitions of the government on air pollution control.
Emission reductions from coal combustion sources	Coal combustion has always been a major air pollution source in Beijing, and the city has continuously promoted end-of-pipe control and energy structure adjustment over the past 20 years. Focusing on power plants, coal-fired boilers and residential coal use, the pollution sources were controlled simultaneously, achieving remarkable progress. Take coal-fired power plants as an example. Beijing has implemented a 'coal-to-gas' policy since 2005 and reduced coal combustion by nearly 11 million tonnes by 2017. High-efficiency terminal treatment facilities were continuously renovated and ultralow emission standards were enforced during this period. In 2017, emissions of $PM_{2.5}$, SO_2 and NO_x were reduced by 97, 98 and 86%, respectively, compared with 20 years earlier resulting in significant environmental and health benefits.
Vehicle emission control	The prevention and control of vehicle pollution have long been a critical task in Beijing's air pollution control. Focusing on new vehicles, in-use vehicles and fuel quality, Beijing has implemented a series of local emission standards and comprehensive control measures, as well as strengthened traffic management and economic incentives continuously. The integrated 'Vehicle-Fuel-Road' framework was developed. More important, a large-scale public transport system has been built to allow gradual formation of a green and low-carbon in-city travel habit by the people. Although the number of vehicles increased threefold in Beijing during the past two decades, the total pollutant emissions decreased remarkably. Compared with 1998, CO, THC, NO_x and $PM_{2.5}$ emissions from the transportation sector were reduced by nearly 1105, 94, 71 and 6 kt in 2017 with a decrease rate of 89, 64, 55 and 81%, respectively. Phasing out older vehicles made the most significant contributions during this period.
Intensive pollution control during 2013–2017	The 'Beijing 2013–2017 Clean Air Action Plan' was the most comprehensive and systematic pollution control programme in Beijing. In 2017, Beijing's annual average $PM_{2.5}$ concentration lowered to 58 $\mu g/m^3$ and decreased by 35.6% compared with 2013, achieving the enhanced air quality goal which was generally considered difficult at home and abroad. Estimation of emission reductions by major control measures found that coal-fired boilers control, clean fuels in residential sectors and optimization of the industrial structure were the top three effective measures. During 2013–2017, the emissions of SO_2, NO_x, VOCs and $PM_{2.5}$ decreased by 83, 43, 42 and 55%, respectively.

(*Continued*)

Table 5.4 (Continued)

Key action areas	Specific measures
Co-ordination between Beijing and its surrounding areas	Besides enhancing local air pollution control, Beijing also actively sought to co-ordinate air pollution control measures with the surrounding areas. At the end of 2013, Beijing was asked to lead the establishment of the Mechanism for Coordinated Prevention and Control of Air Pollution in Beijing-Tianjin-Hebei and surrounding areas with the support of China's State Council. In 2017, the former Ministry of Environmental Protection identified the 28 cities in the Beijing-Tianjin-Hebei and surrounding areas as air pollution transportation channel. Through collaborative planning, unified standards, joint emergency response and information sharing, the air quality of the whole region has significantly improved. The annual average $PM_{2.5}$ concentrations of the Beijing-Tianjin-Hebei and surrounding areas decreased by nearly 25% during 2013–2017.
Future action	***(1) Considering synergistic control of $PM_{2.5}$ and O_3 pollution.***
	(2) Optimizing energy structure and energy efficiency simultaneously for a low-carbon development, to meet both air quality and climate target in future.
	(3) Working on vehicle emission control and transportation structure optimization to build a low-emission, high-efficiency transport system.
	(4) Strengthening the control of non-point source pollution.
	(5) Enhancing the coordination of Beijing-Tianjin-Hebei and surrounding areas.
	(6) Integrating city level environmental goal with 2030 sustainable development goals.

Source: UNEP (2019, pp. 7–14).

To further ground the idea that existing protocols and relatively straightforward measures to curb BC and other SCLPs have been discussed, negotiated and implemented, it is useful to consider that as far back as 2009, Molina et al. highlighted the need to implement strategies for short-lived non-CO_2 GHGs and particles, where existing agreements can be used to accomplish mitigation objectives. Molina et al. noted that BC was estimated to be the second or third largest warming agent, although there is uncertainty in determining its precise RF, but the uncertainty in estimation trended towards research finding that BC emissions impacts were much greater than anticipated. For instance, Flanner et al. reviewed existing BC estimates and based on their Snow, Ice and Aerosol Radiative (SNICAR) model found that 'the global climate response and efficacy of BC/snow forcing' was 'a three fold greater global temperature response than expected from equal forcing by CO_2' (2007, p. 2). Meanwhile, both Quinn et al. (2008) and Ramanathan and Carmichael provided an understanding of the significance of transboundary air pollution due to SLCPs such as BC for the Arctic and Himalayan regions. Quinn et al. referenced the first seasonally averaged forcing and corresponding temperature response estimates focused specifically on the Arctic. Their calculations revealed that climate forcing due to BC, CH_4 and O_3 leads to a positive surface temperature response indicating the need to reduce emissions of these SLCPs within and outside the Arctic. Ramanathan and Carmichael pointed to the gravity of the impacts of BC: 'Anthropogenic sources of black carbon, although distributed globally, are most concentrated in the tropics where solar irradiance is highest. Black carbon is often transported over long distances, mixing with other aerosols along the way. The aerosol mix can form transcontinental plumes of atmospheric brown clouds, with vertical extents of 3 to 5 km. Because of the combination of high absorption, a regional distribution roughly aligned with solar irradiance, and the capacity to form widespread atmospheric brown clouds in a mixture with other aerosols, emissions of black carbon are the second strongest contribution to current global warming, after carbon dioxide emissions. *In the Himalayan region, solar heating from black carbon at high elevations may be just as important as carbon dioxide in the melting of snowpacks and glaciers*' (emphasis added, 2008, p. 221).

In 2007 Cofala et al. used a global version of the Regional Air Pollution Information and Simulation (RAINS) model to estimate anthropogenic emissions of the air pollution precursors SO_2, NO_X, CO, BC, OC and CH_4, and developed two scenarios to constrain the possible range of future emissions. Based on their 2007 projections of future economic development and the air quality legislation, Cofala et al. estimated that global anthropogenic emissions of SO_2 and NO_X would slightly decrease between 2000 and 2030, while for BC and CO, reductions between 20 and 35% were computed, and for CH_4 an increase of about 50% was calculated (2007, p. 8486). However, the reality about the increase in BC emissions from a regional perspective in Asia, is markedly different from the preceding optimistic projection. Back in 2009, Wallack and Ramanathan wrote conclusively that fossil fuel use particularly diesel was 'responsible for about 35 percent of black carbon emissions worldwide'. They pointed out that available technologies that filter out BC diesel particulate such as 'filters on cars and trucks, for example, can reduce black carbon emissions by 90 percent or more with a negligible reduction in fuel economy'. They cited a study by the Clean Air Task Force, a US non-profit environmental research organization, that had estimated that retrofitting one million semitrailer trucks with these filters would yield the same benefits for the climate over 20 years as permanently removing over 165,000 trucks or 5.7 million cars from the road. As per Seedon Wallack and Ramananthan, the remaining 65% of BC emissions are associated with biomass burning including through forest fires, fires started by humans for clearing cropland and the use of organic fuels for cooking, heating and small-scale industry. Their recommendation still holds more than a decade later: 'Cleaner options for the man-made activities exist. The greenest options for households are stoves powered by the sun or by gas from organic waste, but updated designs for biomass-fueled stoves can also substantially cut the amount of black carbon and other pollutants emitted. Crop waste, dung, wood, coal, and charcoal are the cheapest, but also the least efficient and dirtiest fuels, and so households tend to shift away from them as soon as other options become reliably available. Thus, the challenge in lowering black carbon emissions is not convincing people to sacrifice their lifestyles, as it is with convincing people to reduce their carbon dioxide emissions. The challenge is to make other options available' (pp. 107–108).

The problem is that wide spread practice of solid fuel burning in poorly ventilated homes continues with grave health impacts in households and communities predominantly in Africa and Asia. Meanwhile, cities in these regions also face the highest levels of PM pollution as a result of rampant and unplanned growth in diesel intensive transportation and industrial practices. The 2019 World Air Quality jointly released by IQAir and Greenpeace found that a staggering 90% of the 200 cities dealing with the highest levels of $PM_{2.5}$ were in China and India. Bangladesh was listed as the country with the worst $PM_{2.5}$ pollution levels followed by Pakistan, Mongolia, Afghanistan and India (IQAir 2020). The 2020 World Air Quality report found that 84% of all monitored countries observed air quality improvements largely due to lock down measures to slow the transmission COVID-19, but only 24 of the 106 monitored countries met WHO annual guidelines for $PM_{2.5}$ in 2020. In the case of South Asia, the report highlighted: 'South Asia is the most polluted region of the world with Bangladesh, India and Pakistan sharing 42 of the 50 most polluted cities worldwide' (IQAir 2020).

The linkages between PM pollution and fossil fuel use have only been made more conclusive over time. Collaborative research conducted by the Max Planck Institute (Germany), the London School of Hygiene and Tropical Medicine (UK) and the Scripps Institute of Oceanography (US) based on a global model to estimate the climate and public health outcomes attributable to fossil fuel use demonstrated that phaseout of fossil fuels can avoid an excess mortality rate of 3.61 (2.96–4.21) million per year from outdoor air pollution worldwide; it could be up to 5.55 (4.52–6.52) million per year by additionally controlling non-fossil anthropogenic sources. Importantly, the researchers highlighted: 'Globally, fossil-fuel-related emissions account for about 65% of the excess mortality, and 70% of the climate cooling by anthropogenic aerosols. The chemical influence of air pollution on aeolian dust contributes to the aerosol cooling. Because aerosols affect the hydrologic cycle, removing the anthropogenic emissions in the model increases rainfall by 10–70% over densely populated regions in India and 10–30% over northern China, and by 10–40% over Central America, West Africa, and the drought-prone Sahel, thus contributing to water and food security' (2020, p. 7192).

TERI's 2019 scoping study on South Asian air pollution which examined Bangladesh, India, Nepal and Pakistan warned that the shortening of life due to air pollution has a disproportionate effect in South Asia because of the high levels of $PM_{2.5}$ in the region: According to the TERI report, South Asia has one of the highest concentrations of BC emissions from cars and trucks, cooking stoves and industrial facilities, but there is significant heterogeneity around sources of ambient air pollution amongst the four countries. It is noteworthy that the report pointed that 'Industrialization, urbanization and increased energy consumption are major drivers. Biomass burning and other intensified agricultural production practices are also major contributors in South Asia. Investments in air pollution abatement can potentially offer no-regrets options to boost economic development (through cost savings from energy-efficient technologies, improved transportation systems, enhanced urban facilities, reduced health costs and improved overall productivity of workforce as a result of reduction in the number of DALYs) as well as mitigate climate change (several air pollutants such as ozone and hydrofluorocarbons also contribute to global warming)' (2019, p. 3).

The UNEP/WMO 2011 Assessment Report identified a package of control measures to reduce SLCPs that can achieve 90% of total potential emissions reductions for BC, methane and HFCs. Many of these measures involve cost-effective technologies and practices that already exist. The role of controlling PM pollution associated with diesel exhaust needs to be emphasized as it consists primarily of solid carbonaceous particles of BC and OC, with the remaining mass composed of metals, ash and semi-volatile organics and secondary particles such as sulphates and nitrates.

SLCP emission reductions could be achieved with net cost savings. Adopting these measures would have major positive co-benefits for public health, especially in the developing world. As CCAC has underscored, if rapidly implemented, these measures can cut the amount of warming that would occur over the next few decades by as much as 0.6 °C, while avoiding 2.4 million premature deaths from outdoor air pollution annually by 2030, and preventing 52 million tonnes of crop losses per year (CCAC website 2021b). Table 5.5 provides a list of key measures to curb individual SLCPs. The list of measures derived from the CCAC outline a range of activities targeting individual pollutants and key emitting sector, but as noted by CCAC, the list needs to continue to evolve based on updated scientific findings and should not be considered exhaustive.

Table 5.5 List of measures to curb individual SLCPs by key emitting sector.

Methane – *40% emissions reduction potential globally by 2030*	
Agriculture	• Improve manure management and animal feed quality. • Apply intermittent aeration of continuously flooded rice paddies. • Improve animal health and husbandry by combining herd and health management, nutrition and feeding management strategies. • Introduce selective breeding to reduce emission intensity and increase production. • Promote farm-scale anaerobic digestion to control methane emissions from livestock. • Adopt guideline on healthy dietary choices.
Fossil fuels	• Carry out pre-mining degasification and recovery and oxidation of methane from ventilation air from coal mines. • Reduce leakage from long-distance gas transmission and distribution pipelines. • Extend recovery and utilization from gas and oil production. • Recover and use gas and fugitive emissions during oil and natural gas production.
Waste Management	• Separate and treat biodegradable municipal waste, and turn it into compost or bioenergy. • Upgrade wastewater treatment with gas recovery and overflow control. • Improve anaerobic digestion of solid and liquid waste by food industry. • Upgrade primary waste water treatment. • Divert organic waste. • Collect, capture and use landfill gas.

(*Continued*)

Table 5.5 (Continued)

Black carbon – *70% emissions reduction potential globally by 2030*	
Household energy	• Replace traditional cooking to clean-burning modern fuel cookstoves. • Replace traditional cooking and heating with clean-burning biomass stoves. • Eliminate kerosene lamps. • Replace lump coal with coal briquettes for cooking and heating. • Replace wood stove and burners with pellet stoves and boilers.
Industrial production	• Modernize traditional brick kilns to vertical shaft brick kilns. • Modernize coke ovens to recovery ovens.
Transport	• Use diesel particular filters for road and off-road vehicles. • Fast transition to Euro VI/6 vehicles and soot-free buses and trucks. • Eliminate high-emitting diesel vehicles.
Agriculture	• Ban open-field burning of agricultural waste.
Fossil fuels	• Capture and improve oil flaring and gas production.
Waste management	• Ban open burning of municipal waste.
Hydrofluorocarbons (HFCs) – *56% emissions reduction potential by 2050 (upon execution of policies under the Kigali Amendment)*	
Cooling and refrigeration	• Ratify and comply with the control measures of the Kigali Amendment. • Replace high global warming potential hydrofluorocarbons with low or zero global warming potential alternatives, combined with improvements in lifecycle energy efficiency. • Improve insulation materials and building designs to avoid the use of or reduce the need for air conditioners.

Sources: UNEP/WMO (2011), and CCAC website (2021b).

5.5 Conclusion

The WHO Special Report on health and climate change made it clear why health considerations including those related to SLCPs and air pollution are critical for the future. Box 5.3 contains a list of key factors identified by the WHO (2018) that address the nexus between curbing carbon emissions and SCLPs. The policy reality is that while developed countries have been the main historical contributors of anthropogenic GHG emissions, China and India are today the first and third largest aggregate GHG emitters. Key findings of the recently released IEA Global Energy Review 2021 (see Box 5.4) demonstrate aggressive fossil fuel growth despite the COVID pandemic. These findings prompt concern as to a global focus on building back better and greener, and serve to underscore the importance of role of NNSAs including cities and energy/private sector in addressing SLCPs and PM pollution as well as the need to better understand the role of CLRTAP and its Gothenburg Protocol.

Developing countries like India have a limited window of opportunity to recognize the full import of the layering of climate adversities, the need to increase access to clean energy

Box 5.3 Key factors identified by WHO related to curbing carbon and SLCPs.

1

Identify and promote actions to reduce both carbon emissions and air pollution, with specific commitments to reduce emissions of short-lived climate pollutants in Nationally Determined Contributions (NDCs) to the Paris Agreement.

2

Include the health implications of mitigation and adaptation measures in the design of economic and fiscal policies, including carbon pricing and the reform of fossil fuel subsidies.

3

Include the commitments to safeguard health from the UNFCCC and Paris Agreement, in the rulebook for the Paris Agreement, and systematically include health in NDCs. National Adaptation Plans and National Communications to the UNFCCC.

4

Remove existing barriers to investment in health adaptation to climate change, especially for climate-resilient health systems and 'climate-smart' health care facilities.

5

Facilitate and promote the engagement of the health community as trusted, connected and committed advocates for climate action.

6

Mobilize city mayors and other sub-national leaders as champions of intersectoral action to cut carbon emissions, increase resilience and promote health.

7

Systematically track progress in health resulting from climate change mitigation and adaption, and report to the UNFCCC, global health governance processes and the monitoring system for the SDGs.

Source: WHO (2018), Executive Summary.

services and take a more active role in ensuring that measures to curb SLCPs are not segregated from climate responsive action at the global and regional level. A 2018 WRI and Oxfam paper found early and ambitious action to reduce SLCPs to be essential to achieving the goals of the PA and the SDGs and that 'actions to reduce these highly potent pollutants help avoid crossing important thresholds a 1.5°C temperature rise above pre-industrial levels, and potential climatic tipping points, which will affect poor and vulnerable communities first and worst'. As in the case of the UNEP/WMO 2011 Assessment, the 2018 joint report by WRI and Oxfam also pointed out that curbing SLCPs can deliver multiple benefits for health, food security and poverty alleviation. It is noteworthy that the 2018 analysis highlighted that despite 'the importance of reducing SLCPs, actions to mitigate these potent pollutants were often underrepresented in the first nationally determined contributions (NDCs) submitted by Parties to the Paris Agreement', and more importantly underscored the significance of Parties scaling up ambition via new or updated NDCs by 2020 under the PA 'to incorporate and strengthen actions to reduce SLCPs in their NDCs' (Ross et al. 2018, p. 1). The WRI/Oxfam report authored by Ross et al. also included a sample list of actions and policies to support a new or updated NDC is provided in Table 5.6. A comparison with the UNEP/WMO 2011 and

Box 5.4 Key findings of IEA global energy review 2021.

Global energy related CO_2 emissions are heading for their second-largest annual increase ever	Demand for all fossil fuels is set to grow significantly in 2021. Coal demand alone is projected to increase by 60% more than all renewables combined, underpinning a rise in emissions of almost 5%, or 1500 Mt. This expected increase would reverse 80% of the drop in 2020, with emissions ending up just 1.2% (or 400 Mt) below 2019 emissions levels.
Sluggish demand for transport oil is mitigating the rebound in emissions	Despite an expected annual increase of 6.2% in 2021, global oil demand is set to remain around 3% below 2019 levels. Oil use for road transport is not projected to reach pre-COVID levels until the end of 2021. Oil use for aviation is projected to remain 20% below 2019 levels even in December 2021, with annual demand more than 30% lower than in 2019. A full return to pre-crisis oil demand levels would have pushed up CO_2 emissions a further 1.5%, putting them well above 2019 levels.
Global coal demand in 2021 is set to exceed 2019 levels and approach its 2014 peak	Coal demand is on course to rise 4.5% in 2021, with more than 80% of the growth concentrated in Asia. China alone is projected to account for over 50% of global growth. Coal demand in the United States and the European Union is also rebounding, but is still set to remain well below pre-crisis levels. The power sector accounted for only 50% of the drop in coal related emissions in 2020. But the rapid increase in coal-fired generation in Asia means the power sector is expected to account for 80% of the rebound in 2021.
Among fossil fuels, natural gas is on course for the biggest rise relative to 2019 levels	Natural gas demand is set to grow by 3.2% in 2021, propelled by increasing demand in Asia, the Middle East and the Russian Federation (Russia). This is expected to put global demand more than 1% above 2019 levels. In the United States – the world's largest natural gas market – the annual increase in demand is set to amount to less than 20% of the 20 bcm decline in 2020, squeezed by the continued growth of renewables and rising natural gas prices. Nearly three-quarters of the global demand growth in 2021 is from the industry and buildings sectors, while electricity generation from natural gas remains below 2019 levels.
Electricity demand is heading for its fastest growth in more than 10 years	Electricity demand is due to increase by 4.5% in 2021, or over 1 000 TWh. This is almost five times greater than the decline in 2020, cementing electricity's share in final energy demand above 20%. Almost 80% of the projected increase in demand in 2021 is in emerging market and developing economies, with the People's Republic China (China) alone accounting for half of globe growth. Demand in advanced economies remains below 2019 levels.
Renewables remain the success story of the COVID-19 era	Demand for renewables grew by 3% in 2020 and is set to increase across all key sectors – power, heating, industry and transport – in 2021. The power sector leads the way, with its demands for renewables on course to expand by more than 8%, to reach 8300 TWh, the largest year-on-year growth on record in absolute terms.

Source: IEA 2021, *Global Energy Review 2021*, accessed 24 April 2021.

Table 5.6 Sample policies and actions to support a new or updated NDC.

Gas	Sector	Sample policies and actions
Methane	Energy	• Promote the capture and utilization of gas and unintended fugitive emissions during oil and gas production. • Reduce leakage from long-distance natural gas transmission pipelines and distribution systems. • Promote pre-mine degasification and recovery and oxidation of methane from ventilation air from coal mines.
	Agriculture	• Promote the intermittent aeration of continuously flooded rice paddies and provide sufficient support for farmers – particularly, smallholder and women farmers – to adopt locally relevant best practices. • Promote reduction of enteric fermentation in livestock through dietary supplements and shifts (e.g. from a cellulosic to a starch-based diet) with sufficient support for farmers, pastoralists and herders, particularly those in poor and vulnerable communities. • Support farmers to implement livestock anaerobic digestion projects. • Review national dietary guidelines to promote the consumption of less meat and more plant protein.
	Waste	• Recover and utilize methane emitted from waste. • Improve waste and wastewater management/upgrade wastewater treatment with gas recovery and overflow control. • Promote the treatment of biodegradable municipal waste and landfill gas collection. • Reduce food loss and waste.
HFCs	Economy-wide	• Increase the percentage of low-Global Warming Potential (GWP) alternatives in economy-wide uses of HFCs, consistent with the HFC phase-down level under the Kigali Amendment to the Montreal Protocol. • Commit to exceeding the country's current Kigali phase-down schedule.
	Sector-specific	• Provide incentives for companies and consumers to replace high GWP HFC commercial equipment or appliances with low-GWP alternatives. • Adopt similar policies to countries/regions that have more stringent F-gas regulations, such as the EU, with or without delay. • Replace high GWP HFCs with low-impact alternatives in specific classes of appliances and equipment, such as using R-290 instead of HFC-410a in room air conditioners. • Introduce a policy requiring all new high-efficiency cooling equipment to use either a low-GWP HFC or an HFC alternative. • Update public procurement processes to transition away from high GWP HFCs.

(*Continued*)

Table 5.6 (Continued)

Gas	Sector	Sample policies and actions
Black carbon	Transport	• Promote diesel particulate filters for road and off-road vehicles.
		• Develop electromobility strategies and/or introduce a policy or legal framework (and associated incentives) to replace internal combustion engine vehicles with electric vehicles.
		• Eliminate high-emitting vehicles from road and off-road transport and/or public transportation.
		• Develop an integrated and sustainable strategy for transport modes in megacities and/or expand towards a greener and more sustainable public transport system.
	Agriculture	• Ban open-field burning of agricultural waste while ensuring sufficient support for farmers – particularly smallholder and women farmers – to transition to more sustainable growing practices.
	Residential	• Replace coal with charcoal briquettes used in cooking and heating stoves in ways that do not cause financial hardship to poor and vulnerable communities and that support, in particular, women's rights.
		• Introduce clean-burning biomass stoves for cooking and heating in developing countries in ways that respect local preferences, do not cause financial hardship to poor and vulnerable communities and support, in particular, women's rights.
	Industry	• Replace traditional brick kilns with vertical shaft kilns and Hoffman kilns.
		• Replace traditional coke ovens with modern recovery ovens, including improving end-of-pipe abatement measures.

Note: The policies and actions listed in Table may not necessarily enhance the mitigation ambition of a Party's NDC. To constitute enhanced mitigation ambition, the cumulative emissions under the new NDC would need to be lower than under the original NDC. In addition, SLCP mitigation actions should be implemented in a rights-based and gender-just way that respects and responds to community needs and capacity constraints. This is because many SLCP sources (e.g. biomass-based cooking, rice production and livestock rearing) are often linked to poor and vulnerable populations, including smallholder farmers, many of whom are women.
Source: Ross et al. (2018, p. 4).

CCAC recommendations shows a close overlap between the measures which indicates a consistency as to available BC and SLCP mitigation policies over time.

The global community must take stock of the fact that linked action on climate and fossil fuel air pollution is both necessary and urgent in the most populous and polluted countries and cities in the world. The 2017 COP-23s motto – 'Further, Faster Together' has yet to be realized especially when it comes to the linkages between clean energy for all and climate responsive action for the poor. While the previous US administration's withdrawal from the PA made the word 'together' seem somewhat of a non sequitur, exactly how much further, faster and together, ambitious climate agreements will be achieved now with the US re-entry into the PA remains to be seen. Certainly, the Biden Administration has scaled up US commitment by reducing GHG emissions by 50–52% compared to 2005 levels, but it has also clearly signaled that: '. . . countries across the globe must also step up. Given that more than 85 percent of emissions come from beyond U.S. borders, domestic action must go hand in hand with international leadership. All countries – and particularly the major economies – must do more to bend the curve on global emissions so as to keep a 1.5 degree C limit on global average temperature rise within reach' (White House Fact Sheet 2021). It was also noteworthy that the Biden Administration reconvened the Major Economies Forum (MEF) – a group of 17 of the world's largest economies responsible for approximately 80% of global GHGs in a two day Summit (22-23 April 2021). The idea for a major GHG emitters group was formally launched by President Barack Obama as the MEF, and it clearly continues to bring together the world's largest aggregate emitters, including key developing countries such as Brazil, South Africa, India and China (also referred to as BASIC Countries). And, for the first time during the 2021 MEF Summit, there was an explicit reference to 'reducing' SLCPs and 'supporting the most vulnerable' within the context of enhancing MEF targets to reach net zero emissions by 2050 (White House Fact Sheet 2021).

The need to engage more actively on SLCPs such as HFCs has received increased global traction with the Biden Administration signalling its intention to focus on the Kigali Amendment to the Montreal Protocol. What is striking about the US announcement within the MEF Summit convened outside of the PA negotiations is just how extensive the agenda of topics covered, which include the issue area of 'Providing urgent support for vulnerable countries to adapt and build resilience to the climate crisis'. The Biden White House has provided clear acknowledgement of the fact that: 'Communities of color and low-income communities around the country are particularly vulnerable to climate change. Abroad, many vulnerable countries already are facing catastrophic climate impacts. They must build their resilience to the climate crisis now'. In support of this goal, the US administration referenced that it would be working towards measures that included:

- '*Supporting environmental justice and climate resilience.* EPA will fund $1 million in grants/cooperative agreements through the Commission on Environmental Cooperation (CEC) to work with underserved and vulnerable communities, including indigenous communities, in Canada, Mexico, and the United States to prepare them for climate related impacts.
- '*Reducing black carbon by investing in clean cookstoves.* Household energy emissions have a significant impact on the climate, environment, human health, gender, and livelihoods. In addition, the reduction of short-lived climate pollutants, such as methane and black carbon, can in the short term contribute significantly to keeping a 1.5 degree C limit on global average temperature rise within reach. Given the urgent need for tangible, ambitious, and global action, the U.S. government is announcing that it is resuming and strengthening its commitment to the United Nations Foundation's Clean Cooking Alliance. The U.S. Environmental Protection Agency (EPA) will work with the Clean Cooking Alliance, other governments, and partners to reduce emissions from home cooking and heating that contribute to climate change and also directly affect the health and livelihoods of almost 40 percent of the world's population.
- *Mitigating black carbon health impacts in Indigenous Arctic communities.* EPA, working through our partners in the Arctic Council, is pleased to announce the Black Carbon

Health in Indigenous Arctic Communities project to be implemented by the Aleut International Association. Indigenous Arctic communities need tools to understand their exposure to black carbon emissions, to help them identify significant local sources, and to share best practices for preventing and mitigating the health impacts of air pollution and climate. The project will help these communities measure, analyze, and addresses black carbon exposure and strengthen their capacity to develop and promote black carbon mitigation strategies' (emphasis included, White House Fact Sheet 2021).

SLCPs have significantly impacted food, water and public health security in certain areas of the world – Africa and Asia predominantly – which are also widely anticipated to suffer the worst impacts of climatic adversities including drastically warmer surface air temperatures and increased drought and desertification in tandem with worsening water and food insecurity. What the relatively short atmospheric lifetime of SLCPs combined with their strong warming potential essentially translates into is that measures to reduce SLCPs can result in climate, human well-being and development benefits within a short-term horizon.

Policy measures and actions to curb SLCPs such as BC are critical to the overall global climate mitigation challenge in the short term, and more importantly in the case of cities within countries like India and regions like South Asia come with the attendant benefits of reducing negative impacts on public health and food and water security. While action to reduce GHGs such as CO_2 are fundamental to addressing climate change, there is another more immediate series of measures related to the curbing of SLCPs which have not been consistently focused upon at the global and/or regional level within countries where community residents face toxic levels of PM pollution. The economic burden of air pollution and the continued prevalence of unclean air resulting from the inefficient combustion of fossil fuels weighs heavily upon residents of cities in developing countries. Premature death, illness, destroyed livelihoods along with health care systems that can least afford to support the additional burdens of ill health caused by air pollution exposure undercut the already precarious infrastructure in many of the most congested and polluted urban cities in the world.

Curbing toxic air pollution is directly associated with a host of broader development challenges – increased urbanization, economic development, poverty reduction, public health and sustainable transportation – and requires difficult socio-economic and political choices to be made. But adopting occasional and/or a random pastiche of anti-pollution measures that are not context specific and long term is arguably the analogue of an ostrich burying its head in the sand most of the time and peeking out in the midst of a sandstorm. Measures such as sustainable mass transit options, diesel particulate filters, stringent traffic regulations governing trucks and passenger vehicles, monitoring protocols to reduce discharge and dust at construction sites and banning of open-field burning of waste are some ways by which PM air pollution can be curbed. But there is another nature-based fix – urban tree corridors and green spaces – which may seem commonsensical but are extremely challenging to successfully implement in the midst of the most congested and populous cities where sprawling and dense slums sit cheek by jowl with sleek urban enclaves. Urban tree corridors and green spaces can be a way to remove PM pollution. A 2013 study conducted by the US Forest Service and the Davey Institute estimated the amounts of $PM_{2.5}$ removed by urban trees and forests in 10 US cities. The study found the greatest effect of trees on reducing health impacts of $PM_{2.5}$ occurred in New York City due to its relatively larger population size and the moderately high removal rate and reduction of $PM_{2.5}$ concentrations based on tree coverage.

Finding the appropriate mix of responses that can tackle the PM pollution in the midst of urbanization and poverty considerations requires citizenry advocacy and involvement, as well as careful policy planning, implementation and investment. To be clear, urban green spaces are not easy or quick fixes, especially in the context of the most polluted, densely populated cities of the world. Urban tree corridors along congested highways, and green areas in the midst of dense urban metropolises are difficult to implement and maintain in the face of pressing poverty reduction needs and development pressures. Then there are additional concerns about ensuring that urban green spaces be carefully crafted using native and appropriate species that do not compete for already scarce water resources, do not displace human development infrastructure/needs and do not result in ecosystem damage and loss. But finding the will to invest in urban green spaces and invest in clean energy access for all including access to sustainable public transportation will pay dividends in human health for the most polluted cities in the world.

References

Bond, T.C. and Sun, H. (2005). Can reducing black carbon emissions counteract global warming? *Environmental Science and Technology* 39 (16): 5921–5926. https://pubs.acs.org/doi/abs/10.1021/es0480421 (accessed 3 July 2021).

Bond, T.C. et al. (2007). Historical emissions of black and organic carbon aerosol from energy-related combustion, 1895-2000. *Global Biogeochemical Cycles* 21 (2) GB 2018: 1–16. https://agupubs.onlinelibrary.wiley.com/doi/full/10.1029/2006GB002840 (accessed 3 July 2021).

Bond, T.C., Doherty, S.J., Fahey, D.W., Forster, P.M., Berntsen, T.K., DeAngelo, B.J., et al. (2013). Bounding the role of black carbon in the climate system: a scientific assessment. *Journal of Geophysical Research: Atmospheres*. doi:https://doi.org/10.1002/jgrd.50171.

Burney, J. et al. (2016). Getting serious about the new realities of climate change. *Bulletin of the Atomic Scientists*. Sage Publications 69 (4): 49–57. https://journals.sagepub.com/doi/pdf/10.1177/0096340213493882 (accessed 3 July 2021).

Carcaillet, C. et al. (2002). Holocene biomass burning and global dynamics of the carbon cycle. *Chemosphere* 40 (8): 845–863. https://www.sciencedirect.com/science/article/abs/pii/S0045653502003855?via%3Dihub (accessed 4 July 2021).

CCAC (2015). Annual Report: 2015. CCAC. https://www.ccacoalition.org/en/resources/climate-and-clean-air-coalition-annual-report-2014-2015 (accessed 20 August 2021).

CCAC website (2021a). Black Carbon. https://www.ccacoalition.org/en/slcps/black-carbon (accessed 4 July 2021).

CCAC website (2021b). Short Lived Climate Pollutants Solutions. https://www.ccacoalition.org/en/content/short-lived-climate-pollutant-solutions (accessed 14 July 2021).

Cofala, J. et al. (2007). Scenarios of global anthropogenic emissions of air pollutants and methane until 2030. *Atmospheric Environment* 41 (38): 8486–8499, ISSN 1352-2310. https://doi.org/10.1016/j.atmosenv.2007.07.010. and https://www.sciencedirect.com/science/article/pii/S135223100700622X (accessed 10 July 2021).

Dickerson, R.R. et al. (2002). Analysis of black carbon and carbon monoxide observed over the Indian Ocean: implications for emissions and photochemistry. *Journal of Geophysical*

Research 107 (D 19): INX2 16-1–INX@16-11. https://agupubs.onlinelibrary.wiley.com/doi/full/10.1029/2001JD000501 (accessed 11 July 2021).

EMEP website (2021). EMEP Home. https://www.emep.int/ (accessed 1 August 2021).

EPA (2020). The Legacy of EPA's Acid Rain Research. https://www.epa.gov/sciencematters/legacy-epas-acid-rain-research (accessed 3 July 2021).

Europa Lex (2017). Council Decision: (EU) 2017/1757 Amendment to the 1999 Protocol to the 1979 Convention on Long-Range Transboundary Air Pollution to Abate Acidification, Eutrophication and Ground-Level Ozone. https://eur-lex.europa.eu/legal-content/EN/TXT/?uri=celex%3A32017D1757#ntr18-L_2017248EN.01000501-E0018 (accessed 11 July 2021).

Europa-Lex site, Summary of CLRTAP (2021). https://eur-lex.europa.eu/legal-content/EN/TXT/?uri=LEGISSUM%3Al28162 (accessed 22 July 2021).

European Society of Cardiology (2020). Study estimates exposure to air pollution increases COVID-19 deaths by 15% worldwide. https://www.escardio.org/The-ESC/Press-Office/Press-releases/study-estimates-exposure-to-air-pollution-increases-covid-19-deaths-by-15-world (accessed 22 July 2021).

Flanner, M. et al. (2007). Present-day climate forcing and response from black carbon in snow. *Journal of Geophysical Research: Atmospheres* **112**: D11. https://agupubs.onlinelibrary.wiley.com/doi/epdf/10.1029/2006JD008003 (accessed 11 July 2021).

Forster, P., Ramaswamy, V., Artaxo, P. et al. (2007). Changes in atmospheric constituents and in radiative forcing. In: *Climate Change: The Physical Science Basis. Contribution of Working Group I to the Fourth Assessment Report of the Intergovernmental Panel on Climate Change* (ed. S. Solomon, D. Qin, M. Manning, et al.). Cambridge UK: Cambridge University Press https://www.ipcc.ch/site/assets/uploads/2018/02/ar4-wg1-chapter2-1.pdf (accessed 10 July 2021).

Hansen, J. and Nazarenko, L. (2004). Soot climate forcing via snow and ice albedos. *Proceedings of the National Academies of the Sciences (PNAS)* 101 (2): 423–428. https://www.pnas.org/content/pnas/101/2/423.full.pdf (accessed 11 July 2021).

Hansen, J. et al. (2000). Global warming in the twenty-first century: an alternative scenario. *Proceedings of National Academies of the Science (PNAS)* 97 (18): 9875–9880.

HEI (2019). *State of Global Air 2019*. Boston: Health Effects Institute http://stateofglobalair.org/resources (accessed 20 August 2021).

HEI (2020). *State of Global Air 2020*. Boston: Health Effects Institute https://www.stateofglobalair.org/resources?resource_category=All&page=4 (accessed on 20 August 2021).

ICCT (2009). *A policy-relevant summary of black carbon climate science and appropriate emission control strategies*. London: ICCT https://theicct.org/wp-content/uploads/2021/06/BCsummary_dec09.pdf (accessed on 20 August 2021).

Im, E.S. et al. (2017). Deadly heat waves projected in the densely populated agricultural regions of South Asia. *Science Advances* 3 (8): e1603322. https://advances.sciencemag.org/content/3/8/e1603322/tab-pdf (accessed 24 July 2021).

IEA (2021). *Global Energy Review 2021*. Paris: IEA https://www.iea.org/reports/global-energy-review-2021 (accessed on 24 April 2021).

International Institute for Sustainable Development (2019). Protocol Regulating Fine Particulate Matter Enters into Force. https://sdg.iisd.org/news/air-pollution-protocol-regulating-fine-particulate-matter-enters-into-force (accessed 21 July 2021).

IPCC (2014a). *Summary for Policymakers in Climate Change 2014: Mitigation of Climate Change. Contribution of Working Group III to the Fifth Assessment Report of the Intergovernmental Panel on Climate Change*. Cambridge, UK: Cambridge University Press.

https://www.ipcc.ch/site/assets/uploads/2018/02/ipcc_wg3_ar5_summary-for-policymakers.pdf (accessed 12 December 2021).

IPCC (2014b) *Technical Summary in Climate Change 2014: Mitigation of Climate Change.* Contribution of Working Group III to the Fifth Assessment Report of the Intergovernmental Panel on Climate Change. https://www.ipcc.ch/site/assets/uploads/2018/02/ipcc_wg3_ar5_technical-summary.pdf (accessed 12 December 2021).

IPCC (2018). *IPCC, 2018: Global Warming of 1.5°C.* Cambridge UK: Cambridge University Press https://www.ipcc.ch/site/assets/uploads/sites/2/2019/06/SR15_Full_Report_High_Res.pdf (accessed 18 July 2021).

IPCC (2021). *Summary for Policymakers-Climate Change 2021: The Physical Science Basis: Contribution of Working Group I to the Sixth Assessment Report of the Intergovernmental Panel on Climate Change.* Cambridge: Cambridge University Press.

IQ Air (2020). *World Air Quality Report: 2020.* Geneva: IQ Air https://www.iqair.com/world-air-quality-report (accessed 3 July 2021).

Jacobson, M. (2004). Climate response of fossil fuel and biofuel soot accounting for soot's feedback to snwo and sea ice albedo and emissivity. *Journal of Geophysical Research* 109 (D21): https://agupubs.onlinelibrary.wiley.com/doi/full/10.1029/2004JD004945 (accessed 25 July 2021).

Klimont, Z. et al. (2016). Global anthropogenic emissions of particulate matter including black carbon. *Atmospheric Chemistry and Physics Discussions* 17: 8681–8723. https://doi.org/10.5194/acp-17-8681-2017 (accessed 25 July 2021).

Kovats, S. and Hajat, S. (2008). Heat stress and public health: a critical review. *Annual Review of Public Health* 29: 41–55.

Lelieveld, J. et al. (2019). Effects of fossil fuel and total anthropogenic emission removal on public health and climate. *Proceeding of the National Academy of Sciences* 116 (15): 7192–7197. https://www.pnas.org/content/116/15/7192 (accessed 17 July 2021).

Mansharamani, A. and Shrivastava, A. (2020). Black carbon is a threat to Himalayan glaciers. *Down to Earth.* https://www.downtoearth.org.in/blog/climate-change/black-carbon-is-a-threat-to-himalayan-glaciers-74542 (accessed 17 July 2021).

Menon, S. et al. (2010). Black carbon aerosols and the third polar ice cap. *Atmospheric Chemistry and Physics* 10: 4559–4571. https://doi.org/10.5194/acp-10-4559 (accessed 18 July 2021).

Ministry of Science and Technology/Government of India (2021). Black carbon could lead to premature mortality: a study. Press release (30 June 2021). https://www.pib.gov.in/PressReleasePage.aspx?PRID=1731444 (accessed 24 July 2021).

Molina, M. et al. (2009). Reducing abrupt climate change risk using the Montreal Protocol and other regulatory actions to complement cuts in CO_2 emissions. *Proceeding of the National Academy of Sciences* 106 (49): 20616–20621. https://www.pnas.org/content/106/49/20616#ref-47 (accessed 24 July 2021).

Myhre, G., Shindell, D. et al. (2013). Anthropogenic and natural radiative forcing. In: *Climate Change 2013: The Physical Science Basis. Contribution of Working Group I to the Fifth Assessment Report of the Intergovernmental Panel on Climate Change.* Cambridge UK: Cambridge University Press https://www.ipcc.ch/site/assets/uploads/2018/02/WG1AR5_Chapter08_FINAL.pdf (accessed 7 July 2021).

Novakok, T. and Hansen, J. (2004). Black carbon emissions in the United Kingdom during the past four decades: an empirical analysis. *Atmospheric Environment* 38 (25): 4155–4163. https://www.sciencedirect.com/science/article/abs/pii/S1352231004004364?via%3Dihub (accessed 24 July 2021).

Pal, J. and Eltahir, E. (2016). Future temperature in southwest Asia projected to exceed a threshold for human adaptability. *Nature Climate Change* **6**: 197–200. (2016).

IPCC (2007). In: *Climate Change 2007: Impacts, Adaptation and Vulnerability. Contribution of Working Group II to the Fourth Assessment Report of the Intergovernmental Panel on Climate Change, Asia* (ed. Parry, M.L. et al., Cambridge: Cambridge University Press. p. 469–506.

Penner, J.E. et al. (1993). Towards the development of a global inventory for black carbon emission. *Atmospheric Environment* 27 (8): 1277–1295. https://doi.org/10.1016/0960-1686(93)90255-W (accessed 25 July 2021).

Quinn, P.K. et al. (2008). Short-lived pollutants in the Arctic: their climate impact and possible mitigation strategies. *Atmospheric Chemistry and Physics* 8: 1723–1735. https://doi.org/10.5194/acp-8-1723-2008 (accessed 25 July 2021).

Ramanathan, V., Crutzen, P.J. et al. (2002). The Indian Ocean experiment and the Asian brown cloud. *Current Science* 83 (8): 947–955.

Ramanathan, V. and Carmichael, G. (2008). Global and regional climate changes due to black carbon. *Nature Geoscience* **1**: 221–227.

Ramanathan, V. et al. (2016). The Next Front on Climate Change: How to Avoid a Dimmer, Drier World. *Foreign Affairs*. March/April 2016. https://www.foreignaffairs.com/articles/2016-02-16/next-front-climate-change (accessed 12 December 2021).

Raymond, C. et al. (2020). The emergence of heat and humidity too severe for human tolerance. *Science Advances* (8 May 2020) 6 (19): https://www.science.org/doi/10.1126/sciadv.aaw1838 (accessed 23 July 2021).

Ross, K. et al. (2018). *Strengthening Nationally Determined Contributions to Catalyze Actions that Reduce Short Lived Climate Pollutants. WRI and Oxfam Working Paper*. Washington DC: WRI https://files.wri.org/d8/s3fs-public/18_WP_SLCPs_toprint2.pdf (accessed 25 July 2021).

Royal Society (2008). *Ground-Level Ozone in the Twenty-First Century: Future Trends, Impacts and Policy Implications*. London: Royal Society https://royalsociety.org/-/media/Royal_Society_Content/policy/publications/2008/7925.pdf (accessed 11 July 2021).

Seedon Wallack, J. and Ramanathan, V. (2009). The other climate changers: why black carbon and ozone also matter. *Foreign Affairs* 88 (5): 105–113. https://www.foreignaffairs.com/articles/2009-09-01/other-climate-changers (accessed 17 July 2021).

Seiler, W. and Crutzen, P. (1980). Estimates of gross and net fluxes of carbon between the biosphere and the atmosphere from biomass burning. *Climatic Change* 2 (3): 207–247.

Sherwood, S. and Huber, M. (2010). An adaptability limit to climate change due to heat stress. *Proceeding of the National Academy of Sciences* 107 (21): 9552–9555. https://doi.org/10.1073/pnas.0913352107 (accessed 17 July 2021).

Sun, T. et al. (2019). Constraining a historical black carbon emission inventory of the United States for 1960–2000. *Journal of Geophysical Research: Atmospheres* 124 (7): 4000–4025. https://agupubs.onlinelibrary.wiley.com/doi/10.1029/2018JD030201#jgrd55339-bib-0027 (accessed 17 July 2021).

TERI (2019). *Scoping Study for South Asia Air Pollution*. New Delhi: TERI https://assets.publishing.service.gov.uk/media/5cf0f3b0e5274a5eb03386da/TERI_Scoping_Study_final_report_May27_2019.pdf (accessed 7 August 2021).

UK legislation website (1956). Clean Air Act. https://www.legislation.gov.uk/ukpga/Eliz2/4-5/52/enacted (accessed 25 July 2021).

UK Met Office website (2021). Great Smog. https://www.metoffice.gov.uk/weather/learn-about/weather/case-studies/great-smog (accessed 25 July 2021).

UN Treaty Collection website (2021). Convention on Long-range Transboundary Air Pollution: Status of Ratification. https://treaties.un.org/Pages/ViewDetails.aspx?src=TREATY&mtdsg_no=XXVII-1&chapter=27&clang=_en (accessed 18 July 2021).

UNECE (2019). *Protecting the Air We Breathe*. Geneva: UNECE https://unece.org/sites/default/files/2021-06/1914867_E_ECE_EB_AIR_NONE_2019_3_200dpi.pdf (accessed 25 July 2021).

UNEP (2019). *A Review of 20 Years' Air Pollution Control in Beijing*. Nairobi: UNEP https://wedocs.unep.org/bitstream/handle/20.500.11822/27645/airPolCh_EN.pdf?sequence=1&isAllowed=y (accessed 25 July 2021).

UNEP/WMO (2011). *Integrated Assessment of Black Carbon and Tropospheric Ozone: Summary for Decision Makers*. Nairobi: Kenya.

Wang, G., Zhang, R. et al. (2016). Persistent sulfate formation from London Fog to Chinese haze. *PNAS* 113 (48): 13630–13635. https://www.pnas.org/content/pnas/113/48/13630.full.pdf (accessed 5 June 2021).

White House (2021). *Fact Sheet: President Biden's Leaders Summit on Climate*. (23 April 2021). https://www.whitehouse.gov/briefing-room/statements-releases/2021/04/23/fact-sheet-president-bidens-leaders-summit-on-climate (accessed 4 July 2021).

WHO (2018). *COP-24: Special Report – Health and Climate Change*. Geneva: WHO https://apps.who.int/iris/handle/10665/276405 (accessed 22 April 2021).

World Bank (2014). *Reducing Black Carbon Emissions from Diesel Vehicles: Impacts, Control Strategies, and Cost-Benefit Analysis*. Washington DC: World Bank https://openknowledge.worldbank.org/bitstream/handle/10986/17785/864850WP00PUBL0l0report002April2014.pdf?sequence=1&isAllowed=y (accessed 4 July 2021).

Xu, B., Cao, J., Hansen, J. et al. (2009). Black soot and the survival of tibetan glaciers. *PNAS* 105 (52): 22114–22118. https://www.researchgate.net/publication/235608346_Black_soot_and_the_survival_of_Tibetan_glaciers (accessed 1 July 2021).

6

The Nexus between Mitigating Air Pollution and Climate Change is Crucial

Time to Stop Knuckle-dragging, Break Global Policy Silos and Spur NNSAs

6.1 Urgency of Integrated Action on Clean Air and Climate: Reframing and Breaking Silos

The UN and its member states have been pursuing a three-decade-old intergovernmental negotiations quest to address the mitigation and adaptation challenges associated with anthropogenic climate change. The 1988 formal UNGA recognition of climate change has become an irrevocable part of global policy at the UN and within the broader sustainable development community. Yet, climate change has still not been effectively integrated as an overarching goal that cuts across multiple SDGs within the context of the UN's ambitious SDA, whilst the UN's PA does not offer any consistent pathways or partnerships to curb the single largest environmental health risk facing the most populous cities in the world- curbing fossil fuel related air pollution. Anthropogenic climate change has consistently been referenced as the defining challenge of our times by numerous global leaders, ranging from the Pope to the CEO of massive corporations and two consecutive UN SGs, but still global goals and partnership silos persist on the inherently linked global challenges of increasing access to sustainable energy for all and climate change as represented by the largely segregated SDGs 7, 13 and 17.

The UN-led global community including global climate focused celebrities and advocates have convened within the context of countless global and high-level summits meant to galvanize action. In spite of 26 annual COP gatherings, the world's largest aggregate emitters and major fossil fuel energy using private sector actors have not combined together in any comprehensive push towards a net zero carbon future that is aimed at improving the lives who contribute the least per capita in terms of GHG emissions but still suffer the double burden of climate adversities and endemic exposure to pollution. Instead, there is a patchwork of UN and NNSA pledges and partnership promises with unclear and hard to track implementation strategies evidenced within the UN's databases and networks on partnerships. It is time to acknowledge that some of the most robust and ambitious climate goals being adopted in the world today are not those adopted by nation-states but instead for example are being undertaken by individual states in the US such as California, Washington State and New York and, equally importantly by cities across the world. Broadening the dialogue to learn how these states and cities are implementing comprehensive mandates on deep carbonization, while also committing to specific targets that can benefit communities that face the greatest climate related health risks is essential

Air Pollution, Clean Energy and Climate Change, First Edition. Anilla Cherian.

especially when contrasted to decades-old the intergovernmental negotiation silos on clean energy, clean air and climate action.

Ahead of the much-ballyhooed COP-26, the onset of the COVID-19 pandemic meant that the last globally significant climate meeting was the one-day UN Climate Action Summit convened on 23 September 2019. The 2019 Summit was one in a continuum of UN Special Summits on Climate Change convened with an express aim to catalyze global action. While there was a notable lack of scaled-up national climate commitments made at the 2019 Summit, and an outright absence of any concrete global action towards a protocol or set of measures that linked energy related air pollution to climate mitigation, it was the role of impassioned young global citizens like Thunberg demanding action to avert an impending climate crisis that made the absence of key aggregate GHG emitters – US and Chinese leaders – at the 2019 UN Summit even more glaring. The harsh, visceral response from some grown men to a teenage Swedish girl's views on climate change was disturbing. In a CBS News article, Christopher Brito pointed out that Fox News was forced to apologize after conservative commentator Michael Knowles' response to Thunberg's speech at the Summit was that: 'The climate hysteria movement is not about science. If it were about science, it would be led by scientists rather than by politicians and a mentally ill Swedish child who is being exploited by her parents and by the international left (Brito 2019)'. Knowles's callousness and blatant ignorance about Asperger's syndrome which is definitively not a mental illness can be starkly contrasted to Thunberg's courage in likening Asperger's syndrome – a high-functioning development disorder on the autism spectrum – as her advantage. Knowles's mendacity about the lack of scientific leadership on climate change needs to be soundly refuted, as it is untrue that scientists have not and are not leading on climate change as the six consecutive series of IPCC ARs have evidenced.

In fact, further bolstering the global scientific consensus regarding climate change just hours ahead of the opening of the UN 2019 Summit, the UN's Science Advisory Group released a Synthesis Report – 'United in Science' – which assembled the key scientific findings undertaken by major global partner organizations on climate change ranging from the WMO, UNEP, Global Carbon Project, the IPCC, Future Earth, Earth League and the Global Framework for Climate Services. The Report was not just a unified assessment on climate change, it was also an authoritative review of the most up-to-date information on climate change and its impacts on humanity and it highlighted 'the urgent need for the development of concrete actions that halt the worst effects of climate change' (WMO 2019). Its key messages can be found in Figure 6.1 excerpted and it leaves little room for confusion about the scale of the climate crisis, as well as the scale of ambition and political will required to tackle this global crisis.

Parties to the PA had unanimously accepted the 2020 deadline for agreement on critical elements of the PA which had to be postponed due to COVID-19 so the pressure was unmistakably on the world's largest aggregate GHG emitting countries to conclusively demonstrate increased ambition in their NDCs. COP-26 – the first climate COP to be held in the UK – from 1 to 12 November 2021 had been widely anticipated to be the most consequential meeting on climate change since COP-21 in Paris that resulted in the PA. COVID-19 laid waste to the plan to convene a COP in 2020, and expectations especially in terms of key aspects of the PA rule-making, the delivery of climate funding related to adaptation and concerted progress on scaled up commitments were high in the lead up to COP-26. Around 120 global leaders

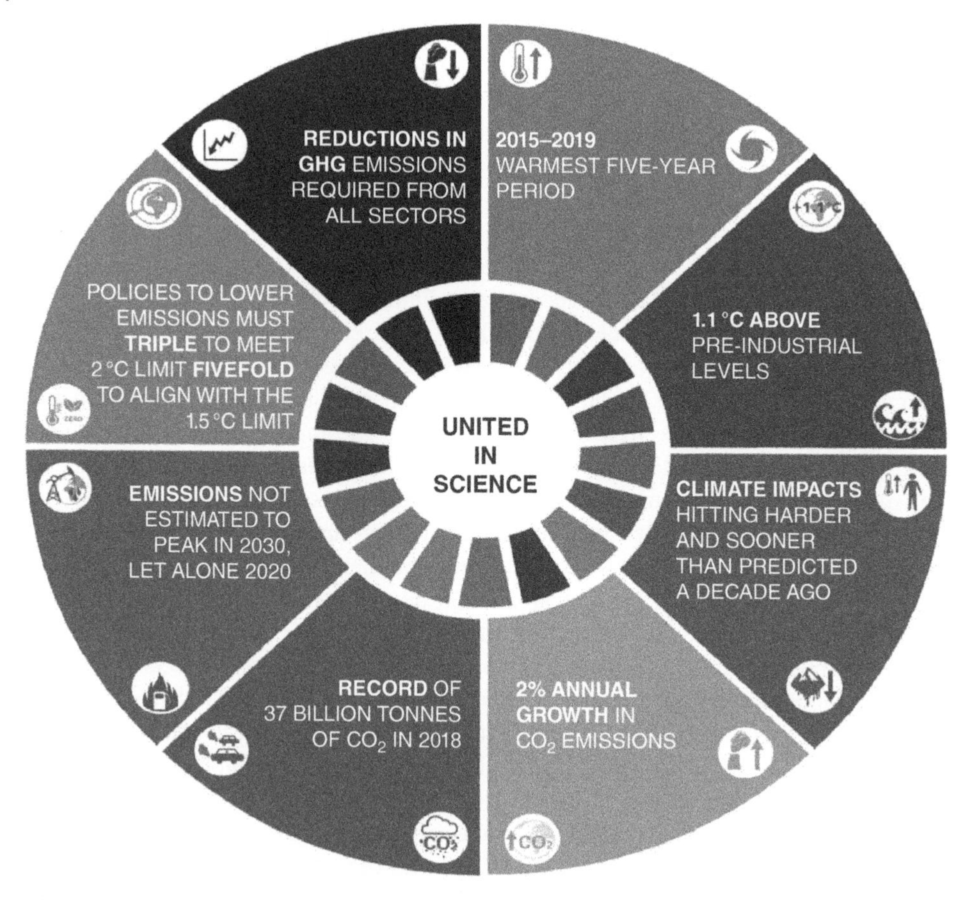

Figure 6.1 Key messages of the 2019 United in Science Synthesis Report. *Source:* WMO (2019).

descended to Glasgow for COP-26 which was billed as the 'last, best chance' to keep the PA's goal of a 1.5^{0}C threshold alive. In his opening address, COP-26 President, Alok Sharma said: *'The science is clear that the window of time we have to keep the goal of 1.5°C alive, and to avoid the worst effects of climate change, is closing fast. But with political will and commitment, we can, and must, deliver an outcome in Glasgow the world can be proud of'* (emphasis included, UKCOP26.org 2021) But the global reality is that the PA is an entirely voluntary driven national commitment exercise for individual countries, and intergovernmental negotiations at COP-26 remained fraught with tension as in the case of previous COPs.

President Biden and Prime Minister Modi representing the world's second and third-largest aggregate GHG emitters respectively were in attendance for the high-level segment of the two-week negotiations, but neither Chinese President Xi nor Russian President Putin chose to attend. India meanwhile had already firmly rejected calls to announce a net zero carbon emissions target a week ahead of COP-26 (Choudhury, 2021), With his own domestic agenda hanging in the balance during COP-26, President Biden stated that the absence of both these leaders was a 'big mistake' and went on to accuse China of 'walking away' (BBC 2021). The stage at COP-26 was set for what would once again prove to be an annual two-week long exercise in textual wrangling and late-night negotiations.

Despite the litany of life-threatening impacts listed by numerous UN and global reports evidenced in the previous chapters and the convening of numerous global climate fora, there are no specific, global policy or partnership modalities that directly target the linked impacts of air pollution, health, access to clean energy for the poor and climate mitigation within the context of the PA or the SDGs 7 and 13 which are largely stand-alone goals. The global implementation challenge which remains is that despite 26 annual COPs, comprehensive time-bound commitments by all major GHG emitters to move to low and zero carbon emissions is still an elusive goal. But, now cities and countries, especially developing countries in Asia and Africa, that are confronted with toxic levels of energy related air pollution have an urgent choice to make within the context of the PA negotiations which is to put curbing SLCPs at the front and centre of their NDCs going forward. Ignoring the multiple public health and clean air, food and water security benefits that can accrue from undertaking robust action and measure to curb SLCPs like BC and O_3 is illogical and imprudent because non-action quite literally increases morbidity and disease burdens as well as lost livelihoods and economic costs. If countries cannot take the lead on synergistic action on increasing access to clean energy and clean air within the context of the PA, then it is time for a framework protocol on curbing SLCPs and improving air quality to be put into place on a basis that is responsive to the needs of the populous and polluted cities in the world. And, this is precisely where India and Indian cities in particular can play a responsive role.

Intergovernmental negotiations have not been able to address the growing dangers of fossil fuel air pollution and it is time for linked action on clean air and clean energy access that can mitigate climate related health risks. More than a decade ago, sociologist, Anthony Giddens – put forward the Giddens Paradox which stated that 'since the dangers posed by global warming aren't tangible, immediate or visible in the course of day-to-day life, many will sit on their hands and do nothing of a concrete nature about them. Yet waiting until such dangers become visible and acute – in the shape of catastrophes that are irrefutably the result of climate change – before being stirred to serious action will be too late' (2009, p. 2). But what is hard to reconcile however is that Giddens specifically discounts the green movement which according to him is largely responsible for pushing for the "precautionary principle," which he argues has been viewed by environmental economists "as so much mumbo jumbo" (p. 49). But the global policy reality is that very early on the 1992 UNFCCC was prescient in invoking the precautionary principle which has been unanimously adopted by all UN member states that are Parties to the UNFCCC. Article 3.3's explicitly recognized that waiting for full scientific certainty should not be used as a reason for postponing climate measures. Article 3.3 of the UNFCCC specifically stated: 'Parties should take precautionary measures to anticipate, prevent or minimize the causes of climate change and mitigate its adverse effects. Where there are threats of serious or irreversible damage, *lack of full scientific certainty should not be used as a reason for postponing such measures,* taking into account that policies and measures to deal with climate should be cost effective so as to ensure global benefits at the lowest possible cost. To achieve this, such policies and measures should take into account different socio-economic contexts, be comprehensive, cover all relevant sources, sinks and reservoirs of greenhouse gases and adaptation, and comprise all economic sectors. Efforts to address climate change may be carried out cooperatively by interested Parties' (emphasis added, (UN/UNFCCC 1992, p. 6).The impetus of precautionary principle when combined with mounting evidence-based warnings that the global window for scaled up climate action is rapidly closing is exactly the impetus for arguing that curbing SLCPs emissions are urgently needed to save human lives.

The continued global knuckle dragging on ambitious low-carbon and zero carbon clean energy pathways that are needed from some of the world's largest aggregate emitters calls into question the logic of convening costly, celebrity focused climate summits. While these summits have resulted in travel-related GHG emissions, these global summits/fora have not resulted in enabling policy framing on behalf of, nor representation by those most likely to suffer as a result of the intersection between fossil fuel related energy pollution, energy poverty and climate adversity.

With the window for scaled-up climate action closing with each passing year, it is quite clear that millennial despair at the non-action or slow pace of climate responsive action by global leaders has reached new heights. Despite the US pull out of the PA, according to a 2019 Washington Post-Kaiser family poll: 'A solid majority of American teenagers are convinced that humans are changing the Earth's climate and believe that it will cause harm to them personally and to other members of their generation. Roughly 1 in 4 have participated in a walkout, attended a rally or written to a public official to express their views on global warming – remarkable levels of activism for a group that has not yet reached voting age' (Kaplan and Guskin 2019). The poll itself is noteworthy as it is the first major survey of teenagers' views of climate change and represents a generational shift in climate activism. In stark contrast to the UN's quest to drum up more ambitious climate action from the world's biggest aggregate GHG emitters, the young climate activist Thunberg cast real doubts that the current batch of global leaders can stop their fiddling while the planet heats up. Her fears were echoed in Propper's bleak assessment of the sustainability community's collective failure to address climate change:

'*Let's be honest.* We are *not going to limit* global warming to 1.5 degrees C. *That target is a bust.* We are already two thirds of the way towards 1.5 degrees, having reached one degree of warming this year . . . *So what of the sustainability community?* I'm pretty sure most of us started out with good intentions. But I think *we have gone about it the wrong way.* We have spread our efforts thinly over a vast array of issues . . . Doing some good here and there, but not conclusively dealing with the problem threatening our existence. Can you name the companies that have cut their absolute carbon footprint while growing their business? These should be our role models.

By trying to tackle everything at once we're diluting our impact, giving too much weight to secondary issues and too little to the really big one. . . *We need all our energy, resources and focus on climate change.* Talk about nothing else, to your board, investors, political connections and customers . . . If we beat climate change, we will automatically make many of the other problems better, and we will have re-established a collective belief that we can act to save our common future' (emphasis added, Propper 2019).

Propper's idea that climate change be viewed as the singular/fundamental organizing principle for both state and non-state actors essentially means that the UN's massive 17 SDGs needs to be reframed on the basis that climate responsive action becomes the principal global goal, and that all other goals are directly integrated and informed by the imperative of climate action. Making climate change the pre-eminent SDG at the UN would certainly be one direct way of breaking through existing policy silos at the UN and of concentrating all resources on addressing the climate crises. But, even doing so, the unpalatable global truth is that extreme exposure to air pollution and climate inequities co-exist and escalate in concert, and hence action to curb SLCPs is essential to any climate, clean air and clean energy equity focused future. Fossil fuel divestment is critical to lowering carbon emissions, but it is also important to recognize the need to invest in clean air and public health. The daily lives of millions of the poorest and most vulnerable communities currently destroyed by the toxic nexus between the lack of access to clean energy and exposure to PM pollution must be factored more directly into the global quest to address climate change.

The SDA's and PA's priority on poverty eradication is predicated on transforming the lives of those most vulnerable, and necessitate that reducing fossil fuel air pollution become an integral part any climate responsive future. Here, the role of the world's largest aggregate emitters are key but so are the responsibilities of fossil fuel consuming NNSAs. The world's second largest aggregate GHG emitter officially signaled on 22 April 2021 – Earth Day – that it would re-join the PA and commit to a more ambitious NDC based on a 50–52% reduction from 2005 levels in economy-wide net GHG pollution in 2030. The goals of a carbon pollution-free power sector by 2035, net zero emissions economy by no later than 2050 for the first time also clearly touched upon the importance of clean energy justice and climate equity via the policy announcement that the US would aim 'to reduce non-CO_2 greenhouse gases, including methane, hydrofluorocarbons and other potent short-lived climate pollutants. Reducing these pollutants delivers fast climate benefits' (US White House 2021). The 2021 US NDC therefore serves as a watershed within the context of the PA given its express linkage to the scaled up ambition in reducing SLCPs. But, the question is what will the other leading aggregate emitters do in terms of carbon pollution and in particular curbing SLCPs and ensuring BC pollution-free measures?

Absent the immediate incentive of the profit bottom line, who really cares about the unsettling, toxic future faced by those who live under polluted hearths and skies? What are countries, companies, investors, including fossil-free divestment and NSA partnerships, and all of us as individuals doing in terms of increasing access to clean energy for the forgotten millions who suffer the daily burden of energy related pollution, poverty and climate insecurities? Those committed to meeting the overarching priority of the UN's SDA cannot gloss over the fact that energy related air pollution – one of the leading causes of global deaths and ill health – has never been consistently within the SDG goal silos. There is simply no ducking around the fact that the UN-led global community is urgently faced with making time-bound verifiable climate and clean air responsive actions that can jointly address poverty reduction and energy related pollution because poverty reduction remains the main priority of the UN-led sustainable development community. The UN SDA and the PA both have poverty eradication as a principal and cross-cutting priority, but as Chapters 2 and 3 demonstrated the agreed global guidance and record on implementing and scaling up partnerships that prioritize poverty reduction within the global policy arenas of climate change and clean energy is confusing at best and contradictory at worst. As Chapters 1, 2 and 3 evidenced, it is simply illogical, impractical and a waste of limited resources and time to have segregated negotiation silos and separate SDGs on sustainable energy and climate change given the central role of energy in improving human well-being, reducing poverty and mitigating climate change.

Arguably, working towards linked action matters most for policy makers in developing countries who not only have a complex arena of policy challenges to respond to, but who can least afford to work on the basis of goal silos on clean air and clean energy. Drawing on Wildavsky (1979) and Weiss (1977), it can be argued that the identification of policy responses to a global challenge like climate change is the result of analytic and research constructs which are embedded in specific sociopolitical or institutional contexts. Consequently, for any policymaker/political leader, the problem of climate change does not exist as an entity distinct from the means of either making or not making sense of the problem. However, from a developing country policymaker's perspective, the intrinsic problem with climate change is that it is not only highly context specific and directly linked to socio-economic development needs and goals but also gravely inequitable in that those who have contributed the least in terms of per capita emissions are the ones who are anticipated to suffer the worst impacts and have the least adaptive resilience to cope with adverse impacts. In his 2002 article, 'Science Policy for Multilateral Environmental Governance', Haas pointed to the wide recognition of the need for science policy to be applied to the management of transboundary and global environmental threats, but also noted: 'Yet science has become extremely politicized. It is often found that good science falls on deaf ears, or is met with bad science, when the politics favor neglecting it' (Haas 2002, p. 8) Building off this point, it is time to call attention to the politics of neglecting the global accumulation of scientific evidence on the health and disease burdens accruing from long-standing exposure to air pollution and ask why there are regional protocols other than CLRTAP?

The record of three decades of intergovernmental climate change negotiations demonstrates that policymakers and diplomats from a number of SIDS and LDCs have long been active in global negotiations as champions for scaled up and more urgent climate action. Despite mounting factual and evidence-based warnings about the devastating multi-sectoral and multi-dimensional impacts of climate change on the poorest and most vulnerable issued by a range of globally relevant and scientifically credible sources, the decades old politically arduous and contested process of intergovernmental climate change negotiations still remains an intractable and seemingly insurmountable global challenge. Reframing climate action to explicitly include the policy nexus between clean energy, clean air and climate is crucial for the future of millions who will suffer inequitably. Climate responsive action therefore cannot, and should not be segregated from increasing access to clean energy for the poorest around the world. It is therefore now both a logical and opportune time to call for the coalescing of SDGs 7 and 13 as a practical and effective means of implementing linked action on sustainable energy for all and climate change.

In arguing for a 'reframing' of climate action within the PA to include curbing SLCPs and public health risks associated with fossil fuel air pollution, it is useful to consider briefly that the broader topic of anthropogenic climate change has a wealth of global scientific consensus behind it, even though this global scientific consensus is consistently downplayed by climate science sceptics. In their 1994 book focused on the resolution of 'intractable policy controversies', M.I.T researchers, Schön and Rein argue that: 'In matters of public policy, disputes are endemic. Whatever the issue may be – the costs of public pensions, the inadequacy of our system of public education, the protection of our natural environment, the causes and remedies of homelessness – the public process of considering and coping with that issue is marked by contention, more or less acrimonious, more or less enduring. We believe, however, that it is critically important to distinguish between two kinds of policy disputes: those that may be settled by reasoned discourse and those that are stubbornly resistant to resolution through the exercise of reason' (1994, p. 3). Schön and Martin use the term 'policy disagreements' to refer to disputes that contending parties are able to resolve by recourse and examination of 'the facts of the situation' which they distinguish from 'policy controversies' that are 'immune to resolution by appeal to the facts' (1994, p. 4). According to Schön and Martin, the way 'facts' are framed by policymakers is what contributes to policy controversies in that parties to a policy controversy employ different strategies of selective attention, and even when parties to a controversy focus on the same facts, they tend to give them different interpretations (1994, pp. 4–5). According to them, resolving policy controversies therefore requires a thoughtful process of policy reframing and reflection: '. . . in order to contribute to the reframing of policy dilemmas, policy makers must be able to reflect on the action frame held by their antagonists. Even to recognize the existence of such dilemmas, policy makers must be able to reflect on their own action frames: they must overcome the blindness induced by their own ways of framing the policy situation in order to see that multiple policy frames represent a nexus of legitimate values in conflict' (1994, p. 187). Here, the reframing of the policy dilemma is not that the scientific underpinnings of anthropogenic climate change needs to be proven at this stage, but that climate and poverty reduction actions within the UN context needs to be more thoughtfully integrated with the clean energy and clean air needs of millions across the world.

Despite growing anguish about climate change across the globe, thoughtful reframing was seen as unlikely when global leaders such as President Trump explicitly called to expunge climate change from the list of topics focused on by the US EPA. In April 2017, the EPA replaced its online climate change section with a holding page that said the content was being updated to 'reflect the agency's new direction under President Donald Trump' and the topic of climate change was longer directly searchable and found only on an archival site. Columbia University's Sabin Center for Climate Change Law had a dedicated and aptly named 'Silencing Science Tracker' which listed over 100 entries – specific examples – of the Trump administration's 'silencing' of climate change related efforts. The Environmental Data and Governance Initiative which formed the basis for some of the work provided by Columbia's tracking tool also analyzed more than 5300 pages on the website of 23 federal agencies and found that the 'Trump administration had removed a quarter of all references to "climate change" from federal government websites since 2016' (Baynes 2019).

Taking a cue from the Trump administration silencing of climate science, on 2 August 2019, Brazilian President Bolsonaro fired the head of a government agency that had revealed Brazil's portion of the Amazon lost more than 1330 square miles of forest cover during the first six months of 2019, a 39% increase over the same period the previous year. The Brazil bureau chief of the *New York Times* noted that Bolsonaro's angry justification of this firing was that 'people within the government were damaging the country's image abroad by disclosing the rate of deforestation', and a week earlier had, 'called concerns about the environment overblown, saying the issue mattered solely to 'vegans, who eat only plants' (Londoño 2019). On 19 August 2019, global media lit up with images of São Paulo, Brazil's largest city, going terrifyingly dark even though it was midafternoon and clouds of ash from forest fires in the Amazon rainforest blocking out the sun. With both the French President Macron and Pope Francis sounding the alarm that 'the lungs of the world' were on fire, on 26 August 2019, G-7 leaders from Canada, France, Germany, Italy, Japan, the UK and the US met in Biarritz, and France announced that the G-7 would release US $22 million to help 'fight' the fires ravaging the Amazon. But, Bolsonaro's initial rejection of the Macron led G-7 'alliance' to 'save' the Amazon treated Brazil 'as if we were a colony or no man's land' which was immediately followed by his grudging acceptance that the personal animus between Macron and himself could be buried reflected an unsettling snapshot. (BBC 2019). Framing the issue of deforestation in the Amazon as a national populist issue that appeases powerful Brazilian agricultural and cattle industries, and consequently any external alliances to save the rainforest as a form of neo-colonialism does not address the thorny issue that the Brazilian administrations of both Temer and Bolsonaro had slashed budgets for environment and support for indigenous and traditional subsistence communities, and that once again it was the poor, vulnerable and marginalized, in this case indigenous communities that ended up suffering (Sullivan 2019).

Climate infused apocalypses resulting from short-sighted political and policy framing at both the global and national level have been playing out in many developing countries. A 2007 edition entitled 'Global Warming', published by a leading Indian journal issued an eerily disturbing article describing the climate change 'apocalypse' that was anticipated to devastate India. R. Chengappa wrote: 'Delhi 2020: They are already calling it the year without winter. From January itself, temperatures across Northern India have soared to 35 degrees Celsius. Much of the Gangetic plain has become transformed into one vast dust bowl. The wheat crop has failed and farmers are committing suicides by droves. Food riots have broken out as the government is no longer able to control the distribution. As disease and death strike with metronomic regularity, the smell of rotting flesh is all pervasive. In a perverse irony, even as the North is sweltering under the heat wave, fierce unseasonal storms lash the West Coast' (2007, p. 38). Chengappa's article written back in 2007 was based on the IPCC's Fourth Assessment Report (AR-4) which found that India would be among the countries worst affected by rising temperatures. Amongst the hellish impacts anticipated was the precipitous decline in water availability where 'half a billion people – or half the country's population would be adversely affected' (2007, p. 43). The article cites then Secretary of the Ministry of Environment and Forests in India- Prodipto Ghosh- as stating that India is 'satisfied' that the IPCC findings are 'scientifically robust and defensible' but that when it comes to responding, India wants the worst polluters to cut back on their GHG emissions. With India's per capita emissions being 25 times less than that of an average American and 15 times less than that of a European in 2007, Chengappa quotes Ghosh: 'India is certainly not responsible for this mess. We are, in fact victims of it. So why expect us to tighten our belts?' (p.43). More than 14 years after the publication of this article, it is time to note that regardless of the per capita inequity associated with GHG emissions, the inefficient combustion of solid fuels within India – the third largest aggregate emitter- cause direct and grave risks to national food and water security and human health.

Recently there has been affirmation of crippling water scarcity in various parts of India, most notably in the Southern Indian city of Chennai – a city rated as one of the safest in India. Chennai has faced consistent water crises after four of its main water reservoirs ran dry and there have been water riots. In 'Parched Lives on the Fringe: How Water Scarcity Has Widened Inequality' it is stated that: 'The bitter struggle for water has become a daily reality for many of Chennai's 9.1 million residents, but vulnerable communities, particularly the 820,000 people living in Chennai's many informal settlements, have suffered the most from the acute water shortage' (Ha 2019). But Chennai serves as only one example of a growing list of megacities – Cape Town and Los Angeles – coping with the devastating multi-sectoral impacts of climatic change. Linkages between water scarcity and climate change have disastrous consequences for the poorest communities within countries and can wreak havoc on already precarious food security and energy access needs. According to the Carbon Disclosure Project (CDP), '18% of the world's population resides in India and they survive on just 4% of the world's freshwater. Rapid population growth and urbanization, a growing agricultural and industrial sector, as well as climate impacts – mainly drought and flooding are putting these limited resources under strain'. Highlighting India as a reminder of just what is at stake – 'no time to lose'- CDP has called for companies to undertake better monitoring via comprehensive risk assessments and management of water resources (CDP 2021). Recognizing the importance of cities in the fight against climate change, the CDP, in its evaluation found that in 2015, 44 cities had set climate targets and the number rose to 88 by 2020, but not a single city in South Asia made the list, while the City of Cape Town was the only city in Africa to do so (CDP website 2021).

Now, it is time to ask whether the challenge of meeting climate mitigation targets should be dependent on the vagaries of political election cycles of individual UN member states and their decades-old predilection for long-drawn out textual outcomes that have not delivered on comprehensive legally binding GHG emission targets nor addressed the human costs of fossil fuel pollution. Continued delays that emanate within the context of global policy silos on clean air, clean energy access and climate change goals are hard to justify and inability of integrating policy and partnership responses to energy pollution, climate change and poverty reduction will consign millions to die. The which non-policy answer lies in the adage that the rich can buy themselves out of trouble, including in this case both climate change and COVID-related stressors. Since energy related pollution and climate adversities do not extract the heaviest health burdens on the rich, those who do suffer at this toxic nexus and whose voices are not well-represented are relegated to living under a regime of extreme climate and energy vulnerabilities and health risks- namely a system of global climate and energy apartheid.

It is time to effectively scale up the role of NNSAs on integrated climate, clean energy and clean air action at the city level holds the key for human health and well-being in many countries across the world, most notably India as discussed in Chapters 4 and 5. Additionally, while the focus of this book has not been on the crucial role of climate finance, it is essential in this concluding chapter to focus briefly on the groundswell of change occurring with the private sector by examining the policy pronouncements made by two different captains of industry – Warren Buffet and Larry Fink. But in so doing, it is relevant to keep in the mind a key question: The growing global trend towards fossil fuel divestment spurred by NNSAs trumps climate denial but does it address the climate, clean energy and clean air apartheid experienced by the global poor?

6.2 NNSA – Private Sector's Shift Away from Fossil Fuels but Where is the Change for the Energy Poor?: Snapshot View of Two Different CEOs on the Risk of Climate Change

There are visible and seismic shifts in NNSAs – private sector – decision-making processes related to clean, low and zero carbon energy sources and technologies, but the question of what fossil-free divestment means for the energy poor and climate vulnerable is extraordinarily difficult to answer. In part this is because there have been for far too long policy silos around clean energy and climate that have impacted frameworks of integrated partnerships that involve NNSAs at all levels – global, regional, national and local. But, 28 years after the first UNGA resolution on climate change was adopted, and close to 25 years after the UNFCCC was adopted at the Rio Earth Summit, a blunt assessment of the risks of climate change and its linkages to clean energy was provided by Warren Buffet (the head of one of the world's largest property and casualty insurance companies) in his shareholder letter on 27 February 2016. The fact that the head of one of the world's largest property and casualty insurance companies focused expressly on climate change for the first time happens to be globally relevant.

Buffet began by pointing out that Berkshire Hathaway Energy which 'invested $16 billion in renewables' owned '7% of the country's wind generation and 6% of its solar generation' in the US and also 'made major commitments to the future development of renewables in support of the Paris Climate Change Conference'. More significantly, he went on to state: 'Our fulfilling those promises will make great sense, both for the environment and for Berkshire's economics' (Berkshire Hathaway Inc. 2016, p. 5). But, it was his focus on climate change provided in response to a sponsor requesting a report on 'the dangers' that climate change presented to insurance operations, as well as his explanation of 'how' these threats are being responded to, that was significant.

Buffet offered a characteristically direct explanation that it is 'highly likely' to him that 'climate change poses a major problem for the planet' and cited his lack of scientific aptitude and the gloomy hand-wringing by so-called experts about Y2K as two reasons why he cannot be 'certain' about climate change. He specifically referenced: 'It would be foolish, however, for me or anyone to demand 100% proof of huge forthcoming damage to the world if that outcome seemed at all possible, and if prompt action had even a small chance of thwarting the danger' (2016, p. 25). Buffet invoked philosophical and biblical precautionary precepts in response to the need to address climate change: 'This issue bears a similarity to Pascal's Wager on the Existence of God. Pascal, it may be recalled, argued that if there were only a tiny probability that God truly existed, it made sense to behave as if He did because the rewards could be infinite whereas the lack of belief risked eternal misery. Likewise, if there is only a 1% chance the planet is heading toward a truly major disaster and delay means passing a point of no return, inaction now is foolhardy. Call this Noah's Law: If an ark may be essential for survival, begin building it today, no matter how cloudless the skies appear' (2016, p. 25). Here, it is useful to point out that Buffet's 2016 reference of precautionary measures was clearly reflected decades prior in the universally adopted UNFCCC's Article 3.3.3 Arguments that posit complete scientific certainty is needed before undertaking practical, cost-effective attempts to integrate sustainable energy for all, and climate mitigation goals are contrary to the logic of this globally agreed upon precautionary principle.

There was another much more disturbing policy reality worth highlighting in Buffet's 2016 shareholder letter that is germane to the broader discussion of who stands to lose the most as a result of bearing the brunt of climatic adversities. Buffet's discussion on the role of precautionary ark building – i.e. planning for floods in midst of sunniest skies – included an extremely clearheaded distinction about who needs to worry versus who can afford to adopt a more blithe, carefree attitude about negative climatic impacts such as SLR. Buffet noted: 'As a citizen, you may understandably find climate change keeping you up nights. *As a homeowner in a low-lying area, you may wish to consider moving. But when you are thinking only as a shareholder of a major insurer, climate change should not be on your list of worries*' (emphasis added, 2016, p. 26). Buffet's differentiation between homeowners in a low-lying area versus shareholders of a major insurer like Berkshire Hathaway within the US context showcased an important dichotomy as to the inequitable burden of climatic adversities that can be extended to a global level. Presumably, those wearing shareholder hats can strike climate change off their list of nightly terrors, but poorer residents living in a heavily leveraged, uninsured homes and/or urban shanty dwellers living precariously in makeshift lean-tos in coastal inundation zones across the world have cause for fear driven by their inability to move. It is precisely these climate vulnerable households and communities who contribute the least in terms of per capita GHG emissions, yet who are encumbered by not having any where to escape to and lacking safety nets that face likely devastation.

Buffet's practical approach was aimed to reassure his shareholders who worry that: 'Berkshire is especially threatened by climate change because we are a huge insurer, covering all sorts of risks'. And here he actually agreed that these worries '. . . might, in fact be warranted if we wrote ten- or twenty-year policies at fixed prices. But insurance policies are customarily written for one year and repriced annually to reflect changing exposures. Increased possibilities of loss translate promptly into increased premiums' (2016, p. 25). And here the bottom line was essentially that Hathaway shareholders could therefore potentially stand to benefit because: *'Up to now,* climate change *has not produced more frequent nor more costly hurricanes* nor other weather related events covered by insurance'. But this reference included a crucially important caveat – the 'up to now' factor. What happens to insurance payouts when adverse climatic impacts are much bigger and more frequent than previously anticipated? When the loss payouts are simply not enough to secure buy-backs given exponentially escalating annual premium prices? And what happens to millions of those who simply do not have the luxury of moving from their low-lying coastal area makeshift homes unlike those who benefit from being insured, are climate resilient and can harness shareholder rewards accruing from speculating on climate risk related insurance?

As it turns out, a series of extreme events in 2020–2021 have proved just how disastrous it was to make the claim that climate change has not until now produced more frequent and more costly hurricanes. Rising temperatures in the Atlantic Ocean spawned so many hurricanes in 2020 that the WMO depleted its annual list of alphabetical names and had to dig into the Greek alphabet. Hurricane Ida was one of the strongest hurricanes to hit New Orleans, Louisiana and other Gulf states in the United States on the 16th anniversary of another devastating hurricane – Katrina – and was anticipated to cost anywhere from '$15 billion to $30 billion in claims', but the figure 'could jump higher, in part because of pandemic pricing that has pushed up the cost of lumber and labor to rebuild' (Scott 2021). In a Forbes article, Leefeldt and Danise revealed: 'While big multinational insurers like Munch Re and Swiss Re have been watching worldwide warming trends for over decade, US property-casualty insurers have for the most part kept their heads down and relied on the huge $825 billion surplus in their war chests to keep them afloat'. Leefeldt and Danise go on to note that this strategy by US insurance companies is 'not working – particularly in the western states plagued by dry, fire-causing heat' and provide evidence of companies shifting responsibility to homeowners with annual premiums 'skyrocketing in hurricane-vulnerable states like Florida and Louisiana' while policy cancellations due to wildfires and hurricanes have caused 'burden shifts to state-sponsored programs like The California Fair Access to Insurance Requirements (FAIR) Plan and Florida's Citizens Property Insurance Corp' (2021) (Leefeldt and Danise 2021).

For those living in cities where the vast majority lack any kind of insurance against extreme climatic adversity and suffer ill health due to air pollution, the situation is clearly much graver. And, now in March 2021, in a volte-face to his 2016 shareholder response letter, Buffet decided to oppose the shareholder proposal submitted by the California Public Employees Retirement System (CalPERS), Federated Hermes, and Caisse de dépôt et placement du Québec (CDPQ) which requested that the board of the company publish an annual assessment addressing how the company manages physical and transitional climate-related risks and opportunities. Robert Eccles in a *Forbes* article offered a succinct piece of advice for Buffet which was that he should change his decision noting: 'This seems like a very modest ask, particularly in light of how companies all over the world are publishing data relevant to climate change, such as using the framework of the Task Force on Climate-related Financial Disclosures (TCFD) and including metrics on greenhouse gas emissions and commitments to be net zero by 2050. Since you own around one quarter of the company's voting shares, this would be a strong and powerful signal that Berkshire Hathaway recognizes the potential impact of climate change on the company. Failure to do so suggests that you do not, an unfortunate mark on your legacy' (2021).

In contrast, on January 2020, it appeared that climate responsive investing (for those who can access asset fund managers) has conclusively arrived – jump-started by a letter by Larry Fink, the head of BlackRock entitled, 'A Fundamental Reshaping of Finance'. The fact that the world's largest asset fund manager was explicit that his company – BlackRock – was going to be putting 'sustainability at the center of our investment approach' was the analogue of a paradigm shift for global investors. A 15 January 2020 article in the *Wall Street Journal* describes just how big of a global imprint BlackRock Inc. has as its 'assets surpassed $7 trillion for the first time as the investment giant reported record-setting flows in 2019 (Lim 2019)'. What set Fink's letter to his peer group of CEOs apart was not just the global significance of his company but his argument that BlackRock is an asset manager that 'invests on behalf of others', and that he was writing 'as an advisor and fiduciary' to clients because as he puts it 'the money we manage is not our own. It belongs to people in dozens of countries trying to finance long-term goals like retirement. And we have a deep responsibility to these institutions and individuals – who are shareholders in your company and thousands of others – to promote long-term value'. His explicit recognition that 'climate change has become a defining factor in companies' long-term prospects' also touched on another key point which was that climate activism by ordinary people resonated in BlackRock's shift towards sustainable, climate responsive investment decision-making. As Fink pointed out, 'Last September, when millions of people took to the streets to demand action on climate change, many of them emphasized the significant and lasting impact that it will have on economic growth and prosperity – a risk that markets to date have been slower to reflect. *But awareness is rapidly changing, and I believe we are on the edge of a fundamental reshaping of finance*' (emphasis included, Fink 2020a).

Fink's blunt truths make it harder for climate deniers who have tried to obfuscate the preponderance of evidence regarding climate risk. BlackRock's conclusion that 'climate risk is investment risk' fundamentally alters investment assumptions, but it also has huge relevance for governmental decision-making related to the policy intersection between highly polluting solid fuel energy reliance and climate vulnerabilities. In his 2020 letter addressed to clients, Fink signalled exactly how sustainability considerations would form the bedrock of

BlackRock's strategies: 'Because sustainable investment options have the potential to offer clients better outcomes, we are making sustainability integral to the way BlackRock manages risk, constructs portfolios, designs products, and engages with companies. We believe that sustainability should be our new standard for investing'. In this 2020 letter, he referenced that BlackRock will exit coal which is 'becoming less and less economically viable, and highly exposed to regulation because of its environmental impacts. With the acceleration of the global energy transition, we do not believe that the long-term economic or investment rationale justifies continued investment in this sector. As a result, we are in the process of removing from our discretionary active investment portfolios the public securities (both debt and equity) of companies that generate more than 25% of their revenues from thermal coal production, which we aim to accomplish by the middle of 2020' (Fink 2020b). This deadline of mid 2020 should be viewed as a tell-tale signifier that private sector investment decisions are moving exponentially faster than governmental decision-making regarding the risks of thermal power reliance. And then in 2021, Fink issued a new letter to his peer CEOs that highlighted the fact that 'conventional wisdom' was that the COVID pandemic crisis 'would divert attention from climate. But just the opposite took place', with the reallocation of capital accelerating even faster than anticipated. 'From January through November 2020, investors in mutual funds and ETFs invested $288 billion globally in sustainable assets, a 96% increase over the whole of 2019. I believe that this is the beginning of a long but rapidly accelerating transition – one that will unfold over many years and reshape asset prices of every type. We know that climate risk is investment risk. But we also believe the climate transition presents a historic investment opportunity' (Fink 2021).

In spite of the shift in climate risk related investing, questions as to investments and actions in curbing SLCPs persist: i) How effectively or when will private sector investment in clean air and clean energy for the poor be seen as historic investment opportunity in addressing global climate change? ii)Will there be yet another separate silo for sustainable investment opportunities that can grow the portfolios of powerful companies while neglecting the clean energy needs of the poor? Fink's 2020 reference in his letter to clients that a sustainable energy transition 'will still take decades' is exactly what is crucially relevant for cities in countries like India trying to simultaneously cope with high levels of energy related air pollution and climatic adversities. Cities, particularly those reeling from toxic exposure to PM pollution, are indeed the loci for linked action on climate change, increasing access to clean air, clean energy and poverty reduction. India, the third largest aggregate GHG emitters, also includes 22 of the top 30 most polluted cities and as Jamrisko pointed out, 'racks up health-care costs and productivity losses from pollution of as much as 8.5 percent of gross domestic product, according to the World Bank' (Jamrisko 2019). The question of how countries like India – where climate vulnerabilities abound and millions lack access to clean energy – can work with the private sector to pursue an energy transition that Fink argues 'cannot leave behind parts of society, or entire countries in developing markets, as we pursue the path to a low-carbon world' remains daunting. Fink made a forceful call for governments to lead the way so that companies/investors can follow, and pointed out that BlackRock was 'a founding member of the Task Force on Climate-related Financial Disclosures (TCFD)', a signatory to the UN's PRI and the 'Vatican's 2019 statement advocating carbon pricing regimes (Fink 2020b)'. And it is precisely cities and households in countries like India where increasing access to clean energy is irrevocably linked to climate action, and which have championed solar and other forms of renewable energy that now urgently need synergistic action. But, delivering on the promise of linked action for millions who suffer at the toxic intersection of energy pollution and climatic adversities requires the elimination of intergovernmental policy silos that segregate UN SDGs on sustainable energy (SDG 7) from climate (SDG 13) and for NNSAs to push towards increasing access to clean energy and clean air access – i.e. being on the side of a climate, energy and poverty reduction nexus.

6.3 Leaning into the Nexus on Clean Energy, Clean Air and Climate Responsive Action Matters

Factoring in climate change is a prevailing trend in the private sector which has seen significant change underway in terms of climate-related engagement and risk assessment over the course of the past few years. But this does not in any way neatly translate or conflate into NNSAs such as major corporations and entities being on the side of clean energy and clean air for the poor. Back in 1999, in his book, *The Carbon War: Dispatches from the end of the Oil Century*, Jeremy Legget traced the early evolution of the intergovernmental negotiations and the role played by oil, gas and coal companies in blocking consensus from 1988 when the UNGA first adopted the climate resolution to 1999 – the post Kyoto era. Legget's arguments are worth recalling in this current post PA adoption era as a backdrop in evaluating progress made towards integrating access to renewable energy for the poor: 'As we approach the end of the hydrocarbon century, the oil companies shuffle for position, patently uncertain of the way forward as their world changes around them, as never before . . . Exxon, Mobil, Texaco and the other residually unrepentant thugs of the corporate world look like continuing to sign the cheques that bankroll the carbon club's crimes against humanity, along with their kindred spirits in the automobile, coal and utility industries. They may well enjoy minor victories along the way. But they have already lost the pivotal battle in the carbon war. The solar revolution is coming. It is now inevitable. The only question left unanswered is, will it come in time' (Legget 2001, p. 328).

Twenty years after Legget penned those words much has changed in terms of renewable energy, but much remains the same in terms of the denial by the pre-eminent cartel of oil and gas producers. The question of whether the clean energy and clean air revolution will come in time for the poorest and most vulnerable amongst us looms larger in light of the COVID-19 pandemic revealing existing health disparities and inequities. According to an Agence France-Presse (AFP) report, at the July 2019 OPEC meeting, the OPEC Secretary General Mohammed Barkindo claimed that, 'Civil society is being misled to believe oil is the cause of climate change'. In a reference to the climate activist inspired school strikes by Swedish teenager Greta Thunberg's 'Fridays for Future' movement, Barkindo mentioned that children of some colleagues at OPEC's headquarters 'are asking us about their future because they see their peers on the streets campaigning against this industry', adding that 'mobilisation' against oil was 'beginning to . . . dictate policies and corporate decisions, including investment in the industry' (AFP 2019). By way of a pointed response, Thunberg and other climate activists were quick to note that it was 'the biggest compliment yet' that the head of the globe's most powerful oil cartel said that their climate campaigns may be the 'greatest threat' to the fossil fuel industry (Watts 2019).

The trend towards low and zero carbon pathways is apparent. But those new to the global discussion on climate change might not be clear about the fact that while Article 4 (Commitments) of the UNFCCC recognizes the 'common but differentiated responsibilities' of all Parties and agrees upon the need for cooperation on technology transfer and the importance of taking into account climate adaption, it also strikingly includes a reference as to oil producing and exporting countries in connection to Article 4.8 and Article 10 (Subsidiary Body of Implementation) to the Convention. More specifically Article 4. 8 outlined the 'specific needs and concerns of developing country Parties arising from the adverse effects of climate change and/or the impact of the implementation of response measures' and goes on to categories of such developing countries ranging from small island countries, to countries with low-lying coastal areas, countries arid and semi-arid areas, countries with areas prone to natural disasters, including countries *'with areas of high urban atmospheric pollution'*. . .and countries *'whose economies are highly dependent on income generated from the production, processing and export, and/or on consumption of fossil fuels and associated energy-intensive products'* amongst others (emphasis added, UN/UNFCCC 1992, p 15). Article 4.10 of the UNFCCC made the case for this latter category of countries even more explicit: 'The Parties shall, in accordance with Article 10, take into consideration in the implementation of the commitments of the Convention the situation of Parties, particularly developing country Parties, with economies that are vulnerable to the adverse effects of the implementation of measures to respond to climate change. *This applies notably to Parties with economies that are highly dependent on income generated from the production, processing and export, and/or consumption of fossil fuels and associated energy-intensive products and/or the use of fossil fuels for which such Parties have serious difficulties in switching to alternatives'* (emphasis added, UN/UNFCCC 1992, p.15) In essence, the very Convention that contains the objective to mitigate climate change conflates the needs of developing countries that are most vulnerable to climate adversities with those that comprise some of the largest oil producing and exporting nations. Interestingly, the needs of countries with areas of highest urban atmospheric pollution have not been addressed in any consistent or comparable manner to date.

Additionally, those captivated by promise of the concept of 'green deals' including the US 'Green New Deal' – introduced as a Bill (H. Res 109) to the US Congress by Representative Ocasio-Cortez – might be surprised to know that back in 2009, in the wake of the globally calamitous financial crisis, then UNEP Executive Director Achim Steiner (current head of UNDP) commissioned a historic report entitled 'Rethinking the Economic Recovery: A Global Green New Deal'. Prepared by Edward Barbier, the report argued that in 2008, the world was confronted with multiple crises – 'fuel, food and financial' and that the result of these combined crises was 'the worst global economic recession since the Great Depression of the 1930s'. The report's finding has a particular resonance in today's world: 'Every 1 per cent fall in growth in developing economies will translate into an additional 20 million people consigned to poverty. Faced with the social and economic consequences of a deepening world recession, it may seem a luxury to consider policies that aim to reduce carbon dependency and environmental degradation. *Such a conclusion is both false and misleading* . . . The multiple crises threatening the world economy today demand the same kind of initiative as shown by Roosevelt's New Deal in the 1930s, but at the global scale and embracing a wider vision. The right mix of policy actions can stimulate recovery and at the same time improve the sustainability of the world economy. If these actions are adopted, over the next few years they will create millions of jobs, improve the livelihoods of the world's poor and channel investments into dynamic economic sectors. A "Global Green New Deal" (GGND) refers to such a timely mix of policies. . . Without this expanded vision, restarting the world economy today will do little to address the imminent threats posed by climate change, energy insecurity, growing freshwater scarcity, deteriorating ecosystems, and above all, worsening global poverty. *To the contrary, it is necessary to reduce carbon dependency and ecological scarcity not just because of environmental concerns but because this is the correct and only way to revitalize the economy on a more sustained basis.*' (emphasis added, UNEP/Barbier 2009, p. 5).

Close to 13 years later, the full-scale implementation of a 'timely mix' of policies related to fostering a GGND is still not-evident, but Barbier's point about reducing fossil fuel subsidies in light of the global crisis caused by the COVID-19 pandemic remains relevant: 'Removal of fossil fuel subsidies eliminates perverse incentives in energy markets and provides an immediate source of financing for low-carbon strategies. Globally around US$300 billion annually, or 0.7 per cent of world GDP, is spent on such subsidies, which are employed mainly to lower the prices of coal, electricity, natural gas and oil products. Most of these subsidies do not benefit the poor but the wealthy, nor do they yield widespread economic benefits . . . Cancelling these subsidies would on their own reduce greenhouse gas emissions globally by as much as 6 per cent and add 0.1 per cent to world GDP. The financial savings could also be redirected to investments in clean energy R&D, renewable energy development and energy conservation, which would further boost economies and employment opportunities. Eliminating fossil fuel subsidies can also benefit low-income economies. For example, energy sector reforms in Botswana, Ghana, Honduras, India, Indonesia, Nepal and Senegal have proven to be effective in leading a transition to more efficient and cleaner fuels that particularly benefit poor households' (2009, p. 10). Barbier also highlighted the need for reducing carbon usage in the transport sector via a mix of innovations such as next generation biofuels and expanding urban public transit and rail networks which have the potential to stimulate growth and create jobs.

The role of the 'green economy' and green jobs associated with environmental and sustainable development has been discussed considerably both within and outside of the UN for quite some time. Decades before the 'Green New Deal' was promoted in the US, in 2008, a joint report by UNEP, International Labor Organization (ILO) and other relevant global organizations provided the first ever comprehensive global assessment entitled 'Green Jobs: Towards Decent Work in a Sustainable Low-Carbon World' which examined the concept of green jobs in a global context. According to this report, green jobs were defined: ' . . . as work in agricultural, manufacturing, research and development (R&D), administrative, and service activities that contribute substantially to preserving or restoring environmental quality. Specifically, but not exclusively, this includes jobs that help to protect ecosystems and biodiversity; reduce energy, materials, and water consumption through high efficiency strategies; de-carbonize the economy; and minimize or altogether avoid generation of all forms of waste and pollution' (p. 3). Although the report forecasted tens of millions of new unemployed people and working poor due to the 2008 global financial crisis, it found that 'compared to fossil-fuel power plants, renewable energy generates more jobs per unit of installed capacity, per unit of power generated and per dollar invested' and also found that 'so far, a small group of countries accounts for the bulk of renewables investments, R&D, and production. Germany, Japan, China, Brazil, and the United States play particularly prominent roles in renewable technology development, and they have so far garnered the bulk of renewables jobs worldwide' (UNEP, ILO et al. 2008, pp. 6 and 8). Today more than ever before, the urgency of climate responsive action demands that the global community look beyond nation-state sponsored SDG silos on clean energy and climate and explore whether NSA-driven global partnerships can efficiently be formulated around the list of identifiable thematic areas or sectors – electricity, agriculture, manufacturing, transportation and buildings, but in all of this the issue of increasing overall access to clean energy and clean air cannot be ignored any longer.

There can be no obfuscating the fact that the shortfall in climate financing and the growing GHG emissions gap remains significant. But increasing access to clean/sustainable, cost-effective energy services happens to be crucial for addressing a wide variety of poverty eradication measures such as reducing infant mortality, improving the lives of women and girls, promoting education, improving access to clean water and enabling productive livelihoods. There are a dizzying array of existing sustainable finance actors, initiatives, clusters/hubs, pledges and new promises that have completely transformed the sustainable finance ecosystem at all levels- global, regional and national. The current reality is that the global financial system is becoming increasingly complex and while it is beyond the immediate purview of this concluding chapter, it is nevertheless important to flag some key developments in relation to the SDGs.

More than three years ago, the OECD Green Finance and Investment flyer summarizing key innovations imperative to achieving the SDGs and PA goals found that '$6.3 trillion in annual, low carbon climate resilient infrastructure will be needed from 2015-2030' to meet these globally agreed upon goals. The OECD was blunt in its assessment that: 'The window to act is narrow – *decisions made on infrastructure investment in developing countries in the next ten years will determine whether the SDGs and Paris Agreement goals will be met*' (emphasis added, OECD 2017, p. 1). But, a July 2017 report jointly prepared by Oil Change International, Friends of the Earth and Sierra Club entitled 'Talk is Cheap: How G20 Governments are Financing Climate Disaster', underscored the point that Group of 20 (G20) governments 'are providing nearly 4 times more public finance to fossil fuels than to clean energy'. The report could not have been more categorical as to its finding: 'Of all public finance for energy provided by G20 institutions and the multilateral development banks between 2013 and 2015, over $71.8 billion annually – or 58 percent – supported fossil fuel production, *while just $18.7 billion annually – or 15 percent – supported clean energy* (including renewable sources such as wind, solar, geothermal, and small hydro). Just over 26 percent of finance went to energy infrastructure categorized as neither clean nor fossil fuel – for example, large hydro dams or transmission infrastructure with no clearly associated energy source' (emphasis added, Oil Change International et al. 2017, Executive Summary, p. 3).

The 2020 'Doubling Back and Doubling Down: G-20 Score Card on Fossil Fuel Funding' published by the International Institute for Sustainable Development (IISD) and Oil Change International issued the following key findings that leaves little room for equivocation as to role of some of the world's largest aggregate emitters when it comes to fossil fuel funding. The key findings include:

- '*G20 governments provided $584 billion annually* (2017–2019 average) via direct budgetary transfers and tax expenditure, price support, public finance, and SOE investment for the production and consumption of fossil fuels at home and abroad. *Governments provided more support to oil and gas production than any other stage of fossil fuel-related activity,* at $277 billion (47% of the total support to fossil fuels).
- *G20 government support has seen a 9% drop relative to the annual 2014–2016 average,* indicating some progress has been made, although around a third of this decrease can be attributed to an average decrease in oil prices. The drop in support does not represent a consistent decline across G20 countries over time. Seven of the G20 countries increased their fossil fuel support: Australia, Canada, China, France, India, Russia, and South Africa. *Progress made between 2014 and 2019 was insufficient: more needs to be done.*
- G20 countries allocated some $170 billion in public money commitments to fossil fuel-intensive sectors in response to the COVID-19 crisis between January 1 and August 12, 2020. This is likely an underestimate due to the dynamic nature of government responses to the COVID-19 crisis and a lack of transparency that doesn't allow for the quantification of many announced policies. Readers can refer to the most up-to-date information at the Energy Policy Tracker (www.energypolicytracker.org). *The support for fossil fuels in response to the COVID-19 crisis indicates that G20 governments are moving in the wrong direction and are likely to undo the little progress made between 2014 and 2019'* (emphasis included, IISD and Oil Change International 2020, p. 4).

The push towards 'naming and shaming' of fossil fuel related companies is a powerful paradigm shift but to be clear, the role of national governments in setting clear policy and regulatory guidelines for clean energy is absolutely crucial. NNSAs in particular energy related private sector entities hold the key to addressing the shift towards cleaner energy and climate mitigation, but political changes to eliminate fossil fuel subsidies by key GHG emitters remain a hurdle. In this regard, it is also worth pointing out that in 2019, the G20 comprising the world's largest aggregate GHG emitters and economies stalled on eliminating fossil fuel subsidies that the G20 itself had agreed to back in 2009. Back in 25 September 2009, G20 leaders issued a Pittsburg meeting statement, citing studies by the OECD and the IEA, agreed that 'eliminating fossil fuel subsidies by 2020 would reduce greenhouse gas emissions in 2050 by ten percent'. The G20 communiqué also emphasized need to provide essential services to '*populations suffering from energy poverty*' (emphasis added) via targeted cash transfers and other mechanisms (G-20 Research Group 2009). But the pace of change towards implementing this 2009 agreement has been extremely slow. More than a decade since this historic 2009 G20 agreement on eliminating fossil fuel subsidies trillions of dollars in government budgets have been spent on subsidies. On the subject of 'energy', 2019 Osaka Summit Leaders' Declaration called attention to the '3E+S' *(Energy Security, Economic Efficiency and Environment + Safety)* in order to transform towards 'affordable, reliable, sustainable and low GHG emissions systems as soon as possible, recognizing that there are different possible national paths to achieve this goal'. It reaffirmed the 'joint commitment on medium term rationalization and phasing-out of Inefficient Fossil Fuel Subsidies that encourage wasteful consumption, while providing targeted support for the poorest', but there was no mention of the SDGs 7 and 13 and no specific details provided as to targeted support (G-20 2019, para. 37).

The 2021 G20 under the Italian presidency focused on three broad, interconnected pillars of action: 'People, Planet, Prosperity'. The final G20 statement that emanated from the 2021 Leaders Summit referenced that climatic impacts 'are being experienced worldwide, particularly by the poorest and most vulnerable', stressed the 'importance of the effective implementation of the global goal on adaptation' and reaffirmed 'the commitment made by developed countries, to the goal of mobilizing jointly USD 100 billion per year by 2020 and annually through 2025 to address the needs of developing countries.' The final statement also for the first time highlighted the role of the methane emissions within the context of curbing overall GHG emissions: 'We acknowledge that methane emissions represent a significant contribution to climate change and recognize, according to national circumstances, *that its reduction can be one of the quickest, most feasible and most cost-effective ways to limit climate change and its impacts.* We welcome the contribution of various institutions, in this regard, and take note of specific initiatives on methane, including the establishment of the International Methane Emissions Observatory (IMEO)' (emphasis added, G-20 2021). While the statement-does not provide clear guidance on the world's largest environmental health risk – air pollution –the significance of the focus on reducing methane emissions, given methane role as both a principal GHG and an SLCP is key to the future climate mitigation strategies of countries like India and China.

There has been a growing trend of companies, cities, not-for-profits and academic institutions divesting funds from fossil fuels companies towards investments in clean, renewable energy in order to accelerate the transition toward a low carbon economy. An example of coalition building amongst private sector entities which showcases the impacts of scaling up like-minded NNSA is the Climate Action 100+, an investor initiative that was launched at the 2017 One Planet Summit: 'To ensure the world's largest corporate greenhouse gas emitters take necessary action on climate change'. More than 320 investors with over $33 trillion in assets collectively under management are engaging companies to:

- 'Curb emissions
- Improve governance and
- Strengthen climate-related financial disclosures'

As per the website, the initiative currently comprises:

- '575 investors
- $54 trillion (in investments managed by investors)
- 167 companies
- 80% of industrial emissions (estimated to be covered by focus companies)'

(Climate Action 100+ website 2021).
But a search of Climate Action 100+ site using the word 'poverty' revealed only one specific link to a 16 July 2020 call issued by the 'Ceres Blueprint for Responsible Policy Engagement on Climate Change' to US companies to establish a corporate governance system that addresses climate change as a systemic risk that can allow for improved alignment with the lobbying practices taken by companies (Ceres Press Release 2020).

It is clear that climate transition risks are being factored in but whether and how this translates into improving the lives of the energy poor and climate vulnerable is unclear. In April 2015, the G20 finance ministers and Central Bank governors asked the Financial Stability Board (FSB) to convene public and private sector participants to review how the financial sector can take account of climate-related issues. As part of its review, the FSB identified the need for better information to support informed investment, lending and insurance underwriting decisions and improve understanding of climate-related risks. To help identify the information needed to assess and price climate-related risks, the FSB established an industry-led task force – the Task Force on Climate-related Financial Disclosures (TCFD)- Box 6.1 provides key takeaways and findings of the TCFD. In 2017, the TCFD released its first report, followed by two others focused on 'the alignment of companies reporting with the TCFD recommendations'. The 2020 report outlined the enormous scope, scale and potential of NNSA's – private sector actors – reporting on climate-related risks:

- 'The number of organizations expressing support for the TCFD has grown more than 85%, reaching over 1,500 organizations globally, including over 1,340 companies with a market capitalization of $12.6 trillion and financial institutions responsible for assets of $150 trillion.
- Similar to the growth in the number of organizations supporting the TCFD, investor demand for companies to report information in line with the TCFD recommendations has also grown dramatically. For example, as part of Climate Action 100+, more than 500 investors with over $47 trillion in assets under management are engaging the world's largest corporate greenhouse gas emitters to strengthen their climate-related disclosures by implementing the TCFD recommendations' (TCFD 2020, p. 1).

Perhaps the most compelling evidence tracing the negative and positive influence of private sector entities and corporations on climate change can be found in a series of reports published by the not-for-profit UK based InfluenceMap. Its 2019, 'Corporate Carbon Policy Footprint: The 50 Most Influential' (2019) identified *50 of the most influential companies* on climate policy globally, *showing the majority (35 companies) were negative* (led by US utility Southern Company, ExxonMobil and Chevron) *and only 15 were positive* (see Box 6.2 below for an excerpt from this report. The InfluenceMap 2020 report meanwhile found that: 'Fossil fuel lobbyists dominating climate policy battles during COVID-19' and tracked '121 instances of corporate and industry lobbying interventions globally that are relevant both to the COVID-19 crisis and the climate emergency from March 1st to July 1st' and found that 'the vast majority of these are associated with fossil fuel value chain companies, including automotive and aviation, as well as, oil, gas and coal production'. More specifically, this 2020 report highlighted that the oil and gas sector 'has been both the most active and the most successful in its lobbying interventions' with '64% of lobbying engagements either completely or most successful, and a further 26% ongoing', whilst automotive sector lobbying was less successful with '59% of its lobbying engagements resolved unsuccessfully or mostly unsuccessfully' and coal 'with 47% of its engagements resolved unsuccessfully or mostly unsuccessfully' (2020, pp. 2–3).

Meanwhile, InfluenceMap's 2021 report, 'Big Tech and Climate Policy' focused on five Big Tech companies (Apple, Alphabet, Amazon, Facebook and Microsoft) that together account for more than 25% of the S&P 500, and that have all grown during the COVID pandemic. The report found:

'While the five Big Tech companies are all highly positive on their own climate program – in terms of minimizing the emissions of their operations, supply chains and products – their climate policy engagement appears ad hoc. It has focused on technical

Box 6.1 Task Force On Climate-Related Financial Disclosures (TCFD): key takeaways and findings.

Key takeaways and findings

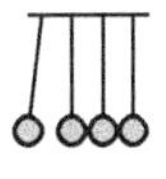

Nearly 60% of the world's 100 largest public companies support the TCFD, report in line with the TCFD recommendations or both. In addition, nearly 700 organizations have become TCFD supporters since the Task Force issued its 2019 status report, an increase of over 85%. The Task Force is encouraged by the growing support for its recommendations and hopes to see similar growth in the percentage of companies disclosing TCFD-aligned information going forward.

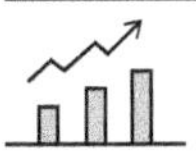

Disclosure of climate-related financial information has increased since 2017, but continuing progress is needed. Disclosure of TCFD-aligned information increased by six percentage points, on average, between 2017 and 2019, and the Task Force applauds the improvements made – both in terms of the number of companies reporting and the quality of such reporting. However, companies' disclosure of the potential *financial* impact of climate change on their businesses and strategies remains low. The Task Force recognizes the challenges associated with making such disclosures but encourages continued efforts and faster progress.

Energy companies and materials and buildings companies lead on disclosure. For fiscal year 2019 reporting, the average level of disclosure across the Task Force's 11 recommended disclosures was 40% for energy companies and 30% for materials and buildings companies.

One in 15 companies reviewed disclosed information on the resilience of its strategy. The review found that the percentage of companies disclosing the resilience of their strategies, taking into consideration different climate-related scenarios, was significantly lower than that of any other recommended disclosure.

Asset manager and asset owner reporting to their clients and beneficiaries, respectively, is likely insufficient. While TCFD-aligned reporting by a sample of asset managers and asset owners increased over the past three years, the Task Force believes reporting by these organizations to their clients and beneficiaries may not be sufficient and that more progress may be needed to ensure clients and beneficiaries have the right information to make financial decisions.

Expert users find the impact of climate change on a company's business and strategy as the 'most useful' for decision-making. Expert users also identified information about a company's material climate-related issues for each *sector* and *geography* and its key metrics as extremely useful for financial decision-making.

Expert users' insights on the most useful information for decision-making may provide a road map for preparers. Companies already disclosing their governance and risk management processes for climate-related issues and working towards full TCFD implementation might consider expert users' relative ranking of specific types of climate-related information – from most useful to least useful – as one factor to consider in prioritizing their efforts.

Source: TCFD (2020, p. 4).

Box 6.2 50 corporations shaping the global climate agenda: key findings from InfluenceMap's 2019 report.

The analysis contained below excerpted from Influence Map's 2019 report is drawn from a universe of the world's 250 largest industrial companies and combines metrics representing (a) corporate climate policy positions (b) its level of engagement (lobbying) and (c) a company's economic and political clout, into a Carbon Policy Footprint Score. This score ranges from −100 (highly influential and climate-oppositional) to +100 (highly influential and climate-positive).

Of the 50 most influential companies who score negatively, the oil/gas sector continues to dominate the list, led by *ExxonMobil, Chevron and BP*. In March 2019 InfluenceMap released "*Big Oil's Real Agenda on Climate Change*," which found that *these three companies along with Shell* and *Total* are spending *hundreds of millions of dollars each year* on sophisticated messaging strategies to *capture the public narrative on climate*. At the same time, they are also *lobbying to control, delay or block climate regulations globally*.

Automotive companies also feature prominently in the list of negative influencers, led by *Fiat Chrysler, Daimler and BMW*. This is the result of a strategy to control and delay the regulatory agenda on vehicle emissions and electric vehicles (EVs). This may now hinder their ability to adapt quickly to any acceleration of emissions and EV rules with a sudden shift in climate politics globally (e.g. in the US following the 2020 elections). Tesla remains the only auto company in the list of 50 who is supportive of climate policy, perhaps not surprisingly given its EV-based business model.

The analysis found lobbying from companies within the coal value-chain to be highly impactful, although in isolated regions globally.

Economically powerful tech companies Microsoft, Facebook and Google remain *outside of the list of the 50 most influential*. They have not translated their climate-positive messaging into strategic, consistent policy engagement.

InfluenceMap analysis continues to show that many strategically influential and positive corporations on climate policy are European, which is likely contributing to a modest but important positive trajectory on climate policy in the region. These consist of utilities pushing for renewables policy (Iberdrola, Enel) and industrials like Royal DSM and Phillips.

Unilever, which has maintained a consistent effort to support a range of climate policy related to the energy system, *is ranked the most influential positive company*' (emphasis added).

Source: InfluenceMap (2019, pp. 2–3).

rules to enable corporate renewable energy procurement that are directly associated with their operations/commitments . . . . Big Tech companies do not appear to be matching their huge economic footprint with corresponding strategic support for strong government action to implement the Paris Agreement. *This represents a lost opportunity for corporate*

Table 6.1 Climate influence of Big Tech.

Company	Key features of climate program
Alphabet	Reported eliminating its entire carbon legacy through carbon offsets as of September 2020; committed to operating on 24/7 carbon-free energy in all data centers and campuses worldwide by 2030 (announced 2020).
Apple	Reported reaching 100% renewable energy use in its data centers in 2018; committed to being carbon neutral across its supply chain (announced 2020)
Microsoft	Goal of being carbon negative by 2030, removing from the environment more carbon than it has ever emitted by 2050, and reaching 100% renewable energy use by the end of 2025 (announced 2020).
Facebook	Goal of buying 100% renewable energy by 2020 (announced 2018); committed to reaching net zero GHG emissions for its value chain by 2030 (announced 2020)
Amazon	Goal of becoming carbon neutral by 2040, including 100% renewable use by 2025 (announced 2019) and converting its delivery fleet to 100% electric vehicles by 2030 (announced 2020)

Source: Influence Map (2021, p. 7).

leadership on climate policy'. Table 6.1 excerpted from the InfluenceMap 2021 report provides an overview of key climate features of the five Big Tech companies surveyed.

Overall, the global private sector's push towards transitioning from fossil fuels appears to be somewhat ad hoc, and even murkier to explain when the question as to what if at all fossil fuel divestment by companies means for those who lack access to clean energy and are coping with toxic levels of exposure to energy related air pollution. But, this is exactly where linked partnership action by NNSAs is necessary not only because of the agreed scope of the SDA and the PA but also because resource constrained national governments in many of the poorest and most vulnerable countries cannot address these developmental challenges without the implementation of NSA-driven partnerships.

6.4 Global Action on Clean Air, Clean Energy and Climate Mitigation Cannot be Implemented in Segregated Silos

A decade after the IEA first pointed out that the window for responsive climate action was rapidly closing, the 2019 United in Science Report highlighted that the 2015 PA's goal to keep global temperature rise this century to 1.50 °C above pre-industrial levels requires that the *current level of climate action policies needs to be roughly tripled for emission reduction* to be in line with the 2 °C goal and increased fivefold for the 1.5 °C goal (emphasis added, WMO 2019). It is evidently clear that fossil fuel energy demand is growing particularly in G20 countries, and in the case of many of the megacities in developing countries, toxic levels of energy related air pollution smother and burden lives of those most vulnerable amongst us. As Jackson et al. noted: 'A robust global economy, insufficient emission reductions in developed countries, and a need for increased energy use in developing countries where per capita emissions remain far below those of wealthier nations will continue to put upward pressure on CO_2 emissions' (Jackson et al. 2019, p. 1).

Undoubtably, more robust, efficient and linked partnership on clean air and climate mitigation including curbing SLCPs are required in the future particularly within the context of developing countries. In addition to the important role of NNSAs, it is clear that energy sector holds the key to solving inherently linked climate change and poverty eradication objectives. There is no getting around the point that mitigating climate change is a major political and developmental challenge precisely because the energy sector as a whole is a crucial sectoral actor in the shift towards sustainable energy at the national and global level. The finding that NDCs need to be urgently scaled up in terms of closing the GHG emissions gaps combined with the finding that NNSA and sub-national actors (cities and local governments) are agents of significant emission reductions informed the discussions in the previous chapters. But, there is now for the first time in recorded history a growing global consensus as evidenced by the previous chapters that it is simply impossible to meet the PA's goals without the active participation of NNSAs. And here, it is important to underscore that the 2018 UNEP Emissions Gap Report clearly notes that it is NSAs (analog to NNSAs) that 'provide important contributions to climate action beyond their quantified emission reductions. They build confidence in governments concerning climate policy and push for more ambitious national goals. They provide space for experimentation or act as orchestrators, coordinating with national governments on climate policy implementation'. In this regard, the report also finds that the scope of the numbers for NSA contributions 'seem impressive, but there is still huge potential for expansion. Based on available data, not even 20 percent of the world population is represented in current national and international initiatives, and many more of the over 500,000 publicly traded companies worldwide still can, and must, act. On the financial side, a record of just over US$74 billion of Green Bonds were issued in the first half of 2018, but this still represents only a very small fraction of the capital markets around the world. The emission reduction potential from NSAs is large, but estimates vary considerably across studies. If international cooperative initiatives are scaled up to their fullest potential, the impact could be considerable compared with current policy: up to 19 $GtCO_2$ e/year by 2030 (range 15–23 $GtCO_2$ e) according to one study. If realized, this would be instrumental in bridging the emissions gap to 2°C pathways' (UNEP 2018, p. 20). But the 2021 UNEP Emissions Gap Report is far more pessimistic in its overall findings at to individual NDC's submitted by Parties to the PA: '. . .new and updated NDCs and announced pledges for 2030 have only limited impact on global emissions and the emissions gap in 2030, reducing projected 2030 emissions by only 7.5 per cent, compared with previous unconditional NDCs, *whereas 30 per cent is needed to limit warming to 2°C and 55 per cent is needed for 1.5°C*. If continued throughout this century, they would result in warming of 2.7°C. *The achievement of the net-zero pledges that an increasing number of countries are committing to would improve the situation, limiting warming to about 2.2°C by the end of the century*. However, the 2030 commitments do not yet set G20 members (accounting for close to 80 per cent of GHG emissions) on a clear path towards net zero. *Moreover, G20 members as a group do not have policies in place to achieve even the NDCs, much less net zero'* (emphasis added, UNEP 2021, p. iv).

The role of NNSAs in sponsoring a shift towards clean air and clean energy in developing countries matters for millions of lives. There is a growing movement towards renewable/

clean energy within the two of world's largest aggregate emitters whose residents face high levels of air pollution as means of meeting climate and clean energy goals. Here, it is useful to recognize that countries like China and India have been making dramatic shifts towards renewable energy. But this shift is still not sufficient in terms of Indian cities for example being able to reduce toxic levels of PM air pollution. Bloomberg NEF which has been tracking for the past 10 years clean energy investments via 'Energy Transition Investment Trends Report' noted the following key highlights for 2020 (see also Figure 6.2):

- 'In 2020, global investment in the low-carbon energy transition totalled $501.3 billion, up from $458.6 billion in 2019 and just $235.4 billion in 2010. This figure includes investment in projects, such as renewable power, energy storage, EV charging infrastructure, hydrogen production and CCS projects – as well as end-user purchases of low-carbon energy devices, such as small scale solar systems, heat pumps and zero-emission vehicles.
- The largest sector in 2020 was renewable energy, which attracted $303.5 billion for new projects and small-scale systems. This was up 2% on 2019, despite COVID-related delays to some deals.
- The second-biggest was electric transport, which saw $139 billion of outlays on new vehicles and charging infrastructure, up 28%. Electric heat got $50.8 billion of investment, up 12%.
- Hydrogen and CCS are small sectors for now, but are expected to grow. In 2020, they received investment of $1.5 billion and $3 billion, respectively down 20% and up 212%.
- Europe and China are currently vying for top position among markets active in energy transition investment' (Bloomberg NEF 2021, p. 1).

Energy transition investment hit $500 billion for the first time in 2020

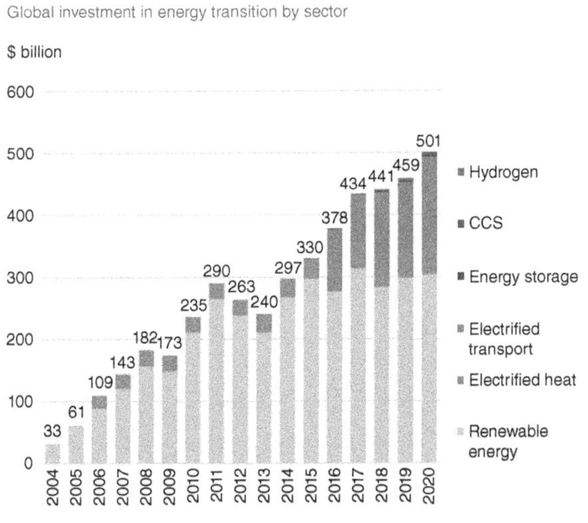

- Our top-level finding is that the world committed a record $501.3 billion to decarbonization in 2020, beating the previous year by 9% despite the economic disruption caused by COVID-19.
- Our analysis shows that companies, governments and households invested $303.5 billion in new renewable energy capacity in 2020, up 2% on the year. They also spent $139 billion on electric vehicles and associated charging infrastructure, up 28% and a new record.
- Other categories also showed strength. Domestic installation of energy-efficient heat pumps came to $50.8 billion, up 12% last year, while that of batteries and other energy storage technologies was $3.6 billion, level with 2019 despite falling unit prices. Global investment in carbon capture and storage rose 212% to $3 billion, and that in green hydrogen was $1.5 billion, down 20% but the second-highest annual number to date.

Figure 6.2 Energy transition investment 2020: Bloomberg NEF report. *Source:* Bloomberg NEF (2021, p. 5).

Renewable energy targets have been referenced in NDCs submitted by parties to the PA, but a lot more remains to be done to close the gap. According to IRENA – the world's only intergovernmental agency focused on renewables, of the 194 Parties to the UNFCCC that submitted NDCs, 145 referred to renewable energy action to mitigate and adapt to climate change, while 109 Parties included some form of quantified target for renewables. Moreover, IRENA's analysis found that 'USD 1.7 trillion would be needed by 2030 to implement renewable energy targets contained in NDCs' and that the renewable energy investment path that will be unleashed as a result will require the mobilization of private investment, stable, consistent and transparent enabling frameworks for renewables. (*IRENA* 2017, p. 2). IRENA's Renewable Energy Report (2020) highlighted how different regions of the world were undertaking renewable energy generation (based on 2018 data) and evidenced that as in past years, Asia's share of global renewable generation continued to increase, reaching 40%, while Europe and North America each have a 20% share, followed by South America (12%) and Eurasia (5%) (IRENA 2020). Table 6.2 provides regional snapshot on renewable energy.

Additionally, IRENA's 'Global Energy Transformation: A Roadmap to 2050' estimated that meeting the PA's objectives would require reducing global energy demand through energy efficiency, increasing the electrification pathway for all end-use sectors, and increasing the share of renewables in the energy matrix, including biofuels. Thus, renewables would need to comprise at least two-thirds of the total final energy supply by 2050. At the same time, the share of renewable energy in the power sector would need to increase from 25% in 2017 to 86% in 2050 (2019, p. 10). This is precisely why global partnerships that are renewable or clean energy sector driven are pivotal because they can get to the heart of the climate change and energy pollution crises. NNSA partnerships such as those coalesced by C-40 and CCAC hold the key because of the global trends towards urbanization and increasing PM pollution within some of the most polluted and populous cities.

Table 6.2 Renewable energy generation highlights by region.

Generation in 2018 (TWh)	Hydro	Wind	Bioenergy	Solar	Geothermal	Marine	Total
Africa	129	14	3	9	5		160
Asia	1719	440	165	293	26	<1	2644
Central America + Caribbean	29	6	6	3	4		48
Eurasia	266	20	3	8	8	<1	306
Europe	578	384	192	132	13	<1	1298
Middle East	19	1	<1	6			27
North America	710	321	81	90	24	<1	1226
Oceania	44	17	4	10	8	<1	84
South America	655	60	69	11	<1	<1	794
World total	**4149**	**1263**	**523**	**562**	**88**	**1**	**6586**

Source: IRENA (2020).

There is an urgent global need to address energy poverty, pollution and climate change in an integrated manner. As evidenced by Chapters 1–3, being energy poor and climate vulnerable has been identified as extracting a heavy human toll for approximately 3 billion people living at the intersection between poverty, climate change and pollution. The disproportionate double burden of marginalization due to poverty and exposure to toxic levels of air pollution cuts across poor communities everywhere and exposes age and gender inequities as a result. As referenced previously, $PM_{2.5}$ which results from various sources ranging from industrial and vehicular emissions, from cooking with solid fuels (wood, crop wastes and dung) as well as the burning of agricultural residues and waste measures one-hundredth the thickness of a human hair and can penetrate deep into human lungs and blood stream. The WHO has highlighted the irreversible damage caused to the lungs of young infants and children and pregnant women from HAP resulting from cooking with solid fuels. Reducing $PM_{2.5}$ emissions is critically important from a human health perspective, but what is often not reflected is that one of the principal components of $PM_{2.5}$ – BC – emitted as a result of the incomplete combustion of solid fuels is known to be an SLCP.

As referenced in the previous chapters, global, regional and national action on reducing SLCPs such as BC is particularly urgent for countries in Asia and Africa. What has often been neglected or less globally recognized is that BC emissions are also directly linked to serious regional climate change impacts including regional rainfall and weather patterns such as the amplification of snow and ice cover patterns in the Himalayas and also the loss of annual regional production levels of rice, wheat and maize. The main problem is that the PA does not address the issue of reducing SLCPs such as BC. It is long over-due that the PA negotiations take stock of the need for SLCPs reductions and implement the recommendations listed earlier by 2018 WHO Special Report and other entities cited in Chapters 4 and 5.

While it is true that NNSAs including the renewable energy sector and sustainable financing sector have contributed to the global quest for sustainability, it would be a mistake to think this motivation is purely altruistic. Put succinctly, for the private sector, there must be some convergence between SDGs and long-term business goals. The future contours and scope of evolution of both the 2030 SDA and the PA will undoubtedly provide NNSA such as the private sector with opportunities to not only be a major partner but also, more importantly, address the larger question of whether or not business can be a positive force to solve global environmental and social issues including most prominently the nexus between clean air, clean energy and climate change. The overall problem, however, is that the future implementation of global partnerships that emanate from the 2030 SDA and PA is clouded by current conceptual confusion and a lack of clarity in terms of GPSD (SDG 17), and there is a demonstrable evidence of silos between sustainable energy and climate change related partnerships related to the separate SDGs 7 and 13 as evidenced in Chapters 1 and 2.

As Chapters 2 and 3 argued, it is time to push back against the fallacy that somehow effectively linked partnership action on low-zero carbon and climate change that puts poverty reduction as the core can be implemented via segregated SDGs that are negotiated and implemented within separate policy silos. But the good news is that globally accessible, integrated clean energy and climate transition partnerships do not need to wait for the moribund textual parsing and political posturing of intergovernmental climate negotiations. Instead the loci for linked action on access to clean energy and climate resiliency for poor and vulnerable communities can and does reside at the local/city government level. Cities are undeniably on the frontline for clean energy and climate justice, as well as climate resiliency solutions. According to UN's, 'The World's Cities in 2018', cities consume more than two-thirds of the world's energy and account for more than 70% of global carbon dioxide emissions. The number of cities with more than 10 million inhabitants – 'megacities' – is projected to rise from 33 in 2018 to 43 in 2030, and Delhi will overtake Tokyo as the world's largest megacity by 2030 as discussed in Chapter 4. Cities are the loci for linked action on clean air and clean energy access because cities are where the climate battle will be largely won or lost. Accordingly, Fink and other high net worth individuals and private sector entities who argue that climate responsive investment should not leave behind 'large parts of society or entire countries in developing markets' need to urgently engage with, and focus on an integrated, implementation nexus between clean energy access, climate justice and community resilience within the bustling yet polluted cities and megacities across the world.

As demonstrated in Chapters 1, 2 and 3, the absence of a fully integrated and target focused UN affiliated partnership framework that can effectively replace existing silos with synergies in terms of the two distinct SDGs 7 and 13 which are inherently linked global challenges is a colossal waste of limited resources and time. Separate silos/SDGs on sustainable energy and climate result in disparate and fragmented partnerships that cannot leverage and scale up and by default defy the very logic of integrated global action that is fundamental to the future implementation of both the SDA and PA.

Despite decades of protracted intergovernmental climate negotiations convened across various global capitals, meeting the objective of the PA by 2030 appears currently unlikely without seriously ramped up actions towards a clean energy transformation. The IEA in its first ever road map for the global energy sector – *'Net Zero by 2050'* – has outlined that '. . . the gap between rhetoric and action needs to close if we are to have a fighting chance of reaching net zero by 2050 and limiting the rise in global temperatures to 1.5^{0}C. Doing so requires nothing short of a total transformation of energy systems that underpin our economies' (2021, p. 3). Notably, the 2021 IEA Road Map to Net Zero is specifically guided by two principles: (i) to promote secure and affordable energy supplies to foster economic growth, and (ii) to ensure that clean energy transitions must be fair and inclusive. In fact, in relation to the second guiding principle, the IEA is quite clear that clean energy transitions are guided by the idea of 'leaving no one behind' and here the report states: 'It is a moral imperative to bring electricity to the hundreds of millions of people who currently are deprived of access to it, the majority of them in Africa' (p. 4). In terms of linked action on clean energy access and climate change, the 2021 IEA Road Map specifically highlights a key point: 'Providing electricity to around 785 million that have no access, and clean cooking solutions to 2.6 billion that lack those options in an integral part of the pathway to net zero. *Emissions reductions have to go hand-in-hand with efforts to ensure energy access for all. This costs about US$ 40 per year, equal to around 1% of average annual energy sector investment, while also bringing major co-benefits from reduced indoor air pollution'* (emphasis added, IEA 2021, p. 17).

The IEA Road map to the goal of net zero emissions by 2050 is predicated on the fact that the energy sector is 'responsible for three quarters of global greenhouse gas emissions today' (p. 13). In the global pathway to net zero identified by the IEA, renewable energy (wind, solar, bioenergy, geothermal and hydro) will comprise 'two-thirds of total energy supplies' and fossil fuels will 'fall from almost four-fifths of total energy supply today to slightly over one-fifth by 2050' (p. 18). In the overall scheme of global energy sector funding, ensuring clean energy access for all is a drop in the proverbial bucket, and yet progress towards this SDG has been so slow. However, the question is exactly how soon and to what extent innovative actions and private sector finance can be directed towards improving the lives of millions who lack access to even basic modern energy services, much less advanced modes of transportation or homes with energy-efficient amenities such as air?

On 2 July 2021, the CEO of the SEforAll Initiative wrote an online article posted on the WEF site about the slow rate of progress on SDG 7 which she notes is 'nowhere near what is required to meet the 2030 target', and called attention to the following facts:

- *'More than 750 million people have no access to electricity and 2.6 billion people continue to lack access to clean cooking technologies.*
- *At the current pace, these numbers will remain above 600 million and 2 billion people respectively in 2030.*
- *Global renewable energy adoption – the second indicator of SDG7 – remains low* despite commendable progress. There's also much room for improvement on energy efficiency – the third and final indicator. *As it stands, only 10% of global energy consumption comes from modern renewable energy sources, while the global annual rate of energy efficiency improvement hovers around the 1% mark versus a 3% target*' (emphasis added, Ogunbiyi 2021).

Given the challenge inherent in implementing the SDGs 7 and 13, there is no question that NNSA especially the energy sector will need to get more actively involved. As Box 6.3 which lists some of the diverse arrays of partnerships initiatives available to private sector based NNSAs seeking to become involved in the implementation of the SDGs reveals, there is no shortage of interest. But as Chapters 1–3 evidenced, the existing level of conceptual, policy and implementation clarity, combined with the absence of a robust, tracking framework that integrates action on SDGs 7 and 13 and the fact that the PA Article 6 mechanism is still not fully defined makes for ample confusion.

Box 6.3 Examples of partnership initiatives for NSAs seeking involvement in implementation of the SDGs.

The UN Global Compact, Global Reporting Initiative (GRI) and World Business Council for Sustainable Development (WBCSD) have developed a platform for understanding how to implement the SDGs (the 'SDG Compass Project') which feature an interactive tool showing how GRI reporting and other corporate reporting intersect with the SDG targets.

The UN-Business Action Hub was developed as a joint effort of the United Nations Global Compact, Global Hand, a Hong-Kong based non-profit specializing in facilitating private sector and NGO connections, and 20 UN entities and aims to foster greater collaboration between the business and UN to advance solutions to global challenges and to support various humanitarian and disaster preparedness and response efforts. On this platform business can learn more about UN entities, their mandates, specific needs and offer programmatic support, in-kind and financial donations, while UN entities can learn more about the specific interests of companies, available resources and engagement opportunities desired by business.

Additionally, both UN and Business can post projects and use the platform to search for and interact with potential partners to scale the impact of their projects.

KPMG and the Global Compact have developed an Industry Matrix that will similarly provide case studies of the opportunities for contributing to achieving the SDGs in different sectors.

Business Fights Poverty is the world's largest network of international development and business professionals. The site provides content on a variety of topics relevant to the 2030 development agenda through different thematic areas, such as inclusive business models, inclusive finance, environment, women, health, nutrition and partnerships.

The SDG Philanthropy Platform is a vehicle to enable partnerships in the global development space and to help them flourish to achieve global development outcomes as the world transitions from the Millennium Development Goals (MDGs) to the SDGs. Philanthropy has adopted a systemic approach to funding and policy work, shifting from fragmented individual projects to long-term collaborative efforts in line with national priorities and the SDGs.

The Global Development Incubator, Inc. (GDI) supports innovative ideas, organizations and initiatives that have the opportunity to create large-scale social change. GDI has supported efforts such as the launch of the Aspen Network of Development Entrepreneurs (ANDE) at the Aspen Institute and the Initiative for Smallholder Finance, among others. GDI has created a number of initiatives specific to the SDGs, including the Global Sustainable Development Goal Partnership and Convergence.

Devex Impact is a global initiative by Devex and USAID in partnership with top international organizations and private industry leaders. Devex Impact is a portal of information on partnerships and existing initiatives for development agencies, companies, global NGOs and other development professionals (over 600 partnerships are indexed and searchable through a database). Devex Impact is led by a Strategic Advisory Council of informed leaders in the development field from the private sector, international NGOs and development agencies.

The Partnering Initiative is an independent not-for-profit organization dedicated to driving cross-sectoral collaboration for a sustainable future. The Partnering Initiative has conducted extensive research on partnerships and developed a holistic approach of integrated programmes and training opportunities to mainstream partnership activity across five levels, from individuals to international policy.

6.5 Conclusion

Poverty reduction is clearly the overarching priority of the UN PA and the SDA. Well-documented, scientifically credible evidence regarding adverse climatic impacts wreaking utter havoc on the poorest and most vulnerable communities has not been framed forcefully and/or acted upon effectively by the rest of the world. The tragedy of the decades old climate and air pollution crises is that the systematic neglect of the policy nexus between the inherently linked pollution, poverty, clean energy and climate challenges has allowed for the global framing of these crises as segregated rather than as integrated challenges. In other words, although there is ample global evidence that demonstrates the direct co-relation/linkages between sustainable energy for the poor, energy related pollution and climate change mitigation, there has been no consistent implementation of integrated global frameworks or modalities that allow for climate, poverty and sustainable energy objectives to be met simultaneously.

The abnegation of responsive action regarding the nexus between energy pollution and climate change experienced by the energy poor and climate vulnerable essentially ends up deciding who gets to live or die. The interlinked climate and pollution apocalypses portend anguish for the poorest communities in SIDS, LDCs and also the largest urban centres in Africa and Asia, so from both the UN and also a global equitable development perspective, there is no time to waste in implementing joint action to curb SLCPs, mitigate pollution and scale up access to clean, cost-effective energy services. The continued existence of separate policy silos and SDGs on sustainable energy for all and climate change should therefore now be publicly acknowledged for what it is – a globally inequitable travesty – one that defies the rationale of equitable human development as well as common sense knowledge that energy is the crucial driver in addressing the intersecting climate change, pollution and poverty eradication concerns of millions across the world.

It is also time to take full stock of the fact that there is a glaring absence of a globally reliable and accessible framework for implementing and verifying partnerships aimed at reducing poverty and energy related air pollution and combating the adverse impacts of climate within UN globally agreed outcomes. This lack of an integrated partnership framework between sustainable energy for all and climate change does not bode well for the future implementation SDGs 7, 13 and 17 despite their shared emphasis on poverty reduction. As Chapters 2 and 3 demonstrate, UN global guidance on sustainable development and climate change related partnerships lacks the necessary policy and programmatic guidance and clarity that would facilitate integrated action on climate change, sustainable energy and poverty and pollution reduction. Having two separate SDGs on sustainable energy (SDG 7) and climate change (SDG 13) within the UN's SDA and an SDG 17 that is ostensibly focused on GPSD yet has no defined process for making linkages with other SDG partnerships makes no practical sense in terms of both financing and implementation. As Chapters 4 and 5 demonstrate, there are a panoply of new and innovative actions being undertaken by a diverse range of NNSA including the energy sector but none that translate effectively into reducing the world's largest environmental health risk.

The gravity of the linked energy pollution, poverty and climate crises is such that continued wastage of limited global time and resources is unmistakably and inequitably felt by those who are least in a position to advocate for themselves in global negotiation assemblages/fora. Segregated UN global goals/policy silos on clean energy and climate are ineffective and inefficient in terms of addressing the linked global challenges of reducing energy related air pollution, mitigating SLCPs and increasing access to low and/or zero-carbon sustainable energy technologies, services and systems for all. The continued existence of silos on sustainable energy and climate also directly impedes synergistic scaled up action, which in turn end up causing implementation challenges for the UN and its member states.

More significantly, developing country member states, particularly the LDCs and SIDS amongst them have little choice but to face the blunt realization of reduced financing and resource constraints in addressing fragmented and disjointed SDGs 7 and 13.

Global partnership efforts on energy and climate change are particularly relevant for meeting poverty reduction objectives which happen to be the stated, overarching priority of both 2015 agreements, and these efforts are germane to the involvement of NSAs including the private sector/business community. Based on an examination of the past record of UN-led global partnership mechanisms, the globally relevant arguments raised previously in this book are that clarity rather continued confusion is urgently needed to:

- replace existing, disjointed silos on sustainable energy and climate change with an integrated partnership nexus within these two universally adopted agreements; and
- implement the proposed global partnership mechanisms envisaged within the 2030 SDGs and the Paris Climate Agreement (Article 6).

There is no glossing over the fact that the future 'success' of the 2015 PA depends entirely on the level of 'ambition' of the voluntary national pledges of climate action required by all Parties – ubiquitously referred to as NDCs. Countries are supposed to submit INDCs every five years with steady increases in their ambitions to curb climate change. But, like all well-intentioned resolutions, it is not the speeches made by climate celebrities at an array of global capitals but the actual, verifiable actions enacted by individual companies, countries, cities, and community residents that matters. It therefore remains to be seen if the next spate of high-level by-invitation-only global fora ostensibly designed to build momentum for new NDCs will result in the scaling up of climate responsive action.

There have been 26 annual COPs convened across the world so far. While, attendance at these COPs has grown each year in terms of NNSAs, the intergovernmental negotiations silo approach has relegated NNSA participation to the outside margins of external events and separated these NNSA events from official COP negotiations fora. An excoriating article written by senior staff from the Center for International Environmental Law (CIEL) entitled 'False solutions prevail over real ambition at COP26' highlighted that the claim made by the UK and the UN that COP-26 would be the most inclusive was exactly the opposite in practice: 'The UK COP presidency and the UNFCCC Secretariat imposed a nearly full ban on civil society presence in the negotiations for the first few days of the conference. In doing so, they breached even the inadequate guidelines that all governments had previously adopted with regards to the transparency of the negotiations, as well as other international standards guaranteeing rights of access and participation in environmental decision-making, including UN processes. UN Special Rapporteurs denounced the situation and issued urgent appeals to the UN Climate Secretariat and the UK presidency to ensure that all voices could be heard.' CIEL's analysis of the failure of COP-26 to deliver on climate justice and equity was blisteringly hard to refute: '*The continued failure of the UN climate process to meet the urgency of the moment will not deter the global movement for climate justice, nor slow the legal and financial momentum for a just transition to a fossil-free future... The fossil fuel era is ending. If global leaders can't summon the courage to help make that happen, then the people will*' (emphasis added, CIEL 2021).

The unvarnished truth is that COP-26 was a failure in terms of ensuring integrated action on clean air and clean energy access aimed at improving the lives of millions who are climate vulnerable and exposed to toxic levels of fossil fuel air pollution. As was the case of COP-25, COP-26 once again failed to even consider, much less deliver on verifiable pathways and protocols for curbing fossil fuel related air pollution including any mechanisms for reducing emissions of BC and PM pollution. As Sarah Woolnough, CEO of Asthma UK and the British Lung Foundation succinctly put it: 'At COP26, whilst world leaders were committed to discussing ways to protect our future from climate change, they failed to address the largest environmental threat to human health we face today. *Welcome commitments to "phase-down" coal, cut methane and end deforestation by 2030 grabbed the headlines, but what was missing was any commitment to improve the quality of the air we all breathe.*' As Woolnough went on to point out, ironically enough, the skirting around the urgency of curbing air pollution happened while COP-26 participants were in Glasgow- the fourth most polluted city in the UK (Air Quality News 2021).

The time to look beyond the confines of word parsing exercises conducted once a year within the context of intergovernmental negotiation silos that segregate access to clean energy for all from climate change goals is long overdue. It is a fallacy to somehow expect that effectively linked global partnership actions on a low to zero carbon energy transition focused on the overarching priority of poverty reduction can be implemented via separated out SDGs and a disjointed and unverifiable global partnership agenda. The good news is that integrated clean air, clean energy and climate transition partnerships do not need to wait for moribund textual and political posturing. Instead, the loci for linked action on access to clean energy, clean air and climate resiliency for frontline and vulnerable communities can and does reside at the local/city and state government level.

References

AFP (2019) Climate campaigners 'greatest threat' to oil sector: OPEC. *AFP*. 2 July 2019. https://www.france24.com/en/20190702-climate-campaigners-greatest-threat-oil-sector-opec (accessed on 7 August 2021)

Air Quality News (2021). Let's clear the air about COP26: It failed to protect our lung health. . .17 December 2021. https://airqualitynews.com/2021/12/17/lets-clear-the-air-about-cop26-it-failed-to-protect-our-lung-health/ (accessed 24 December 2021)

Baynes, C. (2019). Trump administration removes a quarter of all climate change references from government websites. *The Independent* (25 July 2019). https://www.independent.co.uk/news/world/americas/us-politics/trump-climate-change-government-websites-global-warming-a9020461.html (accessed 7 August 2021).

Brito, Christopher (2019) Fox News apologizes after guest calls teen climate activist Greta Thunberg "mentally ill". CBS News (24 September 2019) https://www.cbsnews.com/news/fox-news-greta-thunberg-michael-j-knowles-asperger-syndrome-climate-change-mentally-ill/ (accessed on 18 December 2021).

BBC (2019). Amazon Fires: G-7 to release funds for firefighting planes. (26 August 2020). https://www.bbc.com/news/world-latin-america-49469476 (accessed 18 August 2021).

BBC (2021). As it happened: COP Day 2 and Biden's Anger. *BBC News* (2 November 2021). https://www.bbc.com/news/live/world-59125188 (accessed 5 December 2021).

Bloomberg NEF (2021). Energy Transition Investment Trends. https://assets.bbhub.io/professional/sites/24/Energy-Transition-Investment-Trends_Free-Summary_Jan2021.pdf (accessed 14 August 2021).

Berkshire Hathaway Inc. (2016). Shareholders Letter. (27 February 2016). http://www.berkshirehathaway.com/letters/2015ltr.pdf (accessed 22 June 2021).

Chengappa, R. (2007). Apocalypse Now: The latest UN Panel Report on climate change forecasts that India would be among the countries worst affected by rising temperatures. *India Today* (23 April 2007).

Choudhury, S. (2021). India rejects net-zero emissions target as Modi heads to COP26 climate talks. *CNBC* (29 October 2021). https://www.cnbc.com/2021/10/29/cop26-india-rejects-net-zero-emissions-target-modi-off-to-climate-talks.html (accessed 5 December 2021).

CIEL (2021). False solutions prevail over real ambition at COP26. https://www.ciel.org/false-solutions-prevail-over-real-ambition-at-cop26/.

Ceres Press Release (2020). As US rebuilds economy, companies urged to align lobbying with science based climate action. (16 July 2020). https://www.ceres.org/news-center/press-releases/us-rebuilds-economy-companies-urged-align-lobbying-science-based-climate (accessed 18 August 2021).

CDP website (2021). Cities A List 2020. https://www.cdp.net/en/cities/cities-scores (accessed 7 August 2021).

Climate Action 100+ website (2021). Initiative Snapshot. https://www.climateaction100.org (accessed 7 August 2021).

Eccles, R. (2021). Piece of a climate change advice for Warren Buffet change your vote. (22 March 2021). https://www.forbes.com/sites/bobeccles/2021/03/22/a-piece-of-climate-change-advice-for-warren-buffett-change-your-vote/?sh=cc82e73765fd. (accessed 15 August 2021).

Legget, J. (2001). *The Carbon War: Dispatches from the end of the Oil Century*. New York: Routledge.

Fink, L. (2020a). A Fundamental Reshaping of Finance: Letter to CEOs. https://www.blackrock.com/corporate/investor-relations/2020-larry-fink-ceo-letter (accessed 20 August 2021).

Fink, L. (2020b). Sustainability as BlackRock's New Standard for Investing. https://www.blackrock.com/corporate/investor-relations/2020-blackrock-client-letter (accessed 20 August 2021).

Fink, L. (2021). Larry Fink's 2021 Letter to CEOs. https://www.blackrock.com/corporate/investor-relations/larry-fink-ceo-letter (accessed 20 August 2021).

Giddens, A. (2009). *The Politics of Climate Change*. Malden, MA: Polity.

G-20 Research Group (2009). G-20 Leaders Statement: The Pittsburgh Summit. http://www.g20.utoronto.ca/2009/2009communique0925.html#energy (accessed 20 August 2021).

G-20 (2019). Osaka Leaders Declaration. http://www.g20.utoronto.ca/2019/2019-g20-osaka-leaders-declaration.html (accessed on 20 August 2021).

G-20 (2021). G-20 Rome Leaders' Declaration. https://www.consilium.europa.eu/media/52732/final-final-g20-rome-declaration.pdf (accessed 4 December 2021)

Ha, T. (2019). Parched Lives on the Fringe: How water scarcity has widened inequality in Chennai. *Ecobuisness* (18 September 2019). https://www.eco-business.com/news/parched-lives-on-the-fringe-how-water-scarcity-has-widened-inequality-in-chennai/ (accessed 25 April 2021).

Haas, P. (2002). *Science Policy for Multilateral Environmental Governance*. Tokyo: United Nations University/Institute of Advanced Studies.

IEA (2021). *Net Zero by 2050- A Road Map for the Global Energy Sector*. Paris: IEA/OECD https://iea.blob.core.windows.net/assets/4482cac7-edd6-4c03-b6a2-8e79792d16d9/NetZeroby2050-ARoadmapfortheGlobalEnergySector.pdf (accessed 21 August 2021).

InfluenceMap (2019). Corporate Carbon Policy Footpprint (October 2019). https://influencemap.org/report/Corporate-Climate-Policy-Footpint-2019-the-50-Most-Influential-7d09a06d9c4e602a3d2f5c1ae13301b8 (accessed 27 March 2021). (accessed 20 August 2021).

InfluenceMap (2020). Fossil Fuel Lobbyists are dominating Climate Policy Battles during COVID-19. (20 July 2020). https://influencemap.org/report/Fossil-Fuel-Lobbyists-Are-Dominating-Climate-Policy-Battles-During-COVID-19-a78b11aa1be42aef5d7078d09457603b (accessed 20 August 2021). accessed 27 March 2021.

InfluenceMap (2021). Big Tech and Climate Policy. (January 2020). https://influencemap.org/report/Big-Tech-and-Climate-Policy-afb476c56f217ea0ab351d79096df04a (accessed 20 August 2021).

IRENA (2017). *Untapped Potential for Climate Action.* Abu Dhabi: IRENA https://www.irena.org/-/media/Files/IRENA/Agency/Publication/2017/Nov/IRENA-Untapped-potential-2017-summary.pdf?la=en&hash=97BD94B76DC01A918E5101714E1333AE04E3AEEC (accessed 1 June 2021).

IRENA (2019). *Global Energy Transformation: A Roadmap to 2050.* Abu Dhabi: IRENA.

IRENA (2020). Renewable Energy Statistics 2020. Abu Dhabi: The International Renewable Energy AgencyAbu Dhabi.

IISD and Oil Change International (2020). Doubling Back and Doubling Down: G-20 scorecard on fossil fuel funding. http://priceofoil.org/content/uploads/2020/11/g20-scorecard-report.pdf (accessed 7 August 2021accessed 7 August 2021).

Jackson, R.B., Le Quéré, C. et al. (2019). *Global Energy Growth Is Outpacing Decarbonization.* Canberra Australia: Global Carbon Project, International Project Office.

Jamrisko, M. (2019). The world's dirtiest air is now in India. *Bloomberg* (18 March 2019). https://www.bloomberg.com/news/articles/2019-03-05/the-world-s-dirtiest-air-is-in-india-where-pollution-costs-lives (accessed 30 July 2021).

Kaplan, S. and Guskin, E. (2019). Most American teens are frightened by climate change, poll finds, and about 1 in 4 are taking action. *Washington Post* (16 September 2019). https://www.washingtonpost.com/science/most-american-teens-are-frightened-by-climate-change-poll-finds-and-about-1-in-4-are-takingaction/2019/09/15/1936da1c-d639-11e9-9610-fb56c5522e1c_story.html (accessed 14 August 2021).

Leefeldt, E. and Danise, A. (2021). Homeowners and Insurance Companies will grapple with climate change in 2021. *Forbes Advisor.* https://www.forbes.com/advisor/homeowners-insurance/home-insurance-outlook-2021 (accessed 21 August 2021).

Londoño, E. (2019). Bolsonaro fires head of agency tracking amazon deforestation in Brazil. *New York Times* (2 August 2019). https://www.nytimes.com/2019/08/02/world/americas/bolsonaro-amazon-deforestation-galvao.html?login=email&auth=login-email (accessed 8 August 2021).

OECD (2017) OECD Green Finance and Investment Flyer. https://www.oecd.org/dac/Green%20Finance%20and%20Investment%20Flyer%20DAC%20HLM%202017.pdf (accessed 8 August 2021)

Ogunbiyi, D. (2021). How to end energy poverty and reach net zero emissions post COVID-19. *World Economic Forum* (2 July 2021). https://www.weforum.org/agenda/2021/07/how-to-end-energy-poverty-net-zero-emissions (accessed 15 August 2021).

Oil Change International et al. (2017). *Talk is Cheap: How G20 Governments are Financing Climate Disaster.* http://priceofoil.org/content/uploads/2017/07/talk_is_cheap_G20_report_July2017.pdf (accessed 15 August 2021).

Propper, S. (2019). Sustainability failed: the future is just climate change. *Blog* (20 May 2019). http://priceofoil.org/content/uploads/2017/07/talk_is_cheap_G20_report_July2017.pdf; https://contextsustainability.com/sustainability-failed-the-future-is-just-climate (accessed 15 August 2021).

Schön, D. and Rein, M. (1994). *Frame Reflection: Towards the Resolution of Intractable Policy Controversies*. New York: Harpers Collins.

Scott, A. (2021). Ida's impact likely to be boosted by pandemic pricing. *Reuters* (30 August 2021). https://www.reuters.com/article/us-storm-ida-insurance-idUSKBN2FV1W8 (accessed 31 August 2021).

Sullivan, Z. (2019). Why the Amazon is on fire. *Time Magazine* (26 August 2019). https://time.com/5661162/why-the-amazon-is-on-fire (accessed 14 August 2021).

TCFD (2020). Task Force on Climate-Related Financial Disclosures: 2020 Status Report. https://assets.bbhub.io/company/sites/60/2020/09/2020-TCFD_Status-Report.pdf (accessed 21 August 2021).

Times of India (2018). 14 of the world's 15 most polluted cities in India. (2 May 2018). https://timesofindia.indiatimes.com/city/delhi/14-of-worlds-15-most-polluted-cities-in-india/articleshow/63993356.cms (accessed 30 July 2021).

UKCOP26.org (2021). Around 120 leaders gather at COP26 in Glasgow for 'last,best chance' to keep 1.50C alive. 1 November 2021. https://ukcop26.org/around-120-leaders-gather-at-cop26-in-glasgow-for-last-best-chance-to-keep-1-5-alive/ (accessed 5 December 2021).

US White House (2021). Fact Sheet: President Biden sets 2030 greenhouse gas pollution reduction target aimed at creating good paying union jobs and securing US leadership on clean energy technologies. *White House Briefing Room* (22 April 2021). https://www.whitehouse.gov/briefing-room/statements-releases/2021/04/22/fact-sheet-president-biden-sets-2030-greenhouse-gas-pollution-reduction-target-aimed-at-creating-good-paying-union-jobs-and-securing-u-s-leadership-on-clean-energy-technologies (accessed 20 August 2021).

UN/UNFCCC (1992). *United Nations Framework Convention on Climate Change*. Geneva: UN https://unfccc.int/files/essential_background/background_publications_htmlpdf/application/pdf/conveng.pdf (accessed 1 June 2021).

UNEP/Barbier, E (2009). *Rethinking the Economic Recovery: A Global Green New Deal*. Nairbobi: UNEP.

UNEP, ILO et al. (2008). *Green Jobs: Towards decent work in a sustainable low-carbon world*. Nairobi: UNEP https://www.ilo.org/wcmsp5/groups/public/---dgreports/---dcomm/documents/publication/wcms:098504.pdf (accessed 20 August 2021).

UNEP (2018). *Emissions Gap Report: 2018*. Nairobi: UNEP.

UNEP (2021). *Emissions Gap Report 2021: The Heat Is On – A World of Climate Promises Not Yet Delivered – Executive Summary*. Nairobi: UNEP https://wedocs.unep.org/bitstream/handle/20.500.11822/36991/EGR21_ESEN.pdf (accessed 5 December 2021).

Watts, Jonathan (2019). Biggest compliment yet': Greta Thunberg welcomes oil chief's 'greatest threat' label *The Guardian*. 5 July 2019. https://www.theguardian.com/environment/2019/jul/05/biggest-compliment-yet-greta-thunberg-welcomes-oil-chiefs-greatest-threat-label (accessed on 7 August 2021).

Weiss, C. (1977). Research for policy's sake: the enlightenment function of social research. *Policy Analysis* 3: 531–547.

Wildavsky, A. (1979). *Speaking Truth to Power: The Art and Craft of Policy Analysis*. Boston: Little & Brown.

WHO (2018). *COP-24: Special Report- Health and Climate Change*. Geneva: WHO https://apps.who.int/iris/handle/10665/276405 (accessed 6 June 2021).

WMO (2019). *United in Science: Synthesis Report*. Geneva: WMO https://library.wmo.int/doc_num.php?explnum_id=9937 (accessed 7 November 2021).

Index

Air Pollution, Clean Energy and Climate Change, First Edition. Anilla Cherian.

t

u

y